AIDS

The Biological Basis

The Jones & Bartlett Learning Topics in Biology Series

We are pleased to offer a series of full length textbooks designed specifically for your special topics courses in biology. Our goal is to supply comprehensive texts that will introduce non-science majors to the wonders of biology. With coverage of topics in the news, emerging diseases, and important advances in biotechnology, students will enjoy learning and relating science to current events.

AIDS: The Biological Basis, Sixth Edition
Benjamin S. Weeks, PhD, Adelphi University and Teri Shors, PhD, University of Wisconsin Oshkosh

Human Embryonic Stem Cells, Second Edition
Ann Kiessling, PhD, Harvard Medical School, and Scott C. Anderson

The Microbial Challenge, Third Edition
Robert Krasner, PhD, Providence College, and Teri Shors, PhD, University of Wisconsin Oshkosh

Microbes and Society, Second Edition
Benjamin S. Weeks, PhD, Adelphi University, and I. Edward Alcamo, PhD, formerly of State University of New York at Farmingdale

Plants and People
James D. Mauseth, PhD, University of Texas Austin

Related Titles in Microbiology

20th Century Microbe Hunters
Robert Krasner, PhD, Providence College

AIDS: Science and Society, Seventh Edition
Hung Fan, PhD, University of California Irvine, Ross F. Connor, PhD, University of California Irvine, and Luis P. Villareal, PhD, University of California Irvine

Encounters in Microbiology, Volume 1, Second Edition
Jeffrey C. Pommerville, PhD, Glendale Community College

Encounters in Microbiology, Volume 2
Jeffrey C. Pommerville, PhD, Glendale Community College

Encounters in Virology
Teri Shors, PhD, University of Wisconsin Oshkosh

Fundamentals of Microbiology, Tenth Edition
Jeffrey C. Pommerville, PhD, Glendale Community College

Fundamentals of Microbiology: Body Systems Edition, Second Edition
Jeffrey C. Pommerville, PhD, Glendale Community College

Alcamo's Laboratory Fundamentals of Microbiology, Tenth Edition
Jeffrey C. Pommerville, PhD, Glendale Community College

Principles of Modern Microbiology
Mark Wheelis, PhD, University of California, Davis

Guide to Infectious Diseases by Body System, Second Edition
Jeffrey C. Pommerville, PhD, Glendale Community College

Understanding Viruses, Second Edition
Teri Shors, PhD, University of Wisconsin Oshkosh

AIDS
The Biological Basis

Benjamin S. Weeks
Adelphi University

Teri Shors
University of Wisconsin Oshkosh

JONES & BARTLETT
LEARNING

World Headquarters
Jones & Bartlett Learning
5 Wall Street
Burlington, MA 01803
978-443-5000
info@jblearning.com
www.jblearning.com

Jones & Bartlett Learning books and products are available through most bookstores and online booksellers. To contact Jones & Bartlett Learning directly, call 800-832-0034, fax 978-443-8000, or visit our website, www.jblearning.com.

Substantial discounts on bulk quantities of Jones & Bartlett Learning publications are available to corporations, professional associations, and other qualified organizations. For details and specific discount information, contact the special sales department at Jones & Bartlett Learning via the above contact information or send an email to specialsales@jblearning.com.

Production Credits
Chief Executive Officer: Ty Field
President: James Homer
SVP, Editor-in-Chief: Michael Johnson
Executive Publisher: William Brottmiller
Publisher: Cathy L. Esperti
Senior Acquisitions Editor: Erin O'Connor
Editorial Assistant: Rachel Isaacs
Production Editor: Leah Corrigan
Marketing Manager: Lindsay White
Manufacturing and Inventory Control Supervisor: Amy Bacus
Composition: CAE Solutions Corp.
Cover Design: Michael O'Donnell
Photo Research and Permissions Coordinator: Lauren Miller
Cover Image: © BioMedical/ShutterStock, Inc.
Printing and Binding: Edwards Brothers Malloy
Cover Printing: Edwards Brothers Malloy

Library of Congress Cataloging-in-Publication Data

Weeks, Benjamin S.
     AIDS : the biological basis / Benjamin S. Weeks, Teri Shors.—Sixth edition.
       pages cm
     ISBN 978-1-4496-1488-1
     1. AIDS (Disease) I. Shors, Teri. II. Title.
     RC606.6.A432 2014
     616.97'92—dc23

                                                            2013010986

6048

Printed in the United States of America
17 16 15 14 13   10 9 8 7 6 5 4 3 2 1

To the late Elaine (Motschke) Gross, my mother. *Ich vermisse dich jeden Tag.*

To John Cronn, my undergraduate microbiology mentor, colleague, and friend who opened my eyes to the invisible world of microbes and viruses.

To Robert I. Krasner, Emeritus Professor of Providence College in Rhode Island and author of *The Microbial Challenge*, who shares a similar passion for developing instructional materials for undergraduate microbiology education.

To the hundreds of students I have taught; past and present.

To the HIV/AIDS researchers who persisted for over 30 years to unravel the elegant biology of HIV in order to advance treatment and be cautiously optimistic that a cure or vaccine may one day be attainable.

"We know nothing of what will happen in the future, but by the analogy of experience."
–Abraham Lincoln

Teri Shors

I would like to dedicate this book to my children, Samuel, Jessica, Hayden, and David. Without their support my efforts would not have been possible.

Benjamin Weeks

Contents

9   Hopes and Hurdles Towards an HIV Cure or Vaccine   239

10 AIDS in Perspective 274

Preface

Now, in 2013, three decades after the onset of the HIV epidemic, some experts believe an AIDS-free generation is within reach. The rate of HIV infection has decreased or stabilized in many countries. More than 20 different antiviral drugs are now available to treat HIV-infected individuals, resulting in the improvement and prolonging of life and a reduction in the rates of transmission of the virus. In 2012, Truvada was approved as an HIV prevention drug for use by high-risk individuals and a new, at-home HIV testing kit was approved for over-the-counter purchase. Since the development of anti-HIV drugs, the annual number of deaths due to AIDS has decreased by two-thirds in the United States.

In 2012 and 2013, a few individuals were even *cured* of an HIV infection—their viral loads were reduced to insignificant levels and antiretroviral therapy was stopped (e.g. the "Berlin patient" and the Mississippi baby treated early with anti-HIV drugs). These cases, though still rare, are accelerating research efforts toward the understanding of the mechanism needed to eliminate HIV, including HIV reservoirs from the body.

In the Foreword of this textbook, Dr. Bruce D. Walker suggests, "The HIV epidemic is by no means over." For example, HIV is increasing at a high rate in Eastern Europe and Central Asia. A global commitment of resources will be needed to expand testing, improve treatment and prevention programs, and meet the scientific challenge of developing an HIV vaccine and/or cure. All countries need to improve their efforts to strengthen healthcare of their people. This sixth edition of *AIDS: The Biological Basis* is a continued and important effort to educate people in the effective control of the spread of HIV and AIDS.

Organization and Special Features

This latest edition includes new and vital information for students of HIV and AIDS and provides the educator with more tools to explain the nature of the disease. New figures, tables, boxes, and the learning objectives at the beginning of each chapter reflect the important concepts and discoveries pertaining to the biological basis of HIV. Statistics and rates of infection have been updated and new advances in basic and applied HIV/AIDS research have been added to this sixth edition. The Healthline boxes were moved to the end of each chapter and updated or edited for relevance. Two new features were added to the textbook: **Focus on HIV**, which are short, HIV-related facts located in the margins of the textbook, and **Mapping HIV** boxes that detail the status of the HIV epidemic in different parts of the world. Each chapter contains a Mapping HIV Box, and therefore the HIV status of the entire world is available as follows:

- Chapter 1: United States
- Chapter 2: Western and Central Europe
- Chapter 3: The Caribbean
- Chapter 4: The Middle East and North Africa

- Chapter 5: Africa
- Chapter 6: Eastern Europe and Central Asia
- Chapter 7: Oceania (includes Australia, New Zealand, and Papua New Guinea)
- Chapter 8: Canada
- Chapter 9: Central and South America
- Chapter 10: South and Southeast Asia

Chapter 1 contains a more complete history of HIV and AIDS with respect to early recognition of the disease, a more accurate reflection of the diseases that were observed, based on recent revelations, and a quick overview of the updated epidemiological trends of the HIV epidemic around the globe. Chapter 2 includes new figures along with a full explanation of the nature and role of proteins, nucleic acids, and lipids in viruses and in the process of infection, enhancing the students' understanding of the biological basis of viruses and the mechanisms of infection. This allows a deeper understanding and discussion of the mechanisms of HIV testing, HIV prevention, and drug activity. Chapter 2 also includes a new box titled "Peter Duesberg, Rebel with an Anti-HIV Cause."

Chapter 3 is an overview of the immune system and its response toward HIV infection. It contains new boxes that provide molecular biology information focused on the understanding of the body's defense toward HIV. The new boxes are titled "The Massie Puzzle Piece Hiding on Chromosome 6" and "Shutting the Cellular Door to HIV-1: Research Toward a Cure." Chapter 4 now contains an updated case definition of HIV infection, including criteria for AIDS recognition and diagnosis of AIDS, AIDS dementia, and information on opportunistic infections experienced by AIDS patients. It also includes information about the Millennium Village Project sites in Africa aimed to end mother-to-child HIV transmission.

Chapter 5, which focuses on the spread of HIV among people, now presents updated statistics and rates of HIV infection among various groups and populations. It stresses that HIV patients in developed countries who have the necessary healthcare systems in place can live almost as long as the uninfected population in developed countries if they can adhere to a regime of a cocktail of antiviral drugs that reduces the viral loads to nearly undetectable or very low levels. New topics have been added to Chapter 5 such as:

- Incidence of HIV and AIDS by geographic distribution in the U.S.
- HIV in transgendered people in the U.S.
- HIV and seniors in the U.S.
- HIV in correctional facilities in the U.S.

Chapter 6 focuses on the prevention of HIV transmission, including updated statistics, trends in sexually transmitted diseases (STDs), and rare routes of HIV transmission (e.g. organ transplants and pre-mastication). Sections from Chapter 5 on other transmission methods, improbable transmissions, how AIDS is not transmitted, and the mosquito myth are now included in Chapter 6. New information on decontamination procedures and exposures to HIV in clinical settings were also included. Chapter 7 presents new information on screening recommendations, how HIV testing works, and the drugs used to fight HIV.

Chapter 8 focuses on treating patients for HIV infection and AIDS. It includes a new Table 8.1, The Six Classes of HIV Drugs Licensed by the FDA, and Figure 8.4, a timeline of antiretroviral drug development and therapy. Chapter 9 underwent an extensive revision, including a title change from "An AIDS Vaccine" to "Hope and Hurdles Towards an HIV Cure or Vaccine." This title change reflects the following new topics, related to new research discoveries and studies published in scientific literature:

- Reducing rates of HIV infection
- Eliminating viral reservoirs
- Bone marrow transplants
- Gene therapy

Several new figures were added to aid the student learning of molecular biology concepts needed to understand development of new alternative therapies or a potential cure for HIV.

Chapter 10 is a final overview of the last 30+ years of research that is driving toward the goals of creating an AIDS-free generation. It contains new information on drug assistance programs and figures about investment patterns and how money and resources go towards the AIDS response. A new Table 10.2 on the monthly average wholesale prices for the leading antiviral drugs in the United States (2012) is included.

Like the previous edition, the sixth edition provides thorough chapter outlines, modified Healthline reports, an updated glossary, study questions, and, at the end of each chapter, updated additional reading lists. New information and discussions to look for in the sixth edition include:

1. All of the latest available epidemiological data for worldwide trends, regional trends, and U.S. statistics.
2. Discussions on why Eastern Europe and Central Asia HIV cases are increasing as compared to other parts of the world that are stabilizing.
3. New discoveries on viral persistence and clinical trials in progress to remove reservoirs of HIV from the body using anti-cancer drugs such as vorinostat.
4. Peter H. Duesberg's belief that HIV is not the causative agent of AIDS.
5. Rare cases in which individuals have been "cured."
6. The molecular mechanism of viral control by elite controllers like Bob Massie.
7. Updated case definition of HIV, AIDS, pediatric AIDS, AIDS dementia, and opportunistic infections of AIDS patients, including treatment.
8. Updated statistics on *Pneumocystis* pneumonia and AIDS.
9. Mechanisms and classes of new HIV antiviral drugs developed.
10. Timeline of antiviral drug development.
11. HIV in selected populations; transgender, seniors, correctional institutions.
12. New information on drug assistance programs, policies, and funding for HIV/AIDS research.
13. New information about microbiocides and other methods to prevent HIV transmission.
14. Millennium Project Sites in Africa to end mother-to-child HIV transmission.
15. The impact of the FDA-approved drug Truvada use for high-risk individuals and reducing rates of HIV infection.
16. Rare routes of HIV entry such as organ transplants and pre-mastication.
17. New information on alternative therapies including bone marrow transplants and gene therapy.
18. The vision of an AIDS-free generation and what needs to happen to attain it.
19. The hopes and hurdles towards an HIV cure or vaccine.
20. Updated information on funding for HIV/AIDS relief.
21. New and improved photographs and diagrams.

These updates, extended commentary, and discussions enhance the value of this book for professors and students alike. As future editions evolve to keep abreast of epidemiological patterns and research developments, both professors and students will find that the information here sets the mark for compiling an extensive breadth of knowledge with sufficient detail that permits the reader to learn the basics of AIDS immunopathology, epidemiology, and how AIDS drugs, alternative therapies, and vaccines can work.

Instructor's Resources

The following instructor's resources have been developed by Jones & Bartlett Learning to accompany *AIDS: The Biological Basis, Sixth Edition:*

- *A PowerPoint® Image Bank,* including all figures in the book to which Jones & Bartlett Learning holds the copyright, or has permission to reproduce digitally.
- *PowerPoint Lecture Outlines,* supplying the key concepts and figures found in each chapter as a framework for class lectures. Access to these materials can be gained through your Jones & Bartlett Learning representative.
- Text files of the *Test Bank.*

For more information about these resources, please visit go.jblearning.com/AIDS6e.

Foreword

The HIV epidemic is by no means over. Life-extending drugs have converted what was almost certain death into a chronic, manageable disease for those fortunate enough to have access, but without new advances these medications will have to be taken life-long. Over 8 million people are now on treatment worldwide, but the number in need who still lack access may be twice this or more, and the epidemic is still expanding. Even the most robust economies will be challenged by the costs and personnel required to address this ongoing problem. The major hope for ending the epidemic is the development of an effective vaccine, but this remains elusive. Given the central role that this disease will continue to play on a global level and the personal risks involved for all persons who are sexually active, knowledge of this disease should be considered mandatory.

Obtaining a sound working knowledge of HIV and its consequences has been available through this textbook, *AIDS: The Biological Basis,* which is now being released in an updated sixth edition, edited by Teri Shors. The textbook traces not just the science but also the controversies, both scientific and political, that have accompanied the quest to overcome what has become the defining global health issue of this generation. The book takes the reader from the earliest hints that an epidemic was on the horizon, and moves on to the clinical manifestations and the explanation for these—that HIV is an infection of the immune system and it destroys the very system that is supposed to protect against it. The book also addresses HIV prevention and dispels myths that have led to unnecessary hysteria regarding personal risks.

The success of the scientific response to this epidemic has been impressive, as evidenced by the transformation of this infection from a death sentence into a treatable disease, and this exhilarating path is documented in a compelling way that is likely to inspire a new generation of young students to explore biomedical research as a way to impact the lives of millions. And there remains ample room for them to get involved, as outlined in the parts of the book dedicated to the enormous but as yet unfulfilled effort to develop an effective vaccine. We have seen this disease evolve to one that can be treated, but one that requires lifelong treatment. What we do not yet have is a way to protect uninfected, at-risk persons from acquiring the disease. Undoubtedly the AIDS epidemic will continue to impact the political, social, economic, and scientific discourse well into the future and make the information contained in this book all the more valuable.

As someone who has committed his entire career to attempting to understand the battle between the body and HIV, and why the body usually loses this battle, I can only applaud this effort to increase knowledge among those at risk, which is the vast majority of the world's population. This book should be required reading for anyone wishing to consider themselves educated and ready to address the major challenges of this generation and the next.

Bruce D. Walker, MD
Director,
Ragon Institute of MGH, MIT and Harvard,
Boston, Massachusetts

Acknowledgments

Professor Alcamo, along with those at Jones & Bartlett Learning, envisioned the need for a biologically based book on HIV and AIDS that could be used at the introductory level. Dr. Weeks contributed and revised the textbook after Dr. Alcamo's passing. I am honored and grateful to have been given the opportunity by former Publisher Cathy Sether and Senior Acquisitions Editor Erin O'Connor at Jones & Bartlett Learning to continue Ed Alcamo and Benjamin Week's work by updating and extending the material for this new edition. Erin O'Connor was receptive to new features of the text as it continues to evolve as a resource for students with limited prior knowledge about viruses and their biology.

Some revisions were based on reviews solicited and suggestions recommended by Molly Steinbach. Special thanks to reviewers and Molly Steinbach for this. Peer reviews go beyond basic proofreading. They cast a fresh eye to ensure consistency, as well as finding holes or gaps in the information that need updates or clarity. I am also very grateful to the conscientious editorial assistance of Rachel Isaacs during the preparation of the sixth edition manuscript; Lauren Miller for permissions and photo research; Shellie Newell for copyediting; Linda DeBruyn for proofreading; CAE Solutions, Inc., the compositor; and special thanks to Leah Corrigan, Production Editor, for her patience and guidance through production.

I also extend gratitude to Dr. Bruce D. Walker, Director, Ragon Institute of MGH, MIT and Harvard University for his willingness to create the Foreword for this text despite his incredible schedule. I learned that Dr. Walker and I both belong to the Dr. Bernard Moss "research family." Dr. Bernard Moss, distinguished research investigator and Chief of the Laboratory of Viral Diseases, NIAID, at the NIH has been instrumental in training many virologists and educators. Dr. Walker worked tirelessly to lead research that unraveled why hemophiliac Bob Massie was an elite controller of HIV infection. Walker and Massie's journey into the molecular biology of HIV was featured in a segment of the NOVA program *Surviving AIDS*. *Surviving AIDS* was broadcast on PBS on February 2, 1999. Walker's team of dedicated researchers, with the help of over 3,600 HIV-positive patient volunteers, solved the mystery in 2010 (Box 3.2).

I extend special thanks to those researchers who continue to unravel the biology of HIV to the point at which antiviral drug therapies and powerful interventions were/are being developed. Without their persistent efforts, many more individuals would have died of HIV/AIDS. Now, there is hope for a potential cure and/or vaccine.

Lastly, I express thanks and appreciation to others who have tolerated me during the preparation of this text.

Teri Shors, MS, PhD
Professor
Department of Biology and Microbiology
University of Wisconsin Oshkosh

About the Authors

I. Edward Alcamo (first, second, and third editions)

The late I. Edward Alcamo was the original author of *AIDS: The Biological Basis* (1993). Professor Alcamo envisioned the need for a biologically based book on HIV and AIDS that could be used at the introductory level. As a long-time professor of microbiology at the State University of New York at Farmingdale, he was the author of numerous textbooks, laboratory kits, and educational materials in his field. Dr. Alcamo was the 2000 recipient of the Carski Foundation Distinguished Undergraduate Teaching Award, the highest honor bestowed upon microbiology educators by the American Society for Microbiology. He was educated at Iona College and St. John's University and held a deep belief in the partnership between research scientists and allied health educators. He sought to teach the scientific basis of microbiology in an accessible manner as well as to inspire students with a sense of contemporary relevance. In December 2002, after a six-month illness, Dr. Alcamo died of acute myeloid leukemia.

Benjamin S. Weeks (fourth and fifth editions)

Benjamin S. Weeks contributed to the fourth and fifth editions of *AIDS: The Biological Basis* (2006 and 2010). His training as a scientist began in the mid-1980s. He earned a PhD in 1988 while doing graduate work at the Center of Environmental Health at University of Connecticut. He is currently a Professor of Biology at Adelphi University. Dr. Weeks is involved in teaching, research, and has served as the Director of Undergraduate Academic Affairs. His research continues to focus on how environmental pollutants can cause damage to the nervous and the immune system in humans and how nutritional supplements can mitigate those effects.

Teri Shors (sixth edition)

Teri Shors contributed to this sixth edition of *AIDS: The Biological Basis*. She has taught microbiology and virology at the University of Wisconsin Oshkosh for 15 years. Dr. Shors is a devoted teacher and researcher at the primarily undergraduate level. Dr. Shors' graduate and postgraduate education is virology-based and is reflected in her research. Before teaching at UW Oshkosh, she was a postdoctoral fellow in the Laboratory of Viral Diseases under the direction of Dr. Bernard Moss in the National

Institutes of Allergies and Infectious Diseases (NIAID) at the National Institutes of Health (NIH). Dr. Shors is an author of *Understanding Viruses, Second Edition* and *Encounters in Virology* and a co-author of *The Microbial Challenge, Third Edition.* Additionally, Dr. Shors has contributed to and authored a variety of other texts and scientific papers. Initiative, creativity, humor, networking, using current events and the latest technology in her courses, and leading collaborative, cross-disciplinary studies are all hallmarks of Dr. Shors' talents and make her popular among students in the classroom. She recently developed and taught an online virology course for undergraduates.

In the early 1980s, when the early cases of an unknown acquired immune deficiency syndrome were turning into a mysterious and intractable epidemic, Dr. Alcamo related this parable to his students:

One afternoon, about 350 years ago, in the countryside near London, a clergyman happened to meet Plague.

"Where are you going?" asked the clergyman.

"To London," responded Plague, "To kill a thousand."

They chatted for another few moments, and each went his separate way.

Some time later, they chanced to meet again.

The clergyman said, "I see you decided to show no mercy in London. I heard that 10,000 died there."

"Ah, yes," Plague replied, "But I only killed a thousand..."

"Fear killed the rest."

AIDS: The Biological Basis was developed to help students understand the rapidly developing scientific advances in immunopathology, epidemiology, and treatments for HIV/AIDS in order to turn back the tide on uncertainty and fear.

The AIDS Epidemic

LOOKING AHEAD

June 5, 2011 marked 30 years since the Centers for Disease Control and Prevention (CDC) reported the first cases of AIDS in the United States in its weekly *Morbidity and Mortality Weekly Report (MMWR)*. The chronological milestone provides an opportunity to reflect upon the status of HIV/AIDS in the world today. Since these first five cases were reported, nearly 600,000 people have died in the United States as a result of HIV/AIDS. Worldwide, 30 million people have died of HIV-related causes. There is still no vaccine available to prevent this viral disease. However, there remains hope that this viral disease will be conquered one day. Development of anti-virals to treat patients, along with improved HIV diagnosis methods, have increased the life expectancy of U.S. patients after HIV diagnosis from 10.5 to 22.5 years.

This opening chapter introduces **acquired immune deficiency syndrome (AIDS)** by exploring the development of its **epidemic** in the United States and the world and by describing the research to uncover its cause. On completing the chapter, you should be able to . . .

- Understand the conditions that led public health officials to realize that an epidemic of AIDS was in progress.
- Recognize some broad features of AIDS and characterize the individuals at risk for the disease.
- Summarize the research leading to the isolation and identification of the AIDS virus.
- Conceptualize the transmission of the AIDS virus from chimpanzees to humans.
- Discuss some theories for the origin of the AIDS epidemic and its spread in the United States and the world.
- Note the current magnitude of the AIDS epidemic and describe where it is likely to spread in the future.

The end of this chapter contains a **Healthline Q & A** section that addresses questions you may have as you read and digest the information presented.

INTRODUCTION

Human perception can lead to false understandings; however, when observing new patterns in the world, often all we have are our perceptions. This is the case with our early thoughts about AIDS. Physicians and scientists could not adequately

FIGURE 1.1
A fable tells of six blind men (seen here as blindfolded) invited to examine an elephant and report what they believed it to be. Each touched the elephant at a different part and reported something completely unlike the others. In the early years of the AIDS epidemic, physicians, researchers, and members of the general public saw disease in different ways, much as different blind men perceived the elephant.

Focus on HIV
There were approximately 34.2 million people living with HIV in 2011.

define and describe AIDS because they did not know at what they were looking. This inability to see the "big picture" is best illustrated by the old South Asian fable about six blind men invited to examine an elephant. The first man, falling against the elephant's side, claimed it was a wall. The second, seizing a leg, declared the elephant a tree. Another, grasping a tusk, thought the elephant was a spear. The fourth man, grabbing the trunk, decided he was holding a snake. The fifth, touching the ear, believed it to be a fan. The last blind man, holding the tail, proclaimed the elephant a rope (**Figure 1.1**).

So it was in the early 1980s with AIDS: The elephant was the disease, and the blind men were the scientists and doctors grappling with its emergence in an epidemic. Without a clear vision of AIDS, they were forced to treat the epidemic piecemeal—to try and explain each bit of knowledge as it came forth. Physicians were seeing patients whose immune systems were profoundly suppressed, but they had no idea what was causing this problem. Public health officers were charting the course of the AIDS epidemic without knowing its cause. Manufacturers were searching out new drug therapies, but they were unsure about what they were trying to eliminate. Physicians and scientists did not really know what they were seeing and could not adequately describe what was going on.

Even though physicians and scientists could not identify the causes of what was happening or completely describe all of the features of the disease, it was clear from very early on that a large problem had been presented to public health in the United States. Physicians saw it as a medical problem, economists as a potential disaster to the healthcare system, employers as a threat to the smooth operation of their businesses, healthcare workers as a problem requiring personal protection, and politicians as yet another drain on public funds. Until 1984, there was no clear definition of AIDS. No test was available to confirm a diagnosis and no cure was in sight. Most scientists agreed that a cure would follow only after they found a cause and understood the natural history of the disease.

Despite the early pessimism, however, scientists found the cause of AIDS in a shorter span of time than had been required for any previous infectious disease. So quickly was the cause of AIDS found by scientists that it highlighted the astounding advances made during the 1970s when the ongoing revolution in modern molecular biology began. These advances and rapid identification of the cause of AIDS, however, did not prepare physicians, scientists, patients, and our society for the slow and ongoing progress made in combating AIDS. Indeed, in the early 1980s, when the cause of AIDS was identified, there was a sense that it would soon be a historical side note in the annals of infectious disease. No one was prepared for what has followed.

The Early Years

By 1981, most observers believed that infectious diseases were largely under control in the Western world. There had been no smallpox anywhere since October 1977, and diseases such as typhoid fever, polio, and whooping cough were rarely encountered. Even childhood diseases such as measles, mumps, and rubella appeared to be relics of the past. Indeed, the speed with which an effective therapy for Legionnaires' disease was discovered in 1976 and the success in interrupting the outbreak of toxic shock syndrome in 1978 had strengthened public confidence that scientists could defeat any epidemic.

Moreover, in the early 1980s, American researchers had largely turned away from the negative aspects of microorganisms and were pursuing their practical uses. For instance, bacteria had become mainstays of genetic engineering, which was fast developing as the technology of the future. (A 1981 *Time* magazine article described genetic engineering as "the most powerful and awesome skill acquired by man since the splitting of the atom.") Genetically engineered insulin was already in production. Synthetic vaccines were in the planning stage, and breathtaking developments in DNA technology were in the headlines daily. The mood of the day was optimism.

The First Observations

On June 5, 1981, the CDC published a brief but significant article in **Morbidity and Mortality Weekly Report (MMWR)**. This article, appearing between discussions of dengue fever and measles, described the case reports of five young men treated at three different hospitals in Los Angeles, California. All five men were sexually active and gay, and all five were suffering from *Pneumocystis* pneumonia. By the time of the article's publication, two of the five patients had died from what was reported to be a yeast-like fungal lung infection with *Pneumocystis jirovecii*. The terms **Pneumocystis pneumonia** or **pneumocystosis**, as they first appeared in these reports, remain an accurate way to discuss this infection.

In the CDC article, the writer observed that three of the patients had "*profoundly depressed numbers of thymus-dependent lymphocyte cells and profoundly depressed . . . responses to mitogens and antigens.*" Because **lymphocyte** cells are components of the immune system and because responses to mitogens and antigens are among the important functions of the immune system, it was apparent that the patients' immune systems were under stress. Then, in perhaps the understatement of 1981, the writer noted this: "*The occurrence of pneumocystosis in these five previously healthy individuals with underlying immunodeficiency is unusual.*"

The CDC article of June 5, 1981 is considered a benchmark: the unofficial beginning of the AIDS epidemic in the United States. Statistics were compiled from that

┌ ┐
BOX
1.1
└ ┘

The CDC

The CDC, one of six major agencies of the U.S. Public Health Service, has its headquarters in Atlanta, Georgia. Originally established as the Communicable Disease Center in 1946, the CDC was the first government health organization ever set up to coordinate a national control program against infectious diseases. At first, it was concerned with diseases spread from person to person, from animals to people, or from the environment to humans. Eventually, however, all communicable diseases came under its aegis. Atlanta was selected as the site for the CDC because it was a convenient central point for the study of malaria, which was then common in the South.

In April 1955, two weeks after release of the Salk vaccine for polio, the CDC received reports of six cases of polio in vaccinated children. Two days later, it established the Polio Surveillance Unit and began collecting data on the occurrence of polio and summarizing it for health professionals. More than 80% of vaccine-associated polio cases were related to a single manufacturer, whose vaccine was withdrawn at once. This incident established the role of the CDC in health emergencies, and soon it became a national resource for the development and dissemination of information on communicable diseases. In 1960, the CDC moved its headquarters to a new complex adjoining Emory University. The unassuming appearance of the facility belies its importance.

Reorganized under its current name in 1980 (the words "and Prevention" were added in 1992, but the CDC acronym was retained), the CDC is charged with protecting the public health of the U.S. population by providing leadership and direction in the prevention and control of infectious disease and other preventable conditions, such as cancer. It is concerned with urban rat control, quarantine measures, health education, and the upgrading and licensing of clinical laboratories. The CDC also provides international consultation on disease and participates with other nations in the control and eradication of communicable infections.

Today the CDC employs 10,500 employees and 5,000 contractors, of which 72% of the staff is located at the Headquarters office in Atlanta, Georgia while only 2% are assigned overseas, with the remaining employees located at 11 other locations. Nearly 50% of the workforce holds Master or Doctoral degrees. In 2009, its fiscal budget was $10.1 billion, of which 11% is used for HIV/AIDS, STD, and TB prevention. Its publication *MMWR* is distributed each week to thousands of health professionals. In 2011, the CDC celebrated its 65th anniversary of its founding on July 1, 1946.

date onward, and physicians were alerted to watch for similar cases and report them to their state health departments. The health departments then forwarded the reports to the CDC (**Box 1.1**).

Such reports arrived quickly. On July 3, 1981, another article appeared in the CDC's *MMWR*, this time describing **Kaposi's sarcoma** as well as **pneumocystosis** in 26 young homosexual men. Twenty patients were from New York City, and six lived in California. All had been diagnosed during the previous 30 months. Kaposi's sarcoma is normally a relatively mild skin cancer affecting mainly older men of Mediterranean descent, but in these 26 patients, it was aggressive and deadly, with skin and mucous membrane lesions of blue to violet color (**Figure 1.2**). In 10 of the patients, the Kaposi's sarcoma was accompanied by *Pneumocystis* pneumonia. In several others, doctors reported diseases not normally observed in persons with functional immune systems. This article did not comment on the patients' immune systems. It did, however, encourage physicians to continue reporting similar cases.

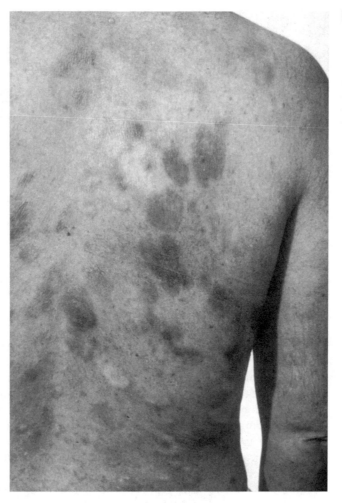

FIGURE 1.2

An AIDS patient displaying the skin lesions of Kaposi's sarcoma on his back. Kaposi's sarcoma is normally a mild skin cancer, but in persons with AIDS, it is aggressive and can be a cause of death. Courtesy of National Cancer Institute.

Because the new disease appeared to be affecting primarily homosexual men, scientists asked what practices might be unique in the homosexual community. They postulated that amyl nitrate, a drug used by some homosexual men to increase sexual pleasure, might be at fault, or perhaps certain disease organisms prevalent in homosexual men might be to blame. One such organism, the cytomegalovirus, was a prime candidate. Another theory was that the immune systems of the homosexual patients were suppressed by a constant barrage of **sexually transmitted diseases** (STDs). In some publications, the disease was given the pejorative name **gay-related immune deficiency** (**GRID**). The idea of a transmissible infectious disease had not yet taken hold.

Then data about further victims began to filter in. Late in 1981, the CDC received reports of heroin addicts and other injection drug users in New York City who were suffering from *Pneumocystis* pneumonia and immune deficiency (**Figure 1.3**). By early 1982, it appeared that heterosexuals could transmit the disease to one another, and within months, the disease surfaced among blood transfusion recipients and hemophiliacs. By June 11, 1982, the CDC had details on 355 cases from five different states (California, Florida, New Jersey, New York, and Texas). The clustering of cases within several groups and the pattern of spread made it apparent that a transmissible agent, probably a virus, was involved.

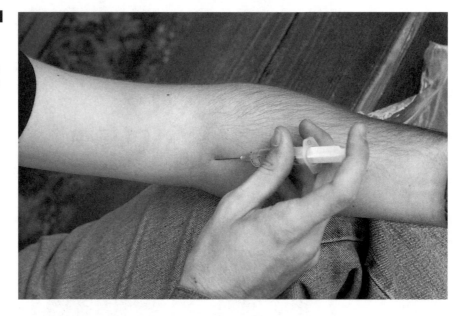

On September 3, 1982, the disease was given a new name. In the *MMWR* of that date, the CDC, acting on a recommendation from biologist Bruce Voeller, noted that "the group of clinical entities, along with its specific immune deficiency, is now called acquired immune deficiency syndrome (AIDS)." The thrust of the article, however, was a warning that a hepatitis B vaccine in use at the time might be a possible agent of transmission and should be used with caution. It was a type of warning that was becoming commonplace whenever blood or blood products were involved.

By early 1983, 16 countries were reporting AIDS cases, and more than 1,000 Americans from 34 states had been diagnosed with the disease (more than half of these patients had died by that time). Also in 1983, physicians in Newark, New Jersey, and New York City reported the first cases of what appeared to be mother-to-child transmission of AIDS. Concern also mounted that the nation's supply of blood in blood banks was contaminated. One public health administrator wrote of a "nightmare time bomb ticking away in the blood supply." Privately, the CDC (Box 1.1) began pressuring blood suppliers to consider ways to protect the public, but without knowing what the agent was and how it was transmitted, blood banks were unable to take meaningful action.

Alarm climbed a notch higher when a woman in California contracted AIDS through sexual intercourse with her husband, who suffered from hemophilia. The woman then passed the disease to her newborn child. It was becoming apparent to the general public that anyone could be at risk. Whatever fantasies Americans had that AIDS was exclusively a disease of homosexual men and injection drug users were clearly over.

The Breakthrough

During the first three years of the AIDS epidemic, the number of diagnosed cases doubled about every six months. To many observers, every six months also brought a new theory of the epidemic's cause. As noted previously here, some scientists initially related the disease to use of amyl nitrate or to immune system depression by a variety of diseases, but these ideas were soon discarded. Certain scientists held to the belief that the agent was the virus that causes African swine fever, as this virus can induce immune suppression in pigs (a temporary alert on undercooked pork was issued); others believed that the agent was a "slow virus"—one that apparently multiplies at an extremely low rate in the body and manifests itself after years or more.

At the fringe were a number of other beliefs. One group maintained that the agent was a viral escapee from a genetics engineering laboratory; another attributed it to a failed biological war against Cuba, and still another suggested that AIDS was due to mutations arising from "secret Soviet Union electromagnetic warfare against the United States." Eventually, however, most scientists focused on a virus as the most likely cause of AIDS. Some opponents were quick to point out, however, that such an assumption had been erroneous before. In the previous decade, they noted, scientists had assumed that viruses were responsible for Legionnaires' disease, Lyme disease, and toxic shock syndrome. In each case, the agent was eventually shown to be a bacterium.

By 1982, two cancer research laboratories were involved in the hunt for an AIDS agent. At the Pasteur Institute in Paris, a group led by Luc Montagnier was intrigued by the possibility that the mysterious AIDS agent could also be the cause of Kaposi's sarcoma. In the United States, at the National Cancer Institute, a team headed by Robert C. Gallo was attempting to prove that AIDS was due to a retrovirus similar to certain other viruses related to cancer. In May 1983, both the Gallo and Montagnier groups published reports hinting that they had found the virus responsible for AIDS. There was a sense of excitement in the medical community.

Then, on April 23, 1984, at a press conference in Washington, D.C., Gallo announced that his group had identified a virus in the blood of 48 persons with AIDS. In addition, he and his colleagues found antibodies to the virus in blood samples from 88% of all those diagnosed as having AIDS. Gallo recommended that the virus be named **human T-cell lymphotropic virus type III (HTLV-III)**. Four articles in the May 4, 1983 issue of the highly respected magazine *Science* carried the details of the research performed by Gallo's group.

Before the articles in *Science* appeared, however, the Montagnier group rushed to call attention to their work on an AIDS virus. In interviews with the international press, Montagnier described a virus that his team had located in the blood of patients with the early manifestations of AIDS. Like the Americans, the French researchers reported evidence of the virus in more than 80% of blood samples from AIDS patients. Montagnier's group suggested that the virus be called **lymphadenopathy-associated virus (LAV)**. The virus appeared to be identical to that isolated by the Americans.

In the months that followed, a rivalry developed between the American and French groups, each claiming to have isolated the virus first. At issue were international awards, research grants, and millions—perhaps billions—of dollars in patent rights for products that might be derived using the AIDS virus, such as an AIDS test. Time did little to sort out the differences between the groups, and the scientific community began referring to the virus as HTLV-III/LAV (in France, it was called LAV/HTLV-III). By February 1985, the American and French teams reported that their viruses were virtually identical. Montagnier and Gallo received much international acclaim (**Figure 1.4**).

Because of legal pressure and the scientific review process, in 1991, Gallo conceded that the virus he identified had originated in tissue samples sent to him by Montagnier. Gallo explained that inadvertent contamination had probably allowed the French virus to enter the tissue samples with which his team was working and

Robert C. Gallo (right) and Luc Montagnier (center), the American and French researchers who performed much of the seminal work leading to identification of the human immunodeficiency virus. In this 1988 photograph, the two researchers are speaking at a news conference in Japan at which they were jointly honored. © Shizuo Kambayashi/AP Photos.

from which the American virus was eventually isolated. However, Gallo insisted that Montagnier's work had relied on techniques developed in Gallo's laboratory in which T-cell cultures were supplemented with interleukin-2, a technique that led directly to the development of blood tests for HIV and the ability to screen donated blood for this virus. Indeed, it was in April 1985 that Gallo's group announced this development of a blood test for AIDS, and by May 1985, the test was perfected and in use. The test was so effective that over the next five years, only four cases of AIDS were linked to transfused blood. The two scientists continued to be in a bitter dispute between the United States and France over royalties from the blood test patent until 1987, when they agreed to share credit for the discovery of HIV. In light of these developments, in 1994, the U.S. government conceded that France's Pasteur Institute deserved a higher percentage of patent royalties from AIDS tests that use the virus, as the Pasteur Institute had gained greater recognition for its role in discovering the virus. Under this arrangement, the French government receives 50% of the royalty monies, the U.S. government 25%, and the World AIDS Foundation 25%. (The World AIDS Foundation funds AIDS research in the developing world.) Consequently, today it is generally agreed that Montagnier's group first identified HIV, although Gallo's group is credited with much of the science that made the discovery by Montagnier possible and, although Montagnier found the virus first, Gallo is credited for showing that it was the cause of AIDS and being the first to grow the virus in culture. The Gallo–Montagnier dispute came to a close in the November 29, 2002 issue of *Science*, in which Gallo and Montagnier coauthored an article in which they both acknowledged the pivotal roles that each had played in the discovery of HIV.

The year 1986 brought a name change for the AIDS virus. That summer, an international commission of prominent virologists and molecular biologists recommended that the name of the virus be changed from HTLV-III/LAV to the human immunodeficiency virus, or HIV. The commission noted that the new name conforms to common nomenclature for related viruses, that it does not incorporate the term

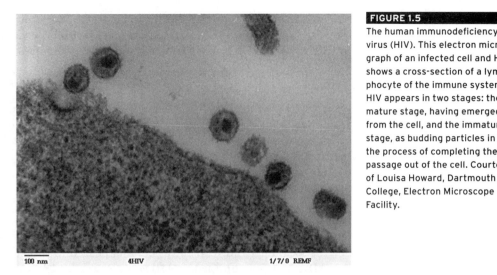

FIGURE 1.5

The human immunodeficiency virus (HIV). This electron micrograph of an infected cell and HIV shows a cross-section of a lymphocyte of the immune system. HIV appears in two stages: the mature stage, having emerged from the cell, and the immature stage, as budding particles in the process of completing their passage out of the cell. Courtesy of Louisa Howard, Dartmouth College, Electron Microscope Facility.

AIDS (because HIV could cause other diseases), that it was chosen without regard to priority of discovery, and that it allows for further names as strains are isolated. Acceptance of the commission's recommendation and confirmation by the scientific community brought one phase of the AIDS saga to an end (**Figure 1.5**).

Another AIDS Virus

No sooner had HIV been identified than another human immunodeficiency virus emerged. Identified in 1985 by Pasteur Institute scientists led by Luc Montagnier, this virus was designated HIV-2, and the original strain began to be referred to as **HIV-1**. **HIV-2** appears to be less pathogenic and progress more slowly, taking longer to damage the immune system after infection; however, although they are different strains, HIV-1 and HIV-2 are the same virus and share many similarities. For example, the modes of HIV-1 and HIV-2 transmission are the same—sexual contact, sharing needles, and so forth. Furthermore, people infected with HIV-2 are subject to the same **opportunistic infections** as those infected with HIV-1, and although they progress to AIDS more slowly, people infected with HIV-2 do get AIDS. Furthermore, HIV-1 and HIV-2 are treated with the same medications and can be detected with the same antibody blood test for CD4 monitoring; however, the efficacy of the medications against HIV-2 is less than that for HIV-1, and some HIV-2 infections do not show cross-reactivity with the HIV-1 anti-CD4 antibody. A specific antibody test for HIV-2 has been developed. The differences between HIV-1 and HIV-2 are as noted before; HIV-2 weakens the immune system more slowly than HIV-1. When compared with HIV-1–infected individuals, patients with HIV-2 are less infectious early in the course of the disease yet more infectious later in the course of disease. These differences are likely due to the entry of HIV-2 into the human population through a different and less common route, thus its discovery in human blood after HIV-1.

Blood tests have shown that HIV-2 is rare in the United States but relatively common in parts of Africa, especially West Africa and countries such as Senegal, Nigeria, Ghana, and the Ivory Coast. Researchers at Harvard University studied 574 women from Dakar, Senegal, and found that five years after infection, one third of the women infected with HIV-1 had progressed to AIDS, whereas none of those infected with HIV-2 had any sign of AIDS. Moreover, the destruction of immune system cells was substantially more extensive in those with HIV-1 than in those with HIV-2. Clearly, HIV-2 takes longer to develop in humans than HIV-1. This is due to the ability of humans to mount a more effective immune response to HIV-2 compared with HIV-1. This possibility has spurred researchers to study closely the interaction between humans and HIV-2.

A Pandemic

In 1985, reports from health officers in Africa indicated that the AIDS epidemic was widespread on that continent. Fearing stigmatization, some African government officials at first sought to distance themselves and deny that AIDS was occurring in their countries; however, the signs of AIDS were unmistakable. Thousands of men and women were affected with severe weight loss, emaciation, and disrupted immune systems, and the epidemic was spreading; indeed, it had become worldwide—a **pandemic**.

Within three years, international health officials recognized that Africa was the continent hit hardest by AIDS. The disease was present not only in homosexual men (more specifically, men who have sex with men) and injection drug users but also in heterosexual men and women, where it appeared to spread by sexual contact.

AIDS was not confined to Africa, however. By 1988, almost 5,000 cases were reported in Europe, and scientists estimated that between 500,000 and one million individuals from European countries were infected with HIV. In Asia, the situation was becoming increasingly severe, and in Australia, AIDS was spreading as fast as in the United States and Europe.

Two years previously, in 1986, the World Health Organization (WHO) had established its Global Program on AIDS. Under the direction of Jonathan Mann, the program began carefully to overcome national sensitivities and gain the confidence of the world's health leaders. Soon it was receiving reports from regional offices and the ministries of health of 175 countries. By 1988, 138 of those countries had reported at least one case of AIDS, and many had more than 1,000 total cases. In 2003, worldwide AIDS cases increased by approximately 4.8 million, and by the year 2004, the total global number of people with AIDS or living with HIV infection rose to 39.4 million, with the highest infection rates seen in sub-Saharan Africa and in South and Southeast Asia. As the leader of the world's health community, the WHO continues to chart the development of the AIDS epidemic all over the globe. The WHO also supports the development of national AIDS prevention and control programs in 120 countries throughout the world. In addition, it collects information about cases and maintains a virus bank because HIV strains from different parts of the world can vary somewhat in their biochemistry. Finally, it makes recommendations on international travel and related issues and serves as a forum for discussing the scientific and practical implications of the AIDS pandemic.

Supplementing the WHO's work is the United Nations Programme on HIV/AIDS (UNAIDS), a joint effort of various subdivisions of the United Nations, including the United Nations Children's Fund. Established in 1995 and currently under the direction of Michel Sidibé, UNAIDS coordinates the UN's efforts in dealing with the AIDS pandemic. Among other activities, the program works with governments of the world to channel resources, identify directions for global research, and implement educational initiatives. Most recently, UNAIDS has become the chief advocate for worldwide action against AIDS. It has brought together 10 organizations of the United Nations system, including the WHO, around a common agenda on AIDS. UNAIDS has established a global database on the rates of HIV infection based on regions and countries. From their database, some good news is emerging. Since 2004 (39.4 million worldwide infections), the global rate of HIV infection appears to be leveling off with 34 million living with HIV worldwide at the end of 2011 (**Figure 1.6**).

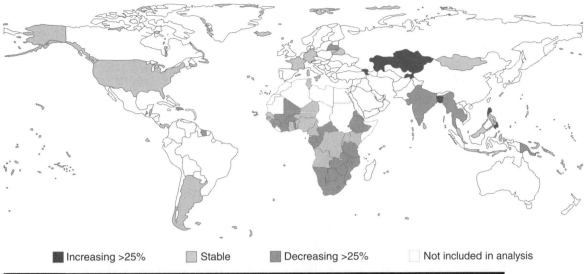

Increasing >25%   Stable   Decreasing >25%   Not included in analysis

FIGURE 1.6

Map illustrating changes in HIV infection 2001-2009. Overall, the AIDS pandemic appears to be stabilizing. In other words, the number of new infections occurring each year is declining. A few areas in the world are hotspots in which the number of HIV infections is increasing dramatically (e.g., Eastern Europe and Central Asia). Modified from 2010 Global Report: UNAIDS Report on the Global AIDS Epidemic 2010 Jointe United Nations Programme on HIV/AIDS (UNAIDS).

Origin of the AIDS Epidemic

Throughout recorded history, epidemics have sprung up to ravage human populations. In the influenza epidemic of 1918–1919, for example, an estimated 20 to 40 million people died worldwide, a number comparable to mortality rates during the infamous Black Death in Europe in the 1300s. Although the source of the influenza epidemic was never found, other outbreaks of disease have defined origins. For instance, when the Spanish arrived in the Americas in 1520, they brought smallpox with them. The Aztec population of Mexico was without any immunity to this disease, and the raging smallpox epidemic that followed reduced that population by half. The sources of some other epidemics are also clear. When the first European navigators reached South Pacific islands in the 1700s, they found the people robust, happy, and well adapted to their environment. The explorers introduced syphilis, tuberculosis, and whooping cough (pertussis), however, and these diseases swept through the population virtually unchecked. Hawaii was struck especially hard. When Captain Cook landed there in 1778, the island's population was about 300,000; by 1860, it had been reduced to fewer than 37,000.

These examples are somewhat reminiscent of the course of AIDS during the twentieth and twenty-first centuries. The disease seemed to burst on the scene during the early 1980s and spread out from central foci in major cities of the United States and Africa and other parts of the world. Setting aside claims that HIV is an entirely new virus or the product of a genetic engineering laboratory, the prevailing hypothesis among scientists is that the virus existed somewhere in nature and emerged at this particular period in history to infect human populations. In this section, we explore where it may have originated and why it came forth at that particular time.

The Origin of HIV

If we assume that HIV existed somewhere in the world prior to its appearance in humans, then the next logical question is, "Where?" During the mid 1980s, Max Essex of Harvard's School of Public Health was among the first to provide evidence that HIV originated in African primates such as monkeys, baboons, and chimpanzees. Essex's research resulted in recovery of a virus called the simian immunodeficiency virus (SIV) from African green monkeys (**Figure 1.7a**). This virus had enough biochemical similarities to HIV to support the notion that the two viruses are related. Essex and his group theorized that SIV crossed the species barrier to humans through a primate bite or scratch.

Further evidence for a primate–human transmission was offered by evidence that HIV-2 crossed the species barrier to humans from the sooty mangabey, a type of monkey belonging to the species *Cercocebus atys* (**Figure 1.7b**). Vanessa Hirsh, a primate researcher at the National Institute of Allergy and Infectious Diseases, established this link in the early 1990s. Hirsh demonstrated that the SIV_{sm}, the SIV strain from the sooty mangabey, has genes nearly identical to those of HIV-2. Research performed by British scientists indicated that SIV_{sm} from the sooty mangabey monkey might have caused six HIV-2 epidemics, which blended together to give the impression of a single West African epidemic.

The research of Essex and Hirsh was expanded and further developed by a host of other scientists, notably the group led by Beatrice Hahn of the University of Alabama at Birmingham. In 1999, Hahn and her coworkers pieced together what was hailed as the best case yet for connecting human HIV to chimpanzees (**Figure 1.7c**). Their research indicates that different subspecies of chimpanzees harbor different strains of HIV-like viruses and that one particular subspecies is the probable source of the HIV-1 (the "original" HIV) that is causing the current pandemic.

(a) (b) (c)

FIGURE 1.7

(a) The African green monkey. The simian immunodeficiency virus (SIV), a virus similar to the human immunodeficiency virus, was recovered from such an animal in the mid 1980s. © Francois Etienne du Plessis/ShutterStock, Inc. (b) The sooty mangabey monkey, believed to be the source of HIV-2 in humans. © Jan van der Hoeven/ShutterStock, Inc. (c) The chimpanzee *Pan troglodytes troglodytes*, which is believed to be the source of HIV-1 in humans. © Timothy E. Goodwin/ShutterStock, Inc.

Before Hahn's work, scientists had identified only three chimpanzees infected with SIVs. The strains of SIV in these animals were designated SIV_{cpz}. Then Hahn's group identified a fourth chimpanzee infected with a strain of SIV_{cpz}. The researchers began an exhaustive and systematic study to examine these SIV strains carefully and the four animals from which they came. The study focused on the nucleic acid of the virus, one of its major chemical components and the substance of which its genes are made. Using sophisticated genetic analyses, Hahn's group compared the nucleic acid content (the genes) of the four SIV_{cpz} strains with that of various samples of HIV obtained from humans.

The results of the research were the key to the origin of HIV: Three of the four SIV_{cpz} strains were found to be closely related (i.e., 70% to 90% genetically identical) to HIV particles isolated from humans. All three strains came from the chimpanzee subspecies *Pan troglodytes troglodytes*. Significantly, the natural range of the chimpanzee *Pan troglodytes troglodytes* coincides precisely with regions of West Africa where the AIDS epidemic has existed for the longest period of time. The fourth SIV_{cpz} strain had a nucleic acid content (genes) much less similar to that of HIV; importantly, it came from the chimpanzee subspecies *Pan troglodytes schweinfurthii* normally found in far-distant East Africa. Hahn's group concluded that chimpanzees living in Gabon, Cameroon, and nearby regions of West Africa were the source of HIV-1 in humans. They surmised that once the virus made its way into humans, it mutated into the existing strains scientists now recognize, while adapting naturally to its new host. We discuss when this crossover presumably happened in the next section.

Hahn's research has several key implications, not the least of which is that it apparently resolves the question of HIV's origin. Moreover, the three SIV strains found in *Pan troglodytes troglodytes* are the presumed forerunners of the three groups of HIV-1 now known to exist. This finding indicates that chimpanzees have been the starting point for at least three independent crossings to humans. It is reasonable to believe that these cross-species transmissions occurred when humans who hunted and butchered chimpanzees for meat were exposed to their blood via a wound or scratch. Moreover, additional transmissions may have occurred and may still be occurring because the hunting and killing of chimpanzees continues in western equatorial Africa.

The significance of the research is further underscored by the observation that chimpanzees in the wild apparently do not develop AIDS, even though the genetic material of humans and chimpanzees is 98.5% identical. Thus, investigators hope to locate a resistance mechanism in the animals that can be applied to humans; however, strengthening the human–chimpanzee link will require finding SIV_{cpz} in wild chimpanzee populations. To do that, a concerted effort will be launched to save these endangered animals for study.

To study the human–primate link further, researchers began to try to determine whether SIV exists in wild baboons and monkeys of varying species. Results reported in 2001 strengthened that link. Investigators led by Beatrice Hahn and Eric Delaporte collected blood samples from 384 wild primates representing 17 species. They found that 18% of the primates harbor SIV antibodies that bind strongly to HIV proteins. The finding indicates that other variants of SIV occur in wild primates and are capable of making the crossover to humans. A collaborative 2010 study led by Preston Marx reported that than more than 40 species of African nonhuman primates were infected by SIV for more than 32,000 years. Thus, for millennia, many subtypes of SIV could continue to pose risks for humans during sporadic encounters as these animals are hunted for meat or kept as pets.

The Jump to Humans

Scientists now generally agree that a strain of SIV made the crossing from chimpanzees to humans and later evolved to the current strains of HIV. Exactly when the crossing occurred is uncertain, but an estimate appears to be provided by analysis of the HIV obtained from the oldest documented case of infection. The analysis shows that the passage to humans probably occurred during the late 1940s or early 1950s.

The research leading to this conclusion was performed by Toufo Zhu of the University of Washington and David Ho of New York's Aaron Diamond AIDS Research Center. The scientists began with the insight that about 1% of HIV's genetic material mutates each year. Then they analyzed HIV fragments obtained from the blood of a Bantu man who lived in what is now Kinshasa in the Democratic Republic of the Congo. The blood had been drawn in 1959, shortly before the man died.

Although evidence of HIV had been observed in the man's blood in 1986 by Max Essex, the methods for studying HIV's genetic material were not then sophisticated enough to gain much information; but in 1998, Zhu and Ho used state-of-the-art technology to multiply the genetic material of HIV available in the blood many millions of times, thus yielding a sufficient supply to work with. Then, using extensive databases and advanced computers, they compared the genetic material from the 1959 HIV to genetic material from current strains of HIV as well as strains of HIV isolated during many intervening years. The comparisons yielded a "family tree" of HIV, showing the progression of changes occurring in its genetic material over a 39-year period. The data indicated that HIV probably entered the human population after World War II had ended, as noted previously here. This estimate is important because it renders unlikely a hypothesis that the AIDS epidemic originated in the 1950s from contaminated lots of polio vaccine (**Box 1.2**).

Moreover, the finding is more than a historical footnote. Comparing the 1959 virus with modern forms yields information on how extensively HIV has evolved over the decades and how much it can be expected to evolve in future years. These data helped in linking HIV to SIV in chimpanzees because Hahn's group could compare the genes of the 1959 virus with those of SIV and search for common features. In addition, identifying when HIV entered the human population puts a time frame on the start of the AIDS epidemic, and by examining social trends of the period, public health officials can understand why the epidemic exploded as rapidly as it did (which we discuss later here).

Finally, knowing about the early form of the virus could conceivably help scientists pinpoint the parts of HIV's genetic material that have changed the least, a finding that would assist vaccine researchers. One problem in HIV vaccine research is how to produce a vaccine that can be effective against the multiple strains of HIV known to exist. A vaccine based on features common between ancestral and modern forms might prove more universal in fighting a global epidemic than a vaccine based on a combination of modern types.

Genetic analyses like this one are used to calculate the dates when strains of an organism mutate and split off from the ancestral strain. At best, this analysis gives an estimate of when the HIV lineage began to diversify, not necessarily when the virus was transmitted to humans. An example of the uncertainty of the data is illustrated by research reported in 2000. A genetic analysis performed by scientists indicates that the M group of HIV (the group that causes the vast majority of AIDS cases) came into existence in 1931, with a 95% confidence interval of 1915 to 1941. Although this

BOX
1.2

Controversy Over the Origin of HIV

The controversy apparently began with a 1992 article in *Rolling Stone* magazine by journalist Tom Curtis and gathered momentum with a 1999 book by science writer Edward Hooper entitled, *The River: A Journey to the Source of HIV and AIDS*. Both Curtis and Hooper theorized that the AIDS epidemic was ignited when an oral polio vaccine was tested in the late 1950s on hundreds of volunteers in what is now the Democratic Republic of the Congo. Hooper contended that the vaccine was contaminated with simian immunodeficiency virus (SIV), which infects monkeys, chimpanzees, and other primates. The contamination occurred, according to Hooper's research, when kidney cells from infected chimpanzees were used to cultivate the poliovirus. In the book, he posits that the earliest cases of AIDS occurred where and when the vaccine was tested.

One of the leaders of the 1950s campaign against polio was Hilary Koprowski from the Wistar Institute in Philadelphia. Answering the hypotheses of Curtis and Hooper, Koprowski maintained that to make the polio vaccine his group had used cells from Asian macaque monkeys rather than chimpanzees. He pointed out that samples of the vaccine were preserved in freezers at the Institute and suggested that they be analyzed.

Three independent laboratories were hired to make the analyses. One laboratory analyzed the DNA in a vaccine sample to determine whether the DNA was from a macaque monkey or a chimpanzee—the results indicated that it was from a macaque monkey. A second laboratory tested the vaccine sample for genes associated with either HIV or SIV—genes for neither virus were found. A third laboratory ran tests duplicating those performed by the other two labs—its results confirmed theirs. Augmenting these results were testimonials from 16 of Koprowski's collaborators who stated that they never worked with chimpanzee cells. Moreover, evidence from genetic analyses indicates an origin of HIV in humans that predates the 1950s. These analyses also show that chimpanzees from the region where Koprowski worked do not harbor the SIV that was the ancestor of HIV.

In his book, Hooper made the point of calling for analysis of the archived vaccine samples, as Koprowski offered; however, when the results were reported, and even though they countered his hypothesis, Hooper refused to retract his theory. In contrast, *Rolling Stone* printed a "clarification" in 1993, and Koprowski won a lawsuit against the magazine claiming defamation of character.

period broadly coincides with the dates deduced in earlier analyses, a slight deviation exists. For this reason, it is important to avoid broad generalizations until the research results have been verified.

Out of Africa

Working on the assumption that HIV made the jump from chimpanzees to Africans somewhere between the 1930s and early 1950s, the next question is this: Why did AIDS become such an explosive pandemic in the following years?

No one knows for sure, but historians point out that the 1960s and 1970s were decades of great turmoil in Africa when the epidemic could have easily gathered momentum. Civil wars were commonplace. National boundaries were redrawn, and population shifts were unprecedented. Demographers note that millions of people in Africa changed from an agrarian lifestyle to an urban lifestyle. It is conceivable that infected people moved from remote areas to the cities and brought the virus to these population centers. Sexual intercourse could then have spread the virus rapidly, especially because those decades saw the emergence of a sexual revolution. Given the roughly 10-year incubation period for the disease, an outbreak in the 1980s appears reasonable. Moreover, a 1988 study indicated that the number of blood samples testing positive for HIV in a city in Zaire increased tenfold in a 10-year span.

Another possibility is that soldiers from Caribbean countries serving as mercenaries in the civil wars could have acquired the virus from remote African populations. The virus may then have made its way to the United States via a Caribbean city. Many people from Haiti, for example, were known to suffer from AIDS early in the epidemic. There is also the possibility that contaminated blood used for transfusions could have been the source of HIV, either by administration to travelers passing through the region or by exportation to other countries.

Most discussions of Africa as the origin of the AIDS virus are bitterly resented by Africans, who do not wish their continent to be thought responsible for the AIDS epidemic. Such discussions have made many African government officials suspicious of foreign scientists and relief organizations and, in some cases, have impeded the flow of information from that continent. To ease tensions, the World Health Assembly passed a 1987 resolution stating that HIV is a "naturally occurring [virus] of undetermined geographic origin." Writing in 1989, Robert Gallo attempted to diffuse the sensitivity further by suggesting that "tracing the origins of the virus to a particular location doesn't imply blame—we don't blame Lyme, Connecticut for Lyme disease."

Entry into the United States

The AIDS virus was in the United States well before the first cases of AIDS were identified in 1981. The case of a Missouri boy points up this fact (**Box 1.3**), as does evidence from blood samples frozen and stored in the late 1970s. During that period, more than 7,000 homosexual men came to public health clinics in San Francisco and New York to participate in a government-sponsored study of hepatitis B. They donated blood samples, and researchers preserved many of the vials for study at a later date. A decade passed before the vials of blood were thawed and tested for evidence of HIV. The results reported in 1987 revealed some interesting patterns: Of 6,700 participants in the San Francisco area, at least 70% had blood samples positive for exposure to HIV; more than 600 of the participants had AIDS, and about 400 of the 600 had already died.

The results provided substantial evidence that HIV was in the United States well before the first cases of AIDS showed up in 1981. Moreover, the results revealed how rapidly the disease was spreading among men who have sex with men: Of blood samples donated in 1978, 3% had evidence of exposure to HIV; of those donated in 1979, 12% showed HIV exposure, and by 1981, 45% were HIV positive. It was clear that the epidemic was gathering steam even before public health agencies recognized its existence.

How the AIDS virus entered the United States is still unknown. One possible source is contaminated blood imported from another country (**Figure 1.8**). Another possibility is sexual contact by Americans with foreigners harboring HIV. Sharing contaminated syringes and needles with infected injection drug users is a third possibility.

Another possible mode of entry into the United States may involve Haiti. Scientists have theorized that cultural exchanges between Zaire and Haiti, both French-speaking countries, offered possibilities for transporting the virus from Africa. American homosexual men vacationing in Haiti could then have acquired the virus and brought it to the United States. This view is supported by the observation that

BOX
1.3

A Much Longer Prologue

Although the "official" beginning of the AIDS epidemic in the United States is considered to be June 1981, there is evidence that AIDS may have made its appearance as early as 1969. The basis for this belief is a report published in October 1987. In the report, researchers related the case of Robert R., a 16-year-old boy from Missouri, who may have died of AIDS-related causes in 1969.

The researchers told of the teenager who came to St. Louis City Hospital in 1968 with substantial swelling of the legs and genital organs and an impaired immune system. A bacterial disease named Chlamydia was diagnosed, and Robert was put on a regimen of antibiotics; however, the drugs did not work. In the ensuing weeks, his muscles wasted away. His lungs filled with fluid (an unmistakable sign of pneumonia), and he died in May 1969. During the subsequent autopsy, pathologists found in his organs the purplish lesions of Kaposi's sarcoma, a type of cancer now known to be related to AIDS.

Unable to locate a microbial cause for Robert's death, doctors froze samples of his blood and tissues, hoping that someday the mystery could be solved. In June 1986, the samples were sent to Tulane University for analysis for HIV. Virologists at the university found that Robert R.'s blood reacted with all nine chemical markers for HIV. There was little question that the virus had been present in the boy's blood.

Is it possible that the AIDS virus may have been in the United States in a less lethal form in the 1960s? Perhaps so, believe some researchers, but they also question where that virus is today. Did all of that less lethal virus simultaneously mutate to a more lethal form? Not likely is the opinion of most. What, then, would explain the more than 10-year lapse before the epidemic's breakout? Thus far there is no answer.

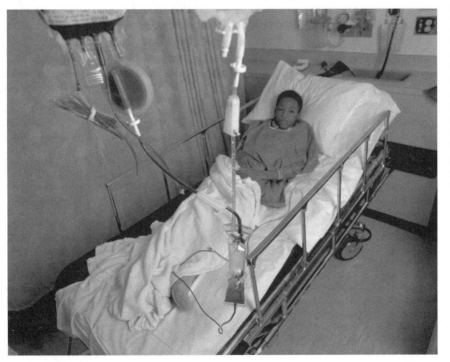

FIGURE 1.8
Although how AIDS entered the United States remains unresolved, one possibility is that the virus was present in contaminated blood imported from abroad. Transfusion of this blood to recipients may have provided a mode of transmission. The young boy pictured here has received hundreds of transfusions since birth to treat a blood disease called Mediterranean anemia.
© Phototake, Inc./Alamy Images.

Mapping HIV

United States

The annual number of new HIV infections has stabilized.

- 1.2 million people are living with HIV.
- 1 in 5 (20%) are unaware of their infection.
- ~50,000 new infections/year.
- ~13 children under the age of 13 were diagnosed in 2009.
- ~60% of new infections occur in gay and bisexual men.
- 20-24 years of age group has the highest number of new diagnoses (6,237 in 2009).
- By race, blacks/African Americans face the most severe burden of HIV (10,800 MSM* new infections in 2009).

- 1.4% of the prison population lives with HIV or AIDS (21,987 in 2008).
- HIV/AIDS diagnosis in seniors (persons over 50 years of age) is increasing.
- Florida has the highest number of people living with HIV: 90,909 in 2009.
- States with the highest total number of people diagnosed with HIV (2009) are: New York, California, Florida.
- States with the least total number of people diagnosed with HIV (2009) are: North Dakota, Wyoming, Montana.

*MSM: gay, bisexual, and other men who have sex with men
Data from: *CDC HIV/AIDS Surveillance Report*, vol. 22, 2010.

AIDS was present in Haitians as well as in American homosexual men in the 1980s; however, AIDS may have been introduced to Haiti by North Americans. The evidence that American men were exposed to AIDS before Haitians bolsters this view. Rather than a source, Haiti may have been a recipient of the virus. This question was finally answered in 2010 by Michael Woroby's collaborative research.

By 2010, technology was available to create viral gene trees to further investigate the hypotheses as to how HIV/AIDS began its world travels into the Americas. Teams led by Woroby conducted a molecular analysis on archived blood samples (collected in 1982–1983) from five Haitian AIDS patients. All of the patients recently immigrated to the United States and were among the first recognized as AIDS victims. Their genetic data suggests that HIV came out of Africa, then to Haiti in 1966, and subsequently to the United States. Most of the circulating viruses can be traced to a viral relative that came from Haiti around 1969. This explains why there were early observations of a high prevalence of AIDS in Haiti. Haiti became the popular sex tourism destination for gay men in the mid-1970s as wells as the stepping-stone that allowed HIV-1 to flourish and spread to the United States and other countries. The current HIV status in the United States is in the **Mapping HIV box.**

The First Decades and Beyond

In late 2000, health officials from the UNAIDS made a startling prediction: They announced that if current trends continued, deaths associated with AIDS could exceed those of the Great Plague of the 1300s (the Black Death) and the catastrophic worldwide influenza epidemic of 1918–1919. AIDS would then become the deadliest contagion in world history. As of 2006, estimates are that nearly 25 million people have died from AIDS since 1985. In 2007, nearly 2.5 million died from AIDS and that annual number seems to have stabilized, suggesting that in the next 5 to 10 years, HIV may very well have claimed more lives than any other single infectious agent.

Thirty-Plus Years of AIDS

In the more than 30 years since *MMWR* carried the first report of pneumocystosis in immune-deficient individuals, the statistics attending the AIDS epidemic have reflected the emergence of a major health crisis. One of the first notable benchmarks was reached in the August 18, 1989 issue of *MMWR,* when the CDC reported that more than 100,000 cases of AIDS had been diagnosed in the United States through July 1989. AIDS-related deaths had reached about 60,000. The CDC also pointed out that the first 50,000 cases of AIDS were reported from 1981 to late 1987 (six years) and the second 50,000 cases between December 1987 and July 1989 (1.5 years). AIDS had become a major cause of illness and death among children and young adults in the United States; in 1988, it ranked 15th among the leading causes of death for Americans.

By 1989, homosexual and bisexual men were still accounting for most reported AIDS cases, but injection drug users, their sexual contacts, and their children represented an increasing proportion of all cases. For example, 63% of AIDS cases occurred in homosexual men before 1985, but only 56% of new cases did in 1989. In contrast, injection drug users accounted for 18% of cases in 1985, but this percentage rose to 23% by 1989. The number of women involved was also rising: Before 1985, only 7% of AIDS cases were in women, but in 1989, the number rose to 11%.

In the January 25, 1991 issue of *MMWR,* the CDC updated its statistics. By that time, more than 100,000 persons had died of AIDS (compared with 100,000 cases in 1989). Among American men aged 25 to 44 years, AIDS had become the second leading cause of death, surpassing heart disease, cancer, and suicide. For women aged 25 to 44, AIDS was projected to be among the top five causes of death. By 1995, the prediction had become reality, and AIDS was the leading cause of death in women in 15 U.S. cities.

Things began to change in 1996, however. That year, new optimism surfaced with the news that protease inhibitors, a class of drugs that complement **azidothymidine (AZT)**, could be used to reduce blood concentrations of HIV to undetectable levels. Also that year, investigators uncovered striking new clues about how HIV binds to host cells—clues that kindled hopes for new therapies, especially because a genetic defect that prohibits binding apparently makes some individuals immune to AIDS. These advances in therapeutic and basic research were augmented by development of a new test—the viral load test—which uses state-of-the-art biotechnology to measure precisely the number of HIV particles in a sample of blood. This direct measurement of infection is far superior to measurements of host cell destruction for gauging a patient's response to drugs or predicting long-term survival. The optimistic mood of 1996 marked a turning point in the battle against AIDS and persuaded the editors of *Science* magazine to name the collection of advances as the Breakthrough of the Year. 1997 was the last year-to-date that HIV/AIDS appeared in the top 15 causes of deaths in Americans.

By December 1999, the AIDS epidemic had been affecting the United States for over two decades. As the world celebrated the closing of the twentieth century, the CDC reported that there had been a total of 733,374 AIDS cases in the United States since the first case were recorded in 1981. Of this total, 81% of cases had occurred in men, 18% in women, and 1% in children. Statisticians noted that the epidemic was

Cases of AIDS diagnosed in newborns in the United States reported quarterly for the 1985–1999 period. The drop-off starting in 1994 (arrow) is related to the detection of HIV in pregnant women and the use of AZT to interrupt the replication of HIV. Modified from CDC, *Morbidity and Mortality Weekly Report*, vol. 48/RR-13, Dec. 10, 1999.

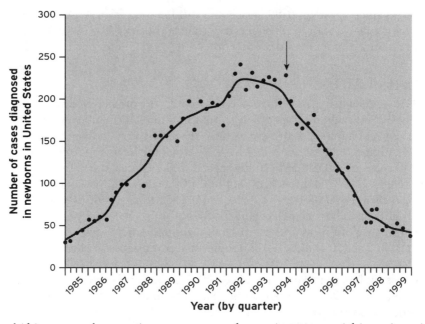

shifting toward a growing percentage of cases in women, African Americans, and Hispanic Americans, and that a decreasing percentage was occurring in homosexual men. Despite this statistic, homosexual men continue to remain the largest single exposure group. In addition to the confirmed cases, almost a million individuals were living with HIV in 1999 but remained undiagnosed. Over 425,000 had died of AIDS-associated illnesses.

In the United States, the good news at the end of the century was the steep decline in AIDS in newborns. AIDS cases in newborns reached a peak in 1992, leveled off during 1993, and then began a dramatic drop-off in 1994 that continues today, as **Figure 1.9** shows. The most apparent reason was improved diagnosis of HIV infection in pregnant women combined with rapid implementation of therapy with AZT to prevent transmission from mother to child. AZT interferes with the replication of HIV in infected cells. Improved treatment of HIV-infected newborns is another notable reason for the decrease because the treatment delays the progression from HIV infection to AIDS.

In the rest of the world at the end of the century, however, the news was much grimmer. Worldwide, the UNAIDS reported in 2007 that over 36 million people were living with HIV or experiencing AIDS and that an astounding 4.1 million had become infected that year. Already, over 25 million individuals have died from AIDS-related causes since the beginning of the pandemic, with roughly 2.4 million deaths in 2007 alone. Roughly 67% of all individuals who are infected with HIV are from sub-Saharan Africa.

As the new century opened, the AIDS crisis in the United States had lessened. To some observers, the disease had practically dropped off the proverbial radar screen. Many people infected with HIV were living almost normal lives thanks to a sophisticated combination of anti-HIV drugs. Moreover, the infection rate had leveled off and even declined for some groups (although it was rising for others), and the public was becoming somewhat complacent about the epidemic. Public health officials warned that, despite the level rate of approximately 35,000 new infections per year in 2006, in the foreseeable future, AIDS will continue to affect homosexual men while becoming more prevalent among poor African-American and Hispanic-American heterosexuals living in the inner cities. Contributing to this prediction is the observation that the rate at which homosexual men contract AIDS has reached a plateau, whereas the rate has risen dramatically among injection drug users, particularly those of minority groups. In New York City, for instance, in a recent year, four of five AIDS cases attributed to sharing needles occurred among African-American and Hispanic-American individuals. Public health officials stress that fighting the spread of AIDS among drug addicts and their sex partners will require significant increases in drug treatment programs as well as education programs that target poor urban areas.

As we have noted, the situation was much different in Africa: AIDS was threatening to be more devastating than civil wars and famines put together. Each day at the turn of the century, between 6,000 and 7,000 people in sub-Saharan Africa were dying of the effects of AIDS, and public health scientists estimated that HIV had infected over 35 million adults. Those numbers remain high, with roughly 5,500 deaths daily in 2007 in sub-Saharan Africa. This death rate is drastically affecting life expectancy in sub-Saharan Africa, and **Figure 1.10** shows some of the alarming projections related to the AIDS epidemic in 2009.

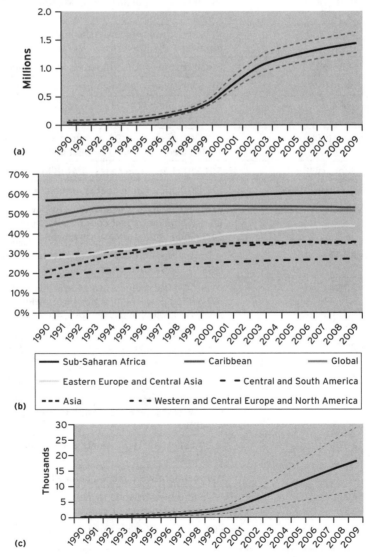

FIGURE 1.10

The devastating long-term effects of HIV/AIDS. (a) The number of people living with HIV in Eastern Europe and Central Asia is increasing. (b) Trends in women living with HIV. (c) The number of children living with HIV in Eastern Europe and Central Asia is increasing. Data from 2010 Global Report: UNAIDS Report on the Global AIDS Epidemic 2010 Jointe United Nations Programme on HIV/AIDS (UNAIDS).

In the early 1980s, AIDS was an unknown entity, shrouded in uncertainty and conjecture, but the disease began to reveal itself through the mid and late 1980s when the scientific base of information expanded (**Table 1.1**). We do not understand major aspects of the interaction of HIV with the infected individual, nor do we fully comprehend the nature of the host response.

TABLE 1.1	A Brief Chronology of the AIDS Epidemic
June 5, 1981	Five cases of *Pneumocystis* pneumonia reported in the CDC's *MMWR*
December 1981	*Pneumocystis* pneumonia reported in New York City drug addicts
June 11, 1982	365 cases of immune deficiency reported from five states in United States
September 3, 1982	The name acquired immune deficiency syndrome (AIDS) coined
Early 1983	16 countries reporting AIDS; 34 states in United States with AIDS cases
April 1983	Research groups led by Gallo and Montagnier independently report discovery of AIDS virus
May 1985	Blood test for AIDS antibody made available
August 1986	AIDS virus named human immunodeficiency virus (HIV)
1987	Azidothymidine (AZT) licensed for use in AIDS patients; AIDS vaccine research ongoing
July 1989	100,000 cases of AIDS in United States; 60,000 deaths in United States from AIDS and complications
January 1991	501,310 cumulative cases of AIDS in United States; 311,381 deaths
October 31, 1995	100,000 cumulative deaths associated with AIDS in United States
1996	Protease inhibitors licensed for use in United States; new discoveries on HIV binding to host cells; new HIV detection tests approved; declining number of cases in newborns
2003	AIDSvax, the first AIDS vaccine tested in a large number of people, fails to show protection
2004	Over 500,000 cumulative deaths due to AIDS in the United States; 4.9 million new HIV infections worldwide; 3.1 million deaths due to AIDS worldwide; approximately 40 million people living with HIV/AIDS worldwide
2006	Over 1.0 million cumulative cases of HIV/AIDS in the United States, and over 5 million total deaths due to AIDS in the United States
2007	MRK-Ad5, an HIV/AIDS vaccine, fails in phase III trials
2009	Gero Hütter and colleagues published the results of a rare bone-marrow transplantation that has apparently cured an HIV-positive man of AIDS
2009	In 33 countries, HIV incidence fell by more than 25% between 2001 and 2009
2010	Approximately 6.6 million people have access to AIDS medications
2012	FDA approves Truvada to use as a preventative measure for people who are at high risk of acquiring HIV through sexual activity
2012	FDA panel backs first rapid, over-the-counter home HIV test

Now, in the first decades of the twenty-first century, our vision of AIDS is clearer, although far from perfectly in focus. The first three decades of research brought hopes of great advances but no effective vaccine. But our society's ability to alter risk-taking behaviors is still limited. In 2012, the U.S. Food and Drug Administration (FDA) approved the antiviral **Truvada** to use as a preventative measure for people who are at high risk of acquiring HIV through sexual activity. HIV testing is one of the most important means to slow new infections. In 2012, a 17-member FDA board unanimously recommended approval of the first, rapid, over-the-counter **OraQuick in-home HIV test** for sale in pharmacy stores and online to anyone age 17 and older. A single use test costs $39.99. It uses a sample of fluid from the mouth for testing. Results are obtained within 20–40 minutes but additional testing should be done in a medical setting to confirm a positive result.

In 2009, the first HIV positive patient was cured of AIDS by a rare bone-marrow transplantation. Two more similar cases of "cured" patients were presented at the International AIDS conference held in Washington D.C. in 2012. Also, in 2012, the cancer drug **vorinostat** was reported to target dormant HIV to come out of hiding or dormancy, potentially allowing antivirals to combat the viral infection.

According to the 2011 *UNAIDS World AIDS Day Report*, we are on the verge of a significant breakthrough in the AIDS response. New HIV infections are falling and more people are starting treatment. The number of people dying of AIDS-related illnesses fell to 1.8 million in 2010 (down from 2.2 million in 2005). Only a few countries do not fit this overall trend. Five of these countries are in Eastern Europe and Central Asia. AIDS incidence has increased by more than 25% between 2001 and 2009 in these countries. This incidence is correlated with intravenous drug use, especially use of homemade heroin.

Wealthier nations must continue to make important scientific and medical strides toward understanding the biology of HIV and preventing AIDS and, second, then deliver this information, medicine, and biotechnology to the regions of the world most affected by HIV. Finally, those populations at particular risk for increasing rates of HIV infection—intravenous drug injectors, the economically underprivileged, and women—must be the focus of education, liberation, medical care, and preventative medicine. Public health officials wonder what hope the poorer nations have and what the economic impact of the pandemic will be. Can the rise of the relative infection rate in women be brought to a plateau or reduced? Can the United States keep HIV from broadening its net? As you read the pages ahead, try to bear in mind that the history of AIDS is being written minute by minute. Answers to some of these questions are probably emerging as you read.

LOOKING BACK

In the early 1980s, AIDS was compared with an elephant being examined by blind men because scientists and public health officials were ignorant of the disease's cause, transmission, symptoms, and other characteristics. Reports beginning in June 1981 made it clear that suppression of the immune system was occurring in homosexual men and that diseases not ordinarily considered dangerous had led to death in many of these men. Similar symptoms were later observed in injection drug users, blood transfusion recipients, hemophiliacs, heterosexual individuals, and newborns.

Many theories purporting to explain the cause of AIDS existed in the early 1980s, but in April 1984, a virus was identified as the agent. The discoverers were members of a group headed by Luc Montagnier of France. Though initially called HTLV-III/LAV, the virus was renamed human immunodeficiency virus (HIV) in 1986 on the recommendation of an international commission. Another AIDS virus called HIV-2 has been identified in people from West Africa. The disease it causes seems to be milder than that caused by the original HIV, now known as HIV-1.

AIDS was recognized as a global problem as early as 1985. The African continent has been particularly hard-hit, with high infection rates in sub-Saharan Africa. Researchers believe that HIV-1 originated in chimpanzees of the subspecies *Pan troglodytes troglodytes* and was probably passed to humans via a scratch or bite from one of these animals. The crossover to humans occurred at least three times, probably in the late 1940s or early 1950s, although some dispute remains about the exact timing. The possibility that HIV was introduced to human populations through tests of polio vaccines has been discounted. Decades of turmoil in Africa encouraged the virus to spread to the remaining world.

By the beginning of the twenty-first century, more than 700,000 cases of AIDS had been reported in the United States, but there was hope of forestalling the epidemic because a cocktail of drugs used since 1996 was shown to decrease the number of deaths associated with AIDS. Moreover, studies indicate that when azidothymodine (AZT) is used in HIV-infected pregnant women, the transmission of HIV to their newborns can be interrupted. Furthermore, the percentage of AIDS cases in sexually active homosexual men was dropping, but the percentage in injection drug users and women was rising. The AIDS elephant is becoming visible after two decades of discoveries, but some of its features, such as its susceptibility to a preventative vaccine, remain obscure.

For the first time in 22 years, the world's largest international AIDS conference convened in Washington D.C. in July 2012. This was possible because the travel ban against HIV-positive people from entering the United States had been lifted by President Barack Obama in 2010. The main worrisome issues confronting the AIDS epidemic included: a global economic recession that which threatened the dollars needed to invest in science, medicine, and education to fight AIDS; getting medication to the poor (including in the United States); treating more pregnant women so they won't spread HIV to their babies; and ways to urge the public and policy makers to pay attention to the disease. The theme of the conference, *Turning the Tide Together*, captured the idea that we now have the potential to change the course of HIV and AIDS if we can scale up intervention and continue the scientific momentum for a cure and vaccine.

Healthline Q & A

Q1 Exactly what is AIDS?

A Defining AIDS can be difficult because AIDS is a complex disease. Basically, AIDS is an infectious disease in which a virus attacks cells of the immune system and thereby renders the body susceptible to microorganisms that otherwise would be held in check. Symptoms of AIDS can be very general (fever, weight loss, night sweats, and diarrhea), or they can be more specific (pneumonia and skin cancer). Multiple stages of AIDS are recognized, beginning with infection by the virus, progressing to "early AIDS," and culminating with full-blown AIDS.

Q2 I've heard that AIDS is caused by a virus. What is that?

A A virus is one of the smallest microorganisms known. It consists of a segment of nucleic acid (such as DNA or RNA) surrounded by a coat of protein and, in some cases, an enclosing membrane. Viruses do not grow, use food, or perform any metabolic processes associated with living things. Inside cells, however, they multiply very efficiently.

Q3 Exactly what is the immune system?

A The immune system is a network of cells and a group of chemical compounds largely responsible for the body's defense against infectious disease, cancer, and other maladies. The system is distributed throughout the body, primarily in the spleen and the lymph nodes of the neck, armpits, and groin. Certain cells of the immune system produce highly specific proteins known as antibodies, which interact with and neutralize microorganisms. Other immune system cells attack microorganisms directly.

REVIEW

This chapter explored the development of the AIDS epidemic during two decades, touching on its origin, method of spread, and magnitude. To test your knowledge of these concepts, consider the following statements. Write T for "true" if a statement is correct as it stands. If the statement is false, change the underlined word or phrase to correct the statement. The correct answers are listed elsewhere in the text.

_____ 1. Although the agent responsible for AIDS was probably in the United States during the 1970s, the "official" beginning of the AIDS epidemic in the United States occurred in <u>1991</u>.

_____ 2. Two diseases that physicians commonly observe in AIDS patients are *Pneumocystis* pneumonia and a form of skin cancer known as <u>adenocarcinoma</u>.

_____ 3. It is now generally recognized that AIDS is caused by a *virus*.

_____ 4. One of the important modes of transmission for the human immunodeficiency virus (HIV) is contaminated <u>food</u>.

_____ 5. AIDS is a disease that occurs in homosexual men as well as in <u>injection drug users</u>.

_____ 6. In France, the research team that confirmed the identity of the AIDS virus was led by <u>Bernard Schwartlander</u>.

_____ 7. The AIDS virus, originally designated <u>HTLV-III/LAV</u>, is now known by the acronym HIV.

_____ 8. Since the introduction of blood tests for AIDS, the number of transfusion-linked cases of AIDS has <u>increased</u> sharply.

_____ 9. Because of its global involvement, the AIDS epidemic is more properly termed an AIDS <u>endemic</u>.

_____ 10. By 1989, the number of cases of AIDS diagnosed in the United States had passed <u>1 million</u>.

_____ 11. A global program to combat the spread of AIDS has been established by the <u>World Health Organization</u>.

_____ 12. One theory relating to the origin of the AIDS virus suggests that it may have existed in African <u>birds</u> before being transmitted to humans.

13. The form of AIDS caused by HIV-2 appears to be <u>deadlier</u> than the form caused by HIV-1.

14. The current research thinking is that epidemics of AIDS due to HIV-2 originated from SIV in the <u>sooty mangabey monkey</u>.

15. At the beginning of the twenty-first century, the AIDS pandemic was particularly severe on the continent of <u>Africa</u>.

FOR ADDITIONAL READING

Chahroudi, A., et al., 2012. "Natural SIV hosts: Showing AIDS the door." *Science* 335:1188–1193.

Gallo, R. C., 2002. "Historical essay: The early years of HIV/AIDS." *Science* 298:1728–1730.

Gilbert, M. T. P., et al., 2007. The emergence of HIV/AIDS in the Americas and beyond. *PNAS* 104:18566–18570.

Hutter G., et al., 2011. "Allogeneic transplantation of CCR5-deficient progenitor cells in a patient with HIV infection: An update after 3 years and the search for patient no. 2." *AIDS* 25:273–274.

Moore, J., 2004. "The puzzling origin of AIDS." *American Scientist* 92:540–547.

Prejean, J., et al., 2011. "Estimated HIV incidence in the United States, 2006–2009." *PLoS One* 6:e17502.

Schwartlander, B., et al. 2000. "AIDS in a new millennium." *Science* 289:64–67.

United Nations. 2012 "UNAIDS." November 8 (http://www.unaids.org/en/dataanalysis/tools/aidsinfo/).

Valdiserri, R. O., 2011. "Thirty years of AIDS in America: A story of infinite hope." *AIDS Educ Prev* 23:479–494.

Wolfe, N. D., et al. 2004. "Naturally acquired simian retrovirus infections in Central African hunters." *The Lancet* 363:932.

Worobey, M., et al., 2010. "Island biogeography reveals deep history of SIV." *Science* 329:1487.

Young, S., 2012. "Talk of 'cure' at historic AIDS conference." July 24 (http://www.cnn.com/2012/07/23/health/hiv-aids-conference/index.html).

2 Viruses and HIV

LOOKING AHEAD

This chapter explores the structure of viruses and how they replicate, focusing on the human immunodeficiency virus (HIV). On completing the chapter, you should be able to . . .

- Understand how viruses are assembled in cells using cellular machinery and building blocks.
- Understand how viruses relate to other microorganisms in structure and replication patterns.
- Describe the general components of a virus and the specific components of HIV.
- Outline the replication process in viruses and explain how the process occurs with HIV.
- Summarize the effect of HIV on the cells of the body's immune system.
- Discuss some general principles of viral inactivation and inhibition, and identify methods for treating viral disease.

The end of this chapter contains a **Healthline Q & A** section that addresses questions you may have as you read and digest the information presented.

INTRODUCTION

Every science has its borderland where the known and visible merge with the unknown and invisible. Startling discoveries often come from this hazy, uncharted realm of speculation, and certain objects loom large. In the borderland of biology, at the fringe of our understanding, are curious and puzzling objects known as **viruses.**

    Scientists have always had an awkard time fitting viruses into the scheme of living things. Viruses do not grow, nor do they move or adapt to their environment. They display few of the biochemical structures or processes we find in living things. Indeed, viruses cannot reproduce unless they are within a host cell. Furthermore, viruses do not produce any metabolic energy; therefore, once inside a host, all of the energy and molecular building blocks used to generate new viruses must also be derived from the host cell.

    In this chapter, therefore, we explore the structure and nature of the cell, without which viruses could not replicate or exist. We also explore the structure and replication patterns of viruses, with emphasis on HIV, the cause of AIDS. We study the properties that make viruses unique and that put them at the threshold between living organisms and inert molecules.

Cell and Molecular Biology

Cells are the basic unit of life. Anything less than a cell is therefore, by definition, not alive. Cells are discrete units and are best first described as small bags of water. Yes, water is the most abundant molecule in cells, and this water is contained by a membrane made of lipids. For organisms like humans, tens of trillions of cells connect to one another to form all of the tissues and organs of an organism, but viruses infect one cell at a time. Thus, although humans have trillions of cells, we must examine how viruses and HIV in particular interact with individual cells. In order to understand this interaction, we need to examine more closely the components of cells.

There are four major components that we call the building blocks of cells. There are proteins, nucleic acids, carbohydrates, and lipids. As noted previously, viruses do not generate any energy, and therefore, cannot synthesize their own molecular building blocks, but rather depend on the host cell to do so. Therefore, the same cellular building blocks are used to build viruses.

Protein

Proteins are important components of cells and viruses. Indeed, all functions and behaviors of cells are mediated by proteins. Proteins do all of the work and conduct all activities of the cell. For example, enzymes are proteins that catalyze the chemical reactions of metabolism and spur the chemical processes that release energy for cell use and then use that energy for building cellular structures, including viruses in infected cells. In addition to promoting the reactions that build viruses, cellular proteins play important structural roles in cells and viruses. For example, proteins are like molecular hands and are able to grab hold of things and recognize them. These "recognition" proteins are essential for viruses to be able to bind (grab like hands) to cell surfaces at the initial stages of infection. The enzymatic and recognition function is based on protein structure, and this structure is based on the subunits that comprise these proteins.

All proteins are polymers of subunits called **amino acids**, whose fundamental structure is shown in **Figure 2.1**. This structure always includes at least one **amino group** (NH_2) and one organic acid or **carboxyl group** (**COOH**). The major elements in amino acids are carbon, hydrogen, oxygen, and nitrogen; in some cases, sulfur is present. Each amino acid has a central carbon between the amino and carboxyl group. Amino acids vary from one another on the basis of which atoms are attached to this central carbon. These atoms are known as the **R-group** and can be as simple as a hydrogen in the case of glycine, involve up to 21 additional atoms in the case of tryptophan, or result in charged R-groups as seen with lysine and glutamic acid (Figure 2.1). There are up to 20 different R-groups on amino acids, which means that there are 20 different amino acids available for use in the synthesis of proteins. The number of amino acids in a single protein may vary from very few (in which case, the small protein is called a peptide) to thousands.

To form a protein from amino acids, a reaction called **dehydration synthesis** takes place. In this reaction, the components of water (−H and −OH) are removed

Hydrophobic amino acids

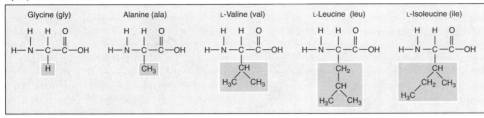

Alcoholic amino acids (hydrophilic)

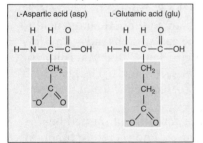

Aromatic amino acids

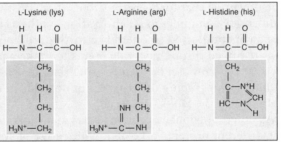

Acidic amino acids (hydrophilic)

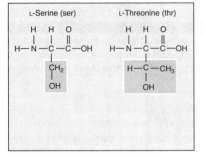

Basic amino acids (hydrophilic)

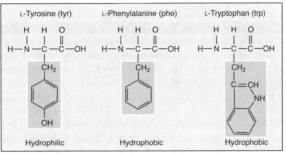

Amides (hydrophilic)   Sulfur-containing amino acids   Imino

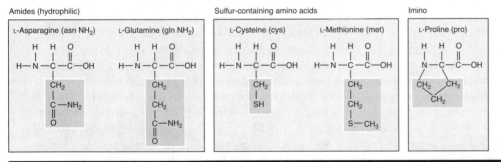

FIGURE 2.1

Structure of the amino acids. All amino acids have the same basic structure, but each varies by the R-group, which is a set of variable atoms attached to the central carbon between the amino (NH_2) and carboxyl (COOH) group of the amino acid. Shaded in this figure, we can see that the R-groups are sometimes charged or have hydrophobic and hydrophilic properties. Interactions between these shaded R-groups give rise to secondary protein structures.

from adjacent amino acid molecules, and a bond forms between the organic acid group of one amino acid and the amino group of the second amino acid. This bond that links two amino acids is referred to as a **peptide bond**. By forming successive peptide bonds, more and more amino acids can be added to the growing chain

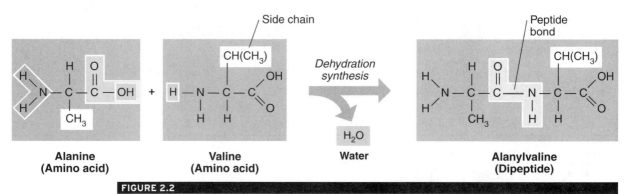

FIGURE 2.2

Formation of a dipeptide. The amino acids alanine and valine are shown, with the differences shaded. The OH group from the acid group of alanine combines with the H from the amino group of valine to form water. The carbon atom of alanine and the nitrogen atom of valine then link together, yielding a peptide bond. The dipeptide in this example is called alanylvaline.

(**Figure 2.2**). In Figure 2.2, the two amino acids alanine and valine react to form a dipeptide in which the two amino acids have become joined by a peptide bond. Examine Figure 2.1 again and note how all of the amino acids can combine through dehydration synthesis and peptide bond formation. Note also that the only difference in the dipeptide that forms is the atoms that constitute the R-groups. In Figure 2.2, the dipeptide can undergo a dehydration synthesis with an additional amino acid to make a tripeptide, and so on and so forth until, in some cases, thousands of amino acids are linked through peptide bonds to make a protein. Thus, an extraordinary variety of proteins can be formed from the 20 available amino acids. The genetic information for placing the correct amino acids in the correct sequence in the protein is provided by the DNA of the cell.

As noted previously, proteins carry out most of the activities of the cell because different proteins take on a wide range of structures, and therefore functions, because of variations in their amino acid sequences. Protein structure can be defined on four levels (**Figure 2.3**). The first or **primary structure** refers to the specific sequence of amino acids in the protein polymer. This sequence is unique to each protein. The R-groups in the amino acids incorporated in the primary sequence are able to interact with one another. For example, sometimes hydrogen, when bound to an oxygen or nitrogen, can have a slight positive charge, and if this occurs on an R-group, as in Figure 2.1, an attraction, known as a hydrogen bond, can form between that R-group and a negative charge on the R-group of another amino acid. Other R-group associations are also possible. These R-group interactions depend on and vary with the primary amino acid sequences and in some cases cause the protein to coil into an **alpha α-helix**. Hydrogen-bonded R-groups, if further away on the primary sequence, can also lead to the formation of a **beta (β)-pleated sheet**. The α-helix, β-pleated sheet and **random coils** are **secondary structures** that arise from the primary sequence (Figure 2.3). The third, or tertiary, structure of proteins results from combinations of two or more secondary structures to give an overall shape to a protein (Figure 2.3). When combinations of α-helices, β-pleated sheets, and random coils combine, the protein can have an overall globular structure, such as that seen with hemoglobin or immunoglobulin. The **tertiary structure** of a protein can also form a long fibrous structure such as that seen in collagen. Globular and fibrous structures are examples of **tertiary protein structure**. Finally, not all proteins can carry out their functions alone. Many proteins must work together in an associated

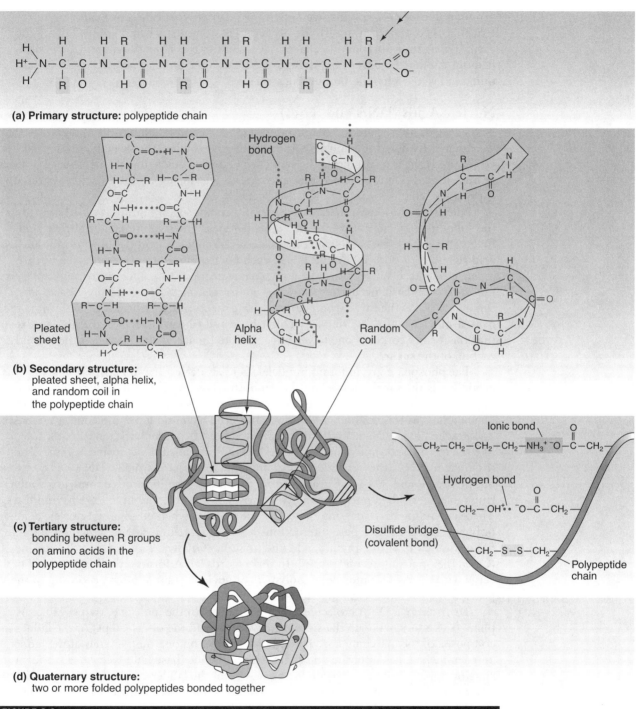

(a) Primary structure: polypeptide chain

(b) Secondary structure:
pleated sheet, alpha helix,
and random coil in
the polypeptide chain

(c) Tertiary structure:
bonding between R groups
on amino acids in the
polypeptide chain

(d) Quaternary structure:
two or more folded polypeptides bonded together

FIGURE 2.3

Primary, secondary, tertiary, and quaternary protein structures. (a) The primary structure of a protein refers
to the sequence of amino acids. (b) As the chain of amino acids gets longer, it takes on a secondary structure.
(c) Combinations of secondary structures give the protein its overall shape, which is known as a tertiary
structure. (d) Some proteins require more than one polypeptide to function, and the polypeptide configuration
of these proteins is known as the quaternary structure.

group, and to these proteins, we assign the term polypeptide. For example, a protein that is composed of multiple polypeptides subunits is hemoglobin. Hemoglobin actually has four polypeptide chains, two pairs of the same polypeptide; therefore, hemoglobin has a **quaternary structure** we call a tetramer, with two α-chain subunits and two β-chain subunits (Figure 2.3).

Nucleic Acids (DNA and RNA)

Nucleic acids are the second major building block component of cells and viruses that we discuss. As noted previously in the section on proteins, the principle role of the DNA and RNA is to prescribe the sequence of amino acids in proteins. Indeed, a piece of nucleic acid (DNA or RNA) that encodes for a sequence of amino acids (a protein) is precisely what is meant by the word "**gene**."

Nucleic acids, like proteins, are generally composed of hundreds of subunits, but the subunits in nucleic acids are called nucleotides. The two important nucleic acid forms that encode for viral proteins are **deoxyribonucleic acid (DNA)** and **ribonucleic acid (RNA)**. For both DNA and RNA, each **nucleotide** subunit is composed of three major elements: a sugar molecule, a phosphate group (PO_4), and a nitrogen-containing molecule with basic properties known as a nitrogenous base, or simply a base. The sugar in DNA is the five-carbon carbohydrate deoxyribose, whereas in RNA, it is the five-carbon carbohydrate ribose. **Deoxyribose** and **ribose** molecules are similar, except that the former contains one fewer oxygen atom (hence, deoxyribose) at the second carbon of the sugar.

The phosphate group found in nucleic acids is derived from phosphoric acid. This group binds the sugars to one another in both RNA and DNA. The chain of alternating sugar and phosphate subunits forms the stem of the nucleic acid molecule, or the "backbone," as it is colloquially known. In both DNA and RNA molecules, the bases are attached to the sugar molecules and extend out from the backbone somewhat like the teeth on a comb. The four bases in DNA are adenine, cytosine, guanine, and thymine; in RNA, they are adenine, guanine, cytosine, and uracil. (DNA has thymine but no uracil, and RNA has uracil but no thymine.) Adenine and guanine are double-ring molecules called **purines**, whereas cytosine, thymine, and uracil are **pyrimidines**. These extending bases can interact with one another when on separate nucleic acid polymer strands. These base interactions follow a matching rule in which an adenine (A) always pairs with a thymine (T) and guanine (G) always interacts with cytosine (C) on the opposing second strand. In this regard, DNA forms a double-stranded molecule with the two strands being polymers of nucleotides that align according to the previously mentioned base-matching rules.

To visualize a DNA molecule, picture a ladder. In the molecule, two sugar phosphate backbones make up the sides of the ladder, and the rungs (steps) of the ladder are composed of the matching base pairs. For example, on one rung or step of the ladder, an adenine base lies opposite and pairs with a thymine base (and vice versa) to form the rung or step. Likewise, the second rung or step of the ladder can be formed by a base pairing of guanine on one strand and cytosine on the other (and vice versa). The ladder is twisted to form a double helix, the so-called spiral staircase shown in **Figure 2.4**. RNA, like DNA, is a polymer of nucleotides, but as mentioned previously here, uses the base uracil instead of thymine, lacks an oxygen at the second carbon of the sugar component, and also remains single stranded; as a general rule, it does not partake in base-pairing like DNA. All cells use DNA as the form of genetic information; however, as we see here, different viruses can use either DNA or RNA as their genetic blueprint. Indeed, the form of nucleic acid that comprises the genetic information (genome) of a

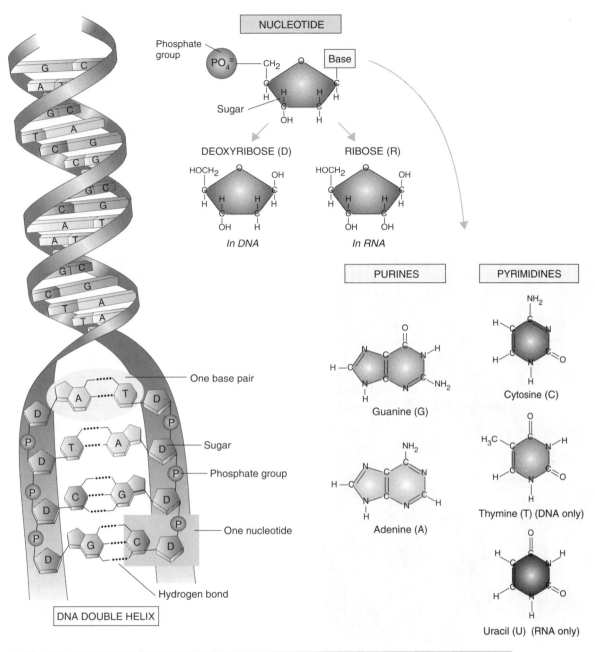

NUCLEOTIDE

Phosphate group

PO₄

Base

Sugar

DEOXYRIBOSE (D)

RIBOSE (R)

In DNA

In RNA

PURINES

PYRIMIDINES

Guanine (G)

Cytosine (C)

Adenine (A)

Thymine (T) (DNA only)

Uracil (U) (RNA only)

One base pair

Sugar

Phosphate group

One nucleotide

Hydrogen bond

DNA DOUBLE HELIX

FIGURE 2.4

Nucleic acid components. The pentose sugars in nucleotides are ribose and deoxyribose. These two small molecules are identical except for the lack of an oxygen atom at one carbon atom in deoxyribose. The nitrogenous bases include adenine and guanine, which are large purine molecules, and thymine, cytosine, and uracil, which are smaller pyrimidine molecules. Note the similarities in the structures of these bases and the differences in their inside groups.

virus greatly affects the nature of the viral replication and cycle and how we study and target the virus with drugs. HIV has an RNA genome.

As noted previously, the sequence of nucleotides in a nucleic acid prescribes the sequence of amino acids in a given protein. We say that a gene is the specific set of nucleotides required to encode for a specific protein. Yes, as a general rule, for each different protein, there is a different gene that encoded for it. The flow of information from the sequence of nucleotides to a sequence of proteins is essential to understand.

In cells, all genetic information is in the form of DNA. The sequence of the bases that compose the nucleotides must be copied in to a single-stranded RNA before the information that prescribes the sequence of amino acids can be accessed. This process of protein synthesis begins with an uncoiling of the DNA double helix and a separation of the two DNA strands. This separation exposes a gene—a region of the DNA molecule that encodes a protein. Using the enzyme **RNA polymerase**, cells synthesize a single-stranded molecule of RNA having bases complementary to the bases along that region of one strand of the DNA (as we discussed previously here). Individual nucleotides yet to be incorporated into the new RNA polymer are free-floating and are available in the cell, and the RNA polymerase is able to match them one at a time to the bases in the DNA template and polymerize these nucleotides into a new polymer as the polymerase moves along the DNA template strand. This production of RNA is called **transcription**. A base code in part of the DNA molecule is transferred to a base code in the RNA molecule. The RNA fragment so constructed is known as **messenger RNA (mRNA)**, shown in **Figure 2.5**. Each single-stranded mRNA molecule eventually passes into the cell's cytoplasm carrying the genetic message. Each three-base sequence on an mRNA molecule is called a codon (a codon will specify an amino acid). The mRNA molecule can thus be considered a series of codons. Once formed, the mRNA molecule moves away from the DNA molecule and is translated into a polymer of amino acids (protein) based on the codon sequence of nucleotides in the mRNA. This **translation** (protein

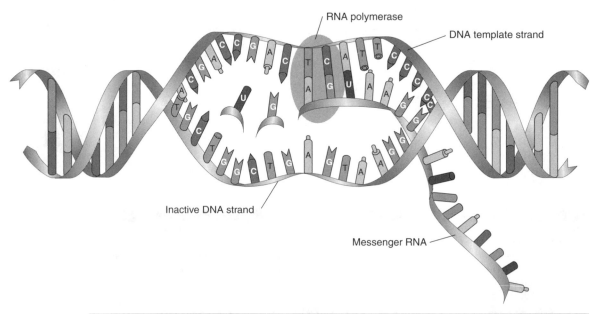

FIGURE 2.5

The transcription process. The enzyme RNA polymerase moves along the template strand of the DNA molecule and synthesizes a complementary molecule of RNA using the base code of DNA as a guide. The mRNA will carry the genetic message of DNA into the cytoplasm, where protein synthesis occurs. Note that the inactive strand of DNA is not transcribed.

synthesis) occurs on the ribosome and is a rather complicated process that we do not need to discuss here.

What you should now understand is that DNA is a polymer code for the synthesis of proteins, and cells must use RNA as an intermediate to translate this code into protein. Viruses are also composed of proteins and nucleic acids, but they do not have the entire set of machinery to make these molecules (nucleic acids and proteins) or to produce the energy required to make these molecules. Therefore, the virus depends on the energy harvested by the host cell and the machinery (ribosomes, etc.) to manufacture the necessary genetic information and proteins that compose the virus.

In addition to making proteins, cells and viruses must replicate their genomic information. As noted, for cells, this is always DNA, and for viruses, it can be either DNA or RNA. Indeed, the replication of the genomic information is an essential step in growth. By cell growth and viral growth we do not mean an increase in size, but rather an increase in number. Because each new virus requires its own genome to encode for the specific viral proteins, the genome must be replicated with the formation of each new viral particle (a *virion*). In order to make a copy of the DNA for newly made cells and viruses, the enzyme DNA polymerase is able to read the original DNA template strands and make duplicates based on base-pair matching rules, as shown in **Figure 2.6**. Individual nucleotides yet to be incorporated into the new DNA polymer strands are available in the cell, and the **DNA polymerases** (two polymerases; one for each strand of the DNA) are able to match them one at a time to the bases in the DNA template (existing strand) and polymerize these nucleotides into a new polymers as the polymerase moves along the DNA template (Figure 2.6).

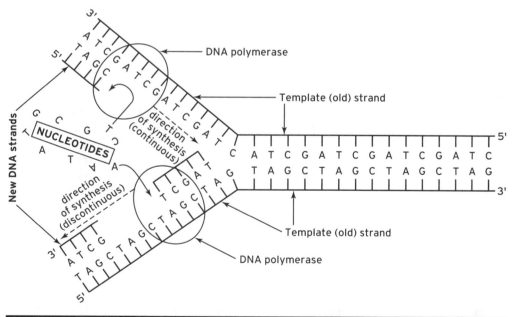

FIGURE 2.6

Replication of DNA is mediated by the enzyme DNA polymerase. DNA has two strands, which serve as templates to be copied. The template DNA is "opened" and the DNA polymerase brings in free-floating nucleotides to match the template and form new strands. Note that the new DNA molecules have an old strand and a new strand.

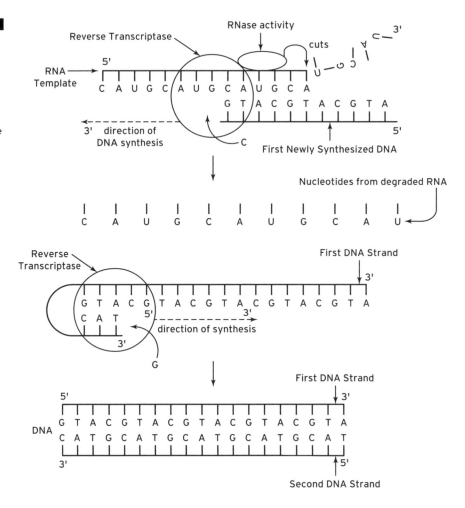

FIGURE 2.7

Reverse transcriptase uses an RNA template to make a double-stranded DNA molecule. As the first DNA strand is made, an RNase subunit cleaves the RNA template into nucleotides, making it possible for the first DNA strand to now serve as a template for second strand synthesis.

Viruses like HIV have an RNA genome, yet these viruses also must make DNA in order to direct the cells to make sufficient levels of viral proteins. To this end, HIV is a **retrovirus**, because like other retroviruses, it has a special polymerase known as **reverse transcriptase**, which can copy a single-stranded RNA into a double-stranded DNA molecule (**Figure 2.7**). The reverse transcriptase is able to read RNA as a template, and to incorporate individual nucleotides. A new DNA polymer strand and this DNA strand are then read by the reverse transcriptase to make the second DNA strand of the double helix (Figure 2.7). The reverse transcriptase has an **RNase** subunit that degrades the original RNA template as the DNA is synthesized.

Structure of Viruses and HIV

Viruses are among the smallest infectious agents that can cause disease in bacteria, plants, animals, and humans (**Figure 2.8**). Indeed, some viruses are so tiny that 10 million laid end to end might stretch across the period at the end of this sentence. Counting these 10 million viruses nonstop at a rate of one per second would consume close to four months.

Using the electron microscope, scientists have been able to magnify viruses many thousands of times and photograph them. These photographs reveal that viruses occur

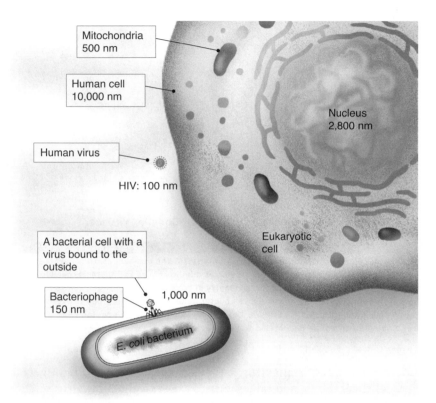

(a)

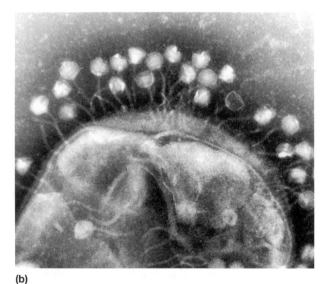

(b)

FIGURE 2.8

(a) A comparison of size relationships among eukaryotic cells, bacterial cells, and viruses. (b) Transmission electron micrograph of bacterial rods isolated from a water sample on a sieve filter. The surface projections on the rods are bacteriophages. Note their size relative to the host bacterium. Courtesy of Dr. Graham Beards.

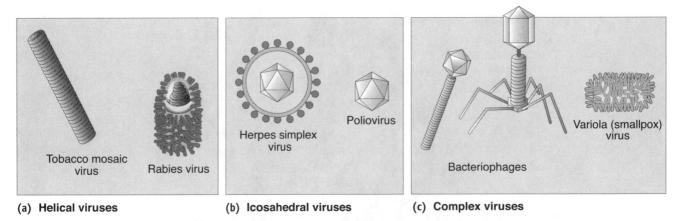

(a) Helical viruses　　　**(b) Icosahedral viruses**　　　**(c) Complex viruses**

FIGURE 2.9

Viruses exhibit numerous variations in shape or symmetry. (a) The nucleocapsid has helical symmetry in the tobacco mosaic and rabies viruses. The helix resembles a tightly coiled spiral. (b) Certain viruses, such as herpesviruses and polioviruses, exhibit icosahedral symmetry in their nucleocapsids. (c) In other viruses, neither helical nor icosahedral symmetry exists exclusively. The bacteriophage, for example, has an extended icosahedral "head" and a helical tail with extended fibers. The variola (smallpox) virus is rectangular-shaped with tubular structures embedded within the membranous envelope.

in a variety of shapes. Certain viruses, for example, appear as a tightly wound coil known as a **helical**. Ebola and rabies viruses are typical helical viruses. Other viruses have the shape of an **icosahedron**, that is, a geometric figure with 20 triangular faces and 12 points. Influenza viruses, herpes simplex viruses, and HIVs can take this form. The shapes of several viruses are depicted in **Figure 2.9.**

All viruses consist of two basic components: a core of nucleic acid (DNA or RNA), called the **genome**, and a surrounding layer of proteins, known as the **capsid**. This capsid can be composed of many copies of a single protein or several proteins, but in either case, each individual protein is known as a **capsomere (Figure 2.10)**. As noted, the nucleic acid of the genome may be either **deoxyribonucleic acid (DNA)** or **ribonucleic acid (RNA)**, but not both. DNA is familiar to most of us as the hereditary material found in the chromosomes and genes of cells. RNA, in contrast, is a nucleic acid used by cells for several purposes, including the production of proteins. The varicella (cause of chickenpox) viruses and herpes simplex viruses contain DNA in their genomes, whereas the measles, mumps, and polioviruses contain RNA. HIV is also an RNA-containing virus. It contains two identical molecules of this type of nucleic acid. Within the HIV capsid, found nestled with the HIV RNA genome, are two copies also of the reverse transcriptase molecules required to convert the RNA genome to DNA upon infection of the host.

The capsid encloses the genome and gives the virus its helical or icosahedral symmetry depending on how the individual capsomeres fit together. The capsid also provides protection for the genome because its protein can resist certain changes in the external environment, such as drying or increasing acidity. This resistance varies among viruses and is relatively low for HIV. The capsid also helps to package the genome in such a way that it can be best delivered to the cell, and in some cases, the capsid proteins serve as attachment sites to the host cell to begin the infection process. Individuals infected with HIV mount an immune or antibody response toward **p24**, a major viral capsid protein. The detection of HIV antibodies in patient serum is the basis of diagnostic tests known as an ELISA and Western blot (**Figure 2.11**). The combination of genome

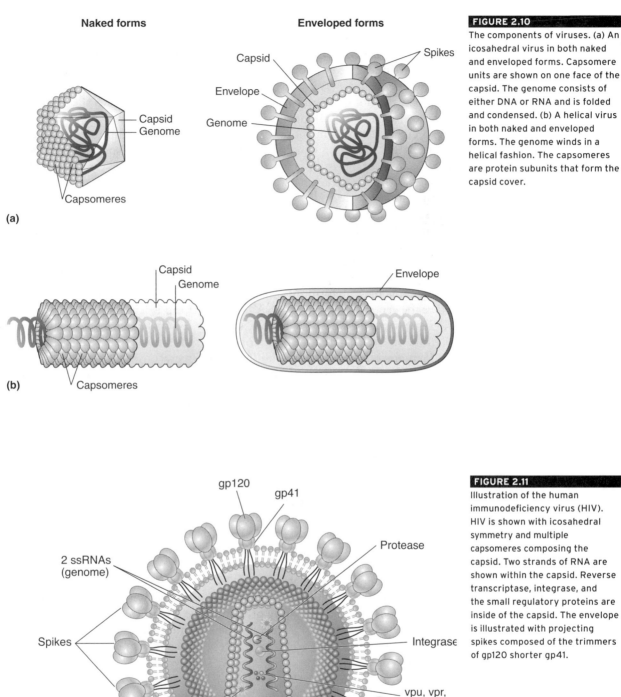

Naked forms

Enveloped forms

Capsid
Genome

Capsomeres

(a)

Spikes

Capsid

Envelope

Genome

Capsid
Genome

Envelope

Capsomeres

(b)

FIGURE 2.10

The components of viruses. (a) An icosahedral virus in both naked and enveloped forms. Capsomere units are shown on one face of the capsid. The genome consists of either DNA or RNA and is folded and condensed. (b) A helical virus in both naked and enveloped forms. The genome winds in a helical fashion. The capsomeres are protein subunits that form the capsid cover.

gp120   gp41

Protease

2 ssRNAs
(genome)

Integrase

Spikes

vpu, vpr,
vif, nef

Matrix

Reverse
transcriptase

p24 (capsid)

FIGURE 2.11

Illustration of the human immunodeficiency virus (HIV). HIV is shown with icosahedral symmetry and multiple capsomeres composing the capsid. Two strands of RNA are shown within the capsid. Reverse transcriptase, integrase, and the small regulatory proteins are inside of the capsid. The envelope is illustrated with projecting spikes composed of the trimmers of gp120 shorter gp41.

and capsid is referred to as the **nucleocapsid**. Special proteins may line the inner surface of the capsid and coat the genome, as we discuss for HIV later here.

Many viruses, including HIV, have an additional layer outside the capsid known as an **envelope**. The envelope is a flexible lipid-bilayer membrane, and because it lays outside the capsid, enveloped proteins interact with host cell through proteins in this envelope and not through capsomeres. This lipid membrane is the same lipid bilayer as seen with human cell membranes. Indeed, viral envelopes are usually derived from the previous host cell membrane, and the virus acquires this membrane as it leaves the host cell and prior to a new infection; however, there are special differences between the host membrane and the viral envelope. Although the lipid nature is identical, viral envelopes contain virus-specified proteins. These viral envelope proteins are encoded by the viral genome and often appear on the envelope as projections known as spikes. These spikes are composed of proteins that assist the union of viruses with cells, as we discuss later. Influenza viruses are well known for their spikes, and researchers have found that the envelope of HIV also has spike proteins (Figure 2.11).

To summarize, viruses are noncellular particles consisting of either DNA or RNA enclosed in a coat of protein that may or may not be enveloped. Essentially, each virus is a genome plus a capsid. In some viruses, an envelope with spikes is also present. In HIV, the genome consists of RNA, and the capsid is icosahedral. Surrounding the capsid of HIV is an envelope with spikes (Figure 2.11). The virus has a diameter of about 0.1 micrometer (one ten-millionth of a meter); 0.1 micrometer can also be expressed as 100 nanometers (one hundred billionths of a meter).

Two features of HIV are worthy of note: First, different HIV particles have highly variable protein spikes. Random **mutation** in the HIV genome results in changes in the amino acid sequence of the protein spikes and consequently causes random changes in the structure of the spikes on different HIV virion. This fact has presented substantial problems for vaccine researchers because a vaccine must take into account all possible forms or mutants of the virus. The second unique feature is the presence of the dual enzyme reverse transcriptase and is discussed previously here. Molecules of reverse transcriptase are found among the strands of RNA in HIV within the nucleocapsid. The function of reverse transcriptase is to synthesize DNA, using the viral RNA as a template, after cell infection has occurred (Figure 2.7). We examine the practical significance of reverse transcriptase later in this chapter. Molecules of the enzymes **integrase** and **protease** are also found beneath the capsid, as we note presently.

Compared with other microorganisms and cells, viruses are structurally very simple. Their chemistry is equally simple. Viruses perform no energy-generating chemical reactions, nor can they synthesize proteins on their own. They consume no food, do not grow, and produce no waste products. Because of their unusual characteristics, many scientists hesitate to refer to viruses as "alive." Viruses replicate, however, and they perform this function efficiently. Replication takes place within a living cell, at the cell's expense.

Replication of Viruses and HIV

The process of viral replication is among the more remarkable of natural phenomena. A virus penetrates a living cell thousands of times its size and uses the chemical compounds and structures of the cell for its own purposes. In doing so, the virus produces hundreds (in some cases, thousands) of copies of itself and leaves behind a poorly functioning or dead cell, which is the basis of pathogenesis. On its own, a virus cannot replicate, but within a cell, the process occurs with extraordinary efficiency.

Such a relationship, in which one organism or virus uses and often destroys another, is known as **parasitism**. For viruses, parasitism is a necessary prerequisite for replication. Later the generalized steps all viruses must go through in order to replicate and generate new progeny are presented. With each step, specific information regarding HIV is given; however, it should be noted that although different viruses have different mechanisms through which they accomplish these steps, all viruses proceed in some way through these steps during their replication cycle.

Adsorption

The first step in viral replication is the union between the virus and its host cell, a step called **adsorption**. Such a union is highly specific, and certain viruses interact only with certain cells. Hepatitis viruses, for instance, unite with liver cells. Influenza viruses unite with respiratory tract cells, and polioviruses unite with brain cells. For HIV, a primary target cell is a cell of the immune system called the **T-lymphocyte** (also known as the T-cell). The basis of this specificity in viral cell targeting is based on the requirement for viral proteins to bind to (recognize) specific proteins found on the cell surface. Because the cells of different tissues possess different proteins on their surface and different viruses have differing surface proteins, the specific recognition of cell surface by viral surface proteins leads to a selective infection of certain cell types by specific viruses as described previously here. For example, the target T-cells for HIV possess on the surface a series of proteins called CD4 molecules (**Box 2.1**). This CD4 molecule is recognized by HIV viral proteins, and thus, these cells can be infected by HIV. Cells that do not have CD4 on the surface are not recognized and therefore not infected by HIV.

Each CD4 molecule contains 433 amino acids. The CD4 molecules are also located on macrophages (microbe-engulfing white blood cells) and on brain cells, which is why HIV can infect macrophages and the brain. Biologists refer to these protein sites of interaction and binding, such as the CD4 molecules, as **receptor** sites. Other sites, called **co-receptors**, are also found on T-lymphocytes, as we see here. It is important to stress that CD4 and the other HIV co-receptors have important normal physiological roles on the T-cell surface during the immune response. These cell surface protein receptors and co-receptors do not exist for the benefit of HIV, but rather, HIV has evolved to be able to recognize these proteins and gain entry to the cell.

The union between the HIV particle and the CD4 receptor site involves a molecule on the spikes of HIV called glycoprotein 120, or **gp120**. This molecule is so named because it is a carbohydrate-containing ("glyco") protein with a molecular weight of 120 kilodaltons (kilo means "thousand;" dalton is a unit of weight equal to the mass of a hydrogen atom). Both CD4 and gp120 molecules play significant roles in the virus–cell interaction. They also have practical importance: CD4 molecules have been investigated for their therapeutic value, and gp120 is used in vaccine preparations.

The gp120 molecule extends out from the viral envelope as a trimer (three gp120 molecules together; Figure 2.11). Each gp120 molecule is a complex structure containing at least five areas (or domains) that can vary in their amino acid content. These are known as the variable domains, V1 through V5, on the gp120 molecule, and again, these variations hamper the development of successful vaccines, as no one vaccine can recognize all variant combinations. Domains 1 to 5 vary for "loop" structure and so sometimes are referred to as V1 loop, V2 loop, and so forth. The CD4 molecule and gp120 appear to interact somewhere within the V1 domain of gp120, and at this point, the virus is adsorbed to the cell surface.

BOX
2.1

On the Trail of CD4

For every lock there is a key, or so it appeared to Steven McDougall, immunologist at the CDC. If HIV is the key, then there must be a lock on the T-lymphocyte where the two fit together. Even if the virus should subtly change its envelope structure (as HIV is known to do), the part that binds with the lymphocyte must remain constant because the binding between the virus and host cell is an essential element in HIV infection. How could McDougall prove the existence of a lock-and-key arrangement, and how could he identify the lock on the T-lymphocyte?

Two articles published in 1985 provided a provocative clue for McDougall. They indicated that when lymphocytes attacked by HIV were treated with highly specific antibodies, the lymphocytes resisted HIV attachment. By reacting with the cells, the antibodies were apparently blocking the binding site. This finding suggested to McDougall that the molecules that identify a lymphocyte cell are also the binding sites of the virus.

To test this hypothesis, McDougall and his colleagues began by "painting" a radioactive substance onto the surface of lymphocytes. Then they exposed the cells to HIV. The radioactive material concentrated where the HIV bound to the cells. Next, they dissolved the lymphocytes with a detergent and mixed the fragments with beads coated with antibodies against HIV. The antibody-coated beads combined with fragments of lymphocytes where HIV was clinging, because HIV antibodies unite specifically with HIV. By analyzing the fragments for radioactivity, the researchers could locate the viruses and, hence, the receptor sites.

The next step was to pass the fragments through a sieve-like gel to separate the fragments according to size. The gel was then placed against special paper sensitive to radioactivity. The researchers analyzed the paper to see where the radioactivity (and, hence, the receptor sites) came to rest. McDougall and his group found that the radioactivity corresponded to molecules weighing 58 kilodaltons. It was clear that one molecule of the lymphocyte had been recognized as the lock by the viral key. Although the molecule's weight was elucidated, little else was known about it at that time. McDougall's group gave the molecule the name **CD4** (CD for "complementary determinant," or "cluster of differentiation," or "cluster designation," depending on one's interpretation and recollection). Now the real work on analysis could begin.

Penetration

The gp120 trimers on the surface of the HIV envelope are associated with a trimer of the shorter **gp41** viral envelope protein that does not extend as far as gp120 from the viral surface (Figure 2.11). The binding between gp120 trimers and CD4 causes a change in the shape of the gp120. This shape change allows the V3 loop of gp120 to interact with additional cell surface **chemokine receptor proteins** known as **CXCR-4** (sometimes called **fusins**). These proteins are the HIV co-receptors mentioned previously here, and as noted, these proteins also play otherwise important roles in the function of the immune system. The interaction between the V3 loop of gp120 and the fusins causes further shape changes, which draws the viral gp41 trimers near to the cell membrane. The gp41 molecule looks somewhat like a harpoon or pointed spring. Once at the membrane and triggered into action by the gp120 attachment to the cell surface, the gp41 molecule darts out (changes shape) and pierces the cell membrane of the macrophage or T-lymphocyte and mediates a fusion between the lipid of the viral envelope and the lipids of the cell membrane, much like two bubbles fusing together.

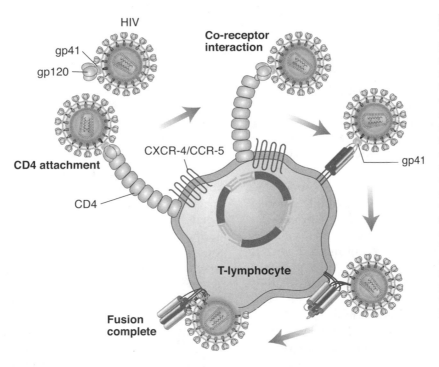

FIGURE 2.12

The union between the human immunodeficiency virus (HIV) and its host T-lymphocyte. The HIV envelope makes contact with the surface of the T-lymphocyte and the gp120 molecule of HIV's spike binds to the CD4 receptor site of the T-lymphocyte. The gp120–CD4 complex then bends down, and a portion of the gp120 molecule binds to the CCR-5 or CXCR-4 co-receptor site. This action releases the gp41 molecule, which pierces the membrane of the T-lymphocyte. Penetration of the HIV nucleocapsid to the T-lymphocyte cytoplasm follows. Adapted from F. Hoffmann-La Roche Ltd., "Blockade of viral peptide helices prevents HIV from fusing with the host cell," Roche Facets Clinical Trials.

This fusion delivers the nucleocapsid of HIV into the cell cytoplasm, and the penetration step is complete.

The observation that penetration requires gp120 V3 interaction with fusin co-receptors bears significance because vaccine-induced antibodies produced against the V3 loop could conceivably prevent HIV's union with T-lymphocytes. Moreover, drugs that destroy the V3 loop could be used to prevent virus–cell union and bring infection to an end. A disquieting note was sounded in 1998, however. Researchers determined the crystalline structure of the gp120 molecule and reported that the amino acid loops of the molecule are shielded from antibodies and drugs by a screen of complex carbohydrates. Moreover, the loops and carbohydrates render the gp120 molecule flexible enough to avoid antibody molecules further, a problem that impacts vaccine development. **Figure 2.12** shows the bonding that takes place.

The CCR-5 co-receptor seems to predominate on macrophages, whereas the CXCR-4 co-receptor works primarily on T-lymphocytes. Although the significance of this observation is not completely understood, it may give insight as to why the infection cycle of HIV proceeds as it does; that is, scientists have noted that early in the infection, macrophages bear the brunt of the HIV infection (the virus is "M-tropic"), whereas later on, the T-lymphocytes are the most heavily involved (the virus is "T-tropic"). Perhaps HIV uses the CCR-5 co-receptor in the initial stages of infection and then shifts its biochemistry and develops the ability to unite with the CXCR-4 co-receptor on the T-lymphocyte surface. The pathology would thus fit the biochemistry; however, as always in science, we must be cautious about drawing hasty conclusions until the experimentation is complete. For example, as of 2002, researchers had identified more than a dozen molecules on host cells that could act as co-receptors. Moreover, they were postulating that HIV may undergo a genetic shift to change its co-receptor allegiance. Still, even at these early stages, the research on co-receptors has provided a clue as to why certain individuals appear to enjoy genetic resistance to HIV (**Box 2.2**).

> **Focus on HIV**
>
> The human chemokine family contains 50 chemokines and 20 chemokine receptors identified to date. They are involved in inflammation, cancer, and HIV-1 infection (e.g., chemokine receptors CCR-5 and CXCR-4).

BOX
2.2

Immune to AIDS

June 1996 was a very hot period for AIDS researchers. During that month, scientists from five different laboratories were working to identify the CCR-5 co-receptor, and within a two-week period, all five crossed the finish line almost simultaneously.

Then a new issue emerged. Biochemist Richard Koup of New York's Aaron Diamond AIDS Research Center had been investigating the mysterious cases of two sexually-active homosexual men who were not infected with HIV despite their high-risk sexual behavior that doubtlessly exposed them to the virus. In all probability, the men should have been infected, but Koup had found that T-lymphocytes from the two men were not attacked by HIV. Could it be that the men's T-lymphocytes lacked co-receptors?

Koup and his colleague Ned Landau accelerated their research from zero to warp speed and began an exhaustive search for a genetic basis for HIV (specifically HIV-1) resistance. They examined the nuclear material from cells of the two men, and within two months, they discovered a gap of 32 missing bases in the DNA of the gene that encodes CCR-5 co-receptors. Apparently, the resulting CCR-5 protein was so badly deformed that the T-lymphocytes destroyed it instead of placing it on their surface as a co-receptor, and without the CCR-5 co-receptor, HIV could not dock on the cell's surface.

The discovery was a bombshell, but it immediately precipitated questions about whether the immunity of the two men was an unusual happenstance of nature or was authentic and prevalent. Thousands of blood samples would be needed to test the theory. It just so happened that 10,000 samples were in the freezer at the laboratory of Stephen O'Brien of the National Cancer Institute. In the late 1970s, O'Brien had observed a gene that protected mice against a leukemia virus by denying entry of the virus to the host cell.

Then, when the AIDS epidemic began in 1981, he was struck by the possibility that such a gene might exist in human cells. Over a 15-year period, he had collected from physicians thousands of blood samples from patients at high risk for contracting HIV. His research had been fruitless, however.

The discovery of the CCR-5 co-receptor put a new spin on O'Brien's search. Now, in 1996, he and his collaborators tested the blood sample for evidence of the defective genes. Their results were surprising: Two copies of the defective genes apparently exist in 1% of Caucasian Americans of Western European descent, and fully 20% have one copy of the gene. Although members of the latter group can be infected by HIV, they remain healthy up to three years longer than those with no copies of the gene (possibly because they have half the number of co-receptors on their T-lymphocytes). Furthermore, the gene seems to be much rarer in Africans, Native Americans, and Asians. This could indicate that the gene originated after the Caucasian line split off from the others. It is, therefore, of rather recent evolutionary origin.

The research has opened several new directions for additional research. For example, understanding the normal gene's activity may lead researchers to methods for mutating it to a defective form to protect patients. Providing defective genes to patients via gene therapy could be used to develop another avenue of treatment, and other researchers are hunting for different HIV-blocking mutations in other groups, mutations that might help explain why some prostitutes in Africa and Thailand have managed to avoid AIDS despite repeated exposure to HIV.

At the very least, these new discoveries remind us that a genetic mutation deemed a "defect" can prove to be a blessing in disguise—that is, it can confer surprising benefits on those lucky enough to inherit it.

Uncoating

After the nucleocapsid is delivered to the cell cytoplasm, the process of uncoating (removing) the genome from the nucleocapsid must begin. The genome must be exposed and the capsid removed if the viral genetic information is to be made accessible to the host cell replication and transcriptional biosynthetic machinery. In the cytoplasm, the protein capsid is stripped away by cellular enzymes (the uncoating phase), and the genome is released. For many viruses (HIV excluded), the viral nucleic acid provides genetic codes for the synthesis of viral parts. Cellular compounds are used in this synthesis. For example, cellular amino acids are used to synthesize viral proteins (such as capsomeres), and raw materials in the cellular cytoplasm are used to build viral nucleic acids. Cellular enzymes are then engaged to fit together the new viral genomes and capsids in the assembly phase. Soon, complete viral particles appear in the cytoplasm (**Figure 2.13**). The process for HIV is somewhat different and more complex, as we discuss later here.

Synthesis of Viral Nucleic Acid and Protein (Biosynthesis)

What we have just explored is a general replication cycle in which viruses use a cell to produce copies of themselves. In biochemical jargon, the process is called the lytic cycle, from lysis, meaning "to break." For HIV, the replication pattern varies slightly.

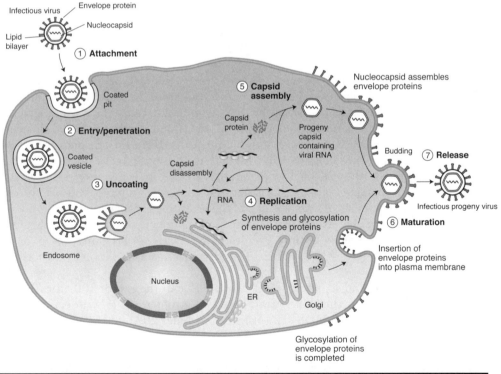

FIGURE 2.13

A generalized schematic of viral replication in a host cell. In this representation, the virus unites with the host cell. Then the capsid and genome penetrate to the cytoplasm of the cell, and the synthesis of viral enzymes begins. Nucleic acid and structural protein molecules for the virus are synthesized next. After assembly, matured, infectious virus particles are released.

The essential difference is that HIV remains in its host T-lymphocyte, becoming part of the cell and programming the synthesis of additional HIV particles during an extended period of time.

As noted in the previous steps, the HIV replication cycle begins with fusion of the virus to the cell's surface membrane, using the interaction between gp120 and gp41 molecules and the CD4 receptors and co-receptors. Penetration of the HIV genome and capsid into the cell cytoplasm follows. Cellular enzymes now remove the capsid, releasing the RNA and two proteins: the p24 protein that coats the HIV genome and the p17 protein that lines the inside of the capsid.

Also released is the dual enzyme known as *reverse transcriptase*. As noted earlier, the reverse transcriptase is able to serve as a DNA polymerase and "reverse transcribe" the RNA genome into a DNA strand, degrading the RNA template as it goes. Following the reverse transcription, the single DNA strand can serve as a template for the reverse transcriptase in the formation of the second DNA strand of the double helix. By using RNA as a template for DNA synthesis, reverse transcriptase reverses the flow of biological information (hence, the enzyme's name, "reverse transcriptase"). Moreover, HIV is known as a retrovirus because of this enzyme's activity ("retro" implies reverse). Within the retrovirus family of viruses, HIV belongs to a subgroup called lentiviruses.

After the double-stranded DNA molecule has been formed, it migrates to the nucleus of its host T-lymphocyte, macrophage, or other cell. The distance the DNA molecule must travel is formidable (up to 20 micrometers), and research reported in 1998 suggests that the molecule may use the actin filaments of the cell's cytoskeleton as an ultramicroscopic highway. Data from experiments using green fluorescent proteins indicate that DNA molecules move along fairly linear paths in bursts somewhat like stop-and-go traffic on a busy street. The entire trip appears to take a few minutes to complete.

On arriving at the nucleus, the DNA molecule is incorporated into the cell's DNA at a random site by a viral enzyme named *integrase*. Researchers indicate that incorporation is most efficient while the cell is dividing because the nuclear envelope has broken down at this time and need not be penetrated. (How the DNA molecule gets through the nuclear envelope of nondividing cells such as macrophages is still uncertain.) The incorporated DNA molecule thus becomes part of one of the cell's 46 chromosomes. Within the nuclear material, the DNA molecule is termed a *provirus* (**Figure 2.14**). A phenomenon like this is not often observed in biology. It is referred to as **lysogeny**, and the cycle of events is called the lysogenic cycle. The process implies that a person carries HIV as a provirus in the T-lymphocytes. With HIV in this **latent** form, the person is perfectly healthy, even though a carrier of HIV. In medical terms, this person does not have AIDS; rather, the individual has a condition known as HIV infection.

Public health specialists at the CDC estimate that at the end of 2009, nearly 1,148,200 individuals in the United States aged 13 or older had HIV infection but were not living with active AIDS. This estimate includes 207,600 (18%) of people whose infections have not been diagnosed. Although in good health, these individuals could transmit the provirus if their infected T-lymphocytes are transferred to another person. Because blood and semen carry T-lymphocytes, it is clear that blood and semen can transfer HIV in its provirus form.

The passage of HIV among healthy individuals is but one consequence of HIV's residing in T-lymphocytes in its DNA form. Another consequence is that the virus escapes body defenses because it is inside a cell. The body's antibodies cannot reach

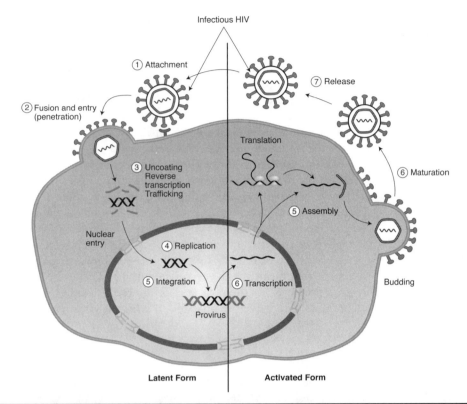

FIGURE 2.14

The replication cycle of human immunodeficiency virus (HIV). (1 and 2) The virus unites with the host cell as the viral envelope contacts the cell surface membrane via the gpl20, gp41, and CD4 molecules. (3) After union, the capsid is stripped away and the RNA of HIV is released. Reverse transcriptase uses the RNA as a template and synthesizes a double-stranded DNA molecule. (4) The new DNA migrates to the cell nucleus and is integrated into one of the host cell's chromosomes as a provirus. (5) The host cell may divide indefinitely with the provirus in position. No infectious particles are made; HIV is in its latent form. (6) The provirus provides the genetic code for the enzymes that will mediate the production of capsid proteins plus RNA genomes. This is the stage of active HIV production. After assembly of viral components (step 5), the new HIV particles move to the cell surface membrane and bud through, thereby acquiring their envelopes (steps 6 and 7). The cycle is now ready to repeat itself. Modified from an illustration by Kate Bishop, MRC National Institute for Medical Research, UK.

the provirus (antibodies do not enter cells), nor can the provirus be attacked by the body's white blood cells (phagocytes) that normally engulf and destroy microorganisms. Moreover, any drug developed against HIV must penetrate living cells and eliminate the provirus without killing the cell, a formidable task. It is also unlikely that a vaccine against AIDS will be able to protect the hundreds of thousands of Americans who carry HIV because they are already infected.

Assembly

Researchers are certain that the DNA provirus continues to encode new HIV particles within T-lymphocytes or other host cells. As new viruses are encoded and synthesized, the injured T-lymphocytes are able to replace themselves and the body defenses continue their response, but the virus eventually overwhelms the cells

The synthesis of HIV components appears to depend on the activity of sections at the end of the nucleic acid molecules of the viral genome. These regions are believed to direct cellular enzymes to form RNA, using the genetic code provided by the proviral DNA. Certain of the RNA strands are used to construct a new generation of HIV genomes, whereas other RNA strands carry genetic instructions for the capsid proteins and reverse transcriptase molecules of the new viruses. During this **active** phase of viral production, genomes, capsid proteins, and enzyme molecules are then assembled at the edge of the cell. Here they form a somewhat circular structure that binds to the cell membrane.

At this point, an enzyme called a *protease* comes into play. Proteases are protein cleaving enzymes. It snips out enzyme molecules and capsid proteins from preliminary protein molecules and then sections the capsid proteins into capsomere segments. The capsomeres unite and form an icosahedral capsid. This icosahedron then collapses to yield a bullet-shaped capsid surrounding the RNA genome and enzymes. Now the virus encloses itself in a portion of the host cell's surface membrane, which will become the envelope. Before forming the envelope, however, the patch of membrane unites with the virus-specified glycoproteins, gp120 and gp41. These glycoproteins then extend from the membrane as spikes. The assembled viral nucleocapsid localizes to just under the membrane of the cells where the gp120 and gp41 have been inserted and the virus prepares for the next step of the replication cycle in which the new viral particle is prepared to leave the host cell in which it was made and go on to infect new cells.

Release

As HIV leaves the cell, it also acquires its envelope. In this process, known as *budding*, the virus coats itself with the cell membrane and pinches off, taking the membrane and spike proteins with it as the envelope. The virus is then released to the extracellular environment (**Figure 2.15**). The cycle of HIV replication is now complete. When the virus combines with a neighboring T-lymphocyte, macrophage, or other cell, the cycle is repeated.

The release of HIV from the cell appears less explosive than for many other types of viruses. How then do the HIV particles destroy the cell? One possibility is that as the virus buds through the cell membrane it tears holes in the membrane. Cytoplasmic leakage through the holes may bring on cell destruction.

Another possibility is that HIV induces T-lymphocytes to cling together in a giant, multinucleated cell mass called a syncytium. Syncytial formation is possible because an HIV-infected cell manufactures gp120 molecules and deposits the molecules on its surface membrane. When an infected cell later encounters a healthy cell, the gp120 molecules of the infected cell bind to the CD4 molecules of the healthy cell. The two cells then continue to fuse with other cells until as many as 500 cells have combined to yield a huge syncytium (**Figure 2.16**). The supply of useful T-lymphocytes is thus depleted, and the efficiency of the immune system is impaired.

The deleterious effects of syncytia were shown in 1998 experiments conducted at the University of Iowa. Researchers led by David R. Soll showed that mobile syncytia can disrupt membranes containing collagen, membranes similar to those found in lymph nodes. The syncytia can also destroy the tissues lining blood cells and open microscopic holes in the blood vessels. These tendencies may explain why AIDS patients often experience disrupted lymph nodes and leaky blood vessels.

Still another possible mechanism of cell destruction is that the provirus is triggered to direct the synthesis of an unusually large number of new HIV particles.

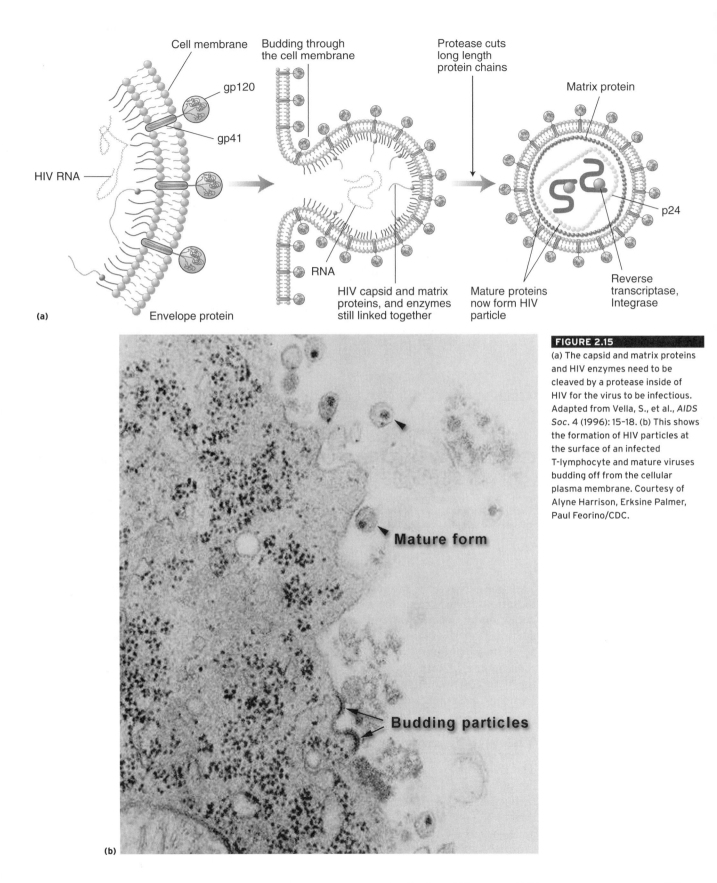

(a) Cell membrane

Budding through the cell membrane

Protease cuts long length protein chains

gp120

gp41

Matrix protein

HIV RNA

p24

Envelope protein

RNA

HIV capsid and matrix proteins, and enzymes still linked together

Mature proteins now form HIV particle

Reverse transcriptase, Integrase

Mature form

Budding particles

(b)

FIGURE 2.15

(a) The capsid and matrix proteins and HIV enzymes need to be cleaved by a protease inside of HIV for the virus to be infectious. Adapted from Vella, S., et al., *AIDS Soc.* 4 (1996): 15–18. (b) This shows the formation of HIV particles at the surface of an infected T-lymphocyte and mature viruses budding off from the cellular plasma membrane. Courtesy of Alyne Harrison, Erksine Palmer, Paul Feorino/CDC.

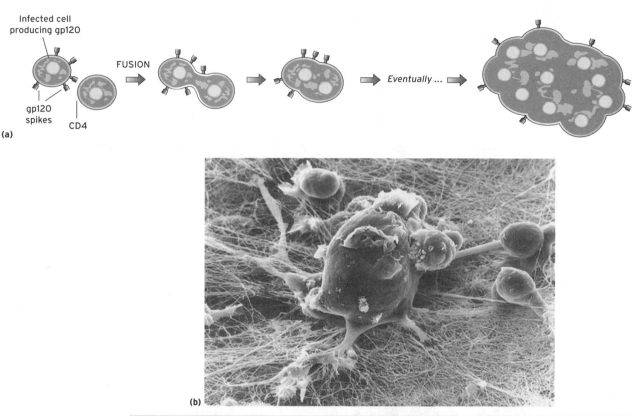

FIGURE 2.16

Syncytium formation. (a) An infected T-lymphocyte displays gp120 molecules on its surface as a result of infection with HIV. It unites with an uninfected T-lymphocyte having CD4 receptor sites on its surface. The union results in a fused T-lymphocyte. Additional uninfected T-lymphocytes unite with this fused cell (shown below the fused cell). The result of multiple fusions is a multinucleated giant cell mass known as a syncytium. (b) An electron micrograph of an HIV-infected syncytium on a bed of collagen fibers. Courtesy of David R. Soll, University of Iowa.

This pulse of activity rapidly uses up cellular components, which leads to cell disintegration. Viruses of many other diseases also induce this destruction. T-lymphocyte depletion by this process and others will soon manifest itself in low immune system efficiency.

Before continuing, we should note that HIV also infects brain cells and depletes their numbers. Brain cell destruction leads to a form of AIDS called HIV encephalopathy. Although studied less thoroughly than T-lymphocyte infection, HIV encephalopathy is an equally serious form of AIDS.

The HIV Genome

The genetic information for the replication of HIV is contained in a genome consisting of 9,747 nucleotides. These nucleotides are known to exist as nine units of genetic activity, or genes (a human cell, in contrast, is currently estimated to have approximately 20,000 to 25,000 genes). HIV is the most complex retrovirus that exists. It contains more genes than all other human retroviruses. When HIV was discovered, some argued that the virus could have been man-made in the laboratory. Because

BOX
2.3

But How Do They Know?

It is generally accepted by the scientific community that HIV is the cause of AIDS. But, critics say, the final proof demands inoculation of pure virus into a human volunteer and development of AIDS in that individual. Such an experiment is ethically unthinkable.

How, then, do scientists associate HIV with AIDS? Consider the following:

- Almost every case of AIDS has occurred in someone who has been shown to harbor HIV.
- In no country where AIDS is present is HIV absent.
- In no country where HIV is present is AIDS absent.
- Blood banks began testing blood for HIV in 1985 and removing HIV-contaminated blood from circulation; the number of transfusion-associated AIDS cases then declined sharply.
- In a long-term study of homosexual men in San Francisco, the incidence of AIDS rose as the rate of HIV infection rose.

- In a study of blood transfusion recipients, 19 recipients developed AIDS; in each of the 19 cases, the donor of the blood was located and found to be positive for HIV.
- In every country of the world studied so far, AIDS has appeared only after HIV has appeared.
- All HIV-positive hemophiliac patients have died from symptoms resembling AIDS; put another way, not a single HIV-negative hemophiliac is known to have died from symptoms resembling AIDS.
- Using highly sophisticated technology, HIV can be located in almost 100% of individuals who have AIDS.
- Laboratory tests show that HIV multiplies in and destroys the very T-lymphocytes whose gradual loss is a signpost of AIDS.

These are but a few of the data pointing to HIV as the AIDS virus. Although arguments may be raised against individual data, the cumulative data provide persuasive evidence that HIV is, indeed, the virus that causes AIDS.

HIV was infecting humans before the 1980s, a time period in which gene technology was not available, the possibility that HIV was concocted in a biology laboratory can be ruled out. In the scientific community it is well accepted that HIV is the cause of AIDS for several reasons (Box 2.3). That being said, there does remain a handful of tenacious dissenters led by Peter H. Duesberg that continue to strongly oppose that HIV causes AIDS (Box 2.4).

Nine Genes

The nine genes of HIV consist of segments of DNA, observed in HIV in its proviral form. Three genes providing genetic information for HIV's structural components are termed *gag, pol,* and *env*. These genes contain the genetic codes for the capsid proteins, viral enzymes, and envelope proteins, respectively. At one point in the genome, the genetic codes for *gag* and *pol* overlap, a situation observed with many genetic codes in nature.

The remaining six genes of HIV appear to be regulatory genes that control such things as penetration of the host cell, uncoating of the HIV genome, production of viral DNA, and integration of the provirus. One regulatory gene is the transactivator gene (*tat*). Research evidence indicates that this gene is responsible for the burst of HIV replication that occurs when infected T-lymphocytes are stimulated. This type of stimulation can be caused by the chemical components of microorganisms or other foreign molecules known as antigens. The *tat* gene is unique in that it is composed of

Peter H. Duesberg, Rebel with an Anti-HIV Cause

Peter H. Duesberg is a professor of molecular and cellular biology at the University of California, Berkeley. Duesberg argued early on during the HIV epidemic that AIDS was caused by recreational drug use and that HIV was a nonpathogenic virus that was along for the ride, like a harmless "passenger virus." He also argued that nitrite inhalants were the cause of Kaposi's sarcoma. Both of these arguments were proven to be incorrect. HIV-positive drug users developed AIDS whereas HIV-negative drug users did not get AIDS, proving recreational drug use was not the cause of AIDS. It is now known that Kaposi's sarcoma is caused by a herpesvirus. Duesberg also argued that the HIV antiviral drug, AZT, caused AIDS. AZT did not cause AIDS in placebo-controlled studies. In 1996, scientist Barry Bloom challenged Duesberg to inject himself with a culture of HIV to prove it was harmless. Duesberg did not accept the challenge.

In 2009, Duesberg was the lead author of an article that was not peer-reviewed titled, "HIV-AIDS hypothesis is out of touch with South African AIDS—a new perspective," published in the journal *Medical Hypotheses*. The article was permanently withdrawn in 2010 with a statement from the journal editor that the controversial opinions about the causes of AIDS "could potentially be damaging to global public health." The article was republished in 2011 after peer review in a revised form titled, "AIDS since 1984: No evidence for a new, viral epidemic—not even in Africa," in the *Italian Journal of Anatomy and Embryology*. Duesberg referred to it as "a new victory in our long quest for a scientific theory of AIDS."

Cornell University HIV researcher John Moore, who had filed a complaint against the publisher of the 2009 Duesberg article, continues to disagree with Duesberg but also believes that Duesberg's latest article will not have a significant effect or impact on politics and public health in South Africa like Duesberg's controversial opinions did in the 2000s.

two separate segments of DNA within the genome (**Figure 2.17**). Mutant strains of HIV that lack this gene react far less actively when antigens stimulate the T-lymphocytes. The gene apparently encodes a protein that increases the expression of HIV genes, thus leading to increased synthesis of new viruses.

Another regulatory gene is the "*r*egulator of *e*xpression of *v*iral protein" gene (***rev***). This gene, also composed of two separate DNA segments, enables the provirus to encode regulatory proteins or viral components, depending on proviral needs at the time. When the provirus lies dormant, for example, *rev* encodes proteins that prevent viral replication (the infection remains silent), but when a pulse of viral reproduction is to take place, *rev* encodes proteins to be used for new viruses (the infection is activated). In the latter case, *rev* apparently acts by increasing the efficiency with which the genetic code in the proviral DNA is translated to make proteins for producing new viruses. After replication is under way, *rev* may interact with *tat* to produce the slow and controlled level of viral production that characterizes HIV.

Still another regulatory gene is the "*n*egative regulatory *f*actor" gene (***nef***). This gene encodes a protein that remains near the nuclear membrane and enhances the movement of genetic messages into the cytoplasm of the host cell. From this position, the protein can also influence biochemical messages to other genes within the nucleus. Apparently these messages profoundly suppress all further gene expression, leading to a dormant provirus. A high concentration of proteins encoded by *rev* apparently can suppress *nef* activity and thereby switch on viral replication.

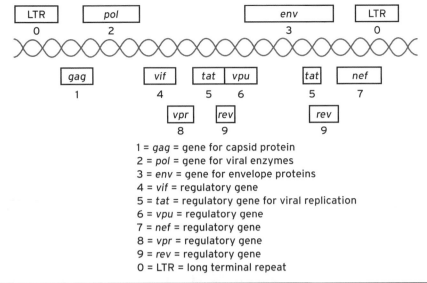

1 = *gag* = gene for capsid protein
2 = *pol* = gene for viral enzymes
3 = *env* = gene for envelope proteins
4 = *vif* = regulatory gene
5 = *tat* = regulatory gene for viral replication
6 = *vpu* = regulatory gene
7 = *nef* = regulatory gene
8 = *vpr* = regulatory gene
9 = *rev* = regulatory gene
0 = LTR = long terminal repeat

FIGURE 2.17

A diagrammatic representation of the genome of the human immunodeficiency virus, displaying nine known genes. The genes have been numbered because portions of two genes (*tat* and *rev*) occur in separate segments. The long terminal repeat (LTR) segments at the ends of the genome are not genes, but are shown for orientation purposes. Activities encoded by the genes are summarized in the text.

The function of another regulatory gene is to encode *v*iral *p*rotein *R*, and hence, it is called **vpr**. This protein assists the transport of viral DNA into the nucleus, where it will integrate into a chromosome. The protein also appears to assist the assembly and release of new HIV particles from the cytoplasm of the host cell.

The regulatory gene **vif** is necessary for the reverse transcription of RNA to DNA. The protein encoded by this gene acts at the end of the process and may function in the correct winding of the new proviral DNA molecule.

Still another regulatory gene called **vpu** encodes *v*iral *p*rotein *U*. This protein breaks down the CD4 receptor protein normally produced within the cytoplasm before it is transported to the cell surface. Unless destroyed, this CD4 protein would inhibit the budding of HIV out of the cell. The activity of the *vpu* gene is thus essential to continued HIV production.

The genome of HIV also contains at its opposite ends two stretches of DNA called *long terminal repeats (LTRs)*. These LTRs are not genes as such, but they apparently direct host cell enzymes to transcribe the biochemical information of proviral DNA into RNA during viral replication. This activity is performed by defining the initiation site for RNA production. Thus, the LTRs play a key role in the activation of the process leading to new HIV particles.

Biochemical and genetic analyses have made it clear that the physiology of HIV is quite complex. The elaborate set of genetic controls reflects a well-adapted virus that carries out its activities in a carefully coordinated manner. In the practical sense, knowledge of these activities can provide possible bases for drugs to control HIV replication.

Three Groups of HIV

Analysis of the HIV genome has made it clear that permanent genetic changes called *mutations* occur often during the replication of HIV. Indeed, one

Genetic relatedness among strains of HIV. The M, N, and O groups are shown originating from a common ancestor. Note that group N viruses are closely related to a strain of SIV and that both may have originated from a common ancestor. HIV-2 is distant from the three strains of HIV-1 showing that it is poorly related to HIV-1. However, it is closely related to an SIV strain. The various lettered subgroups of HIV-1 are indicated as originating from group M HIV.

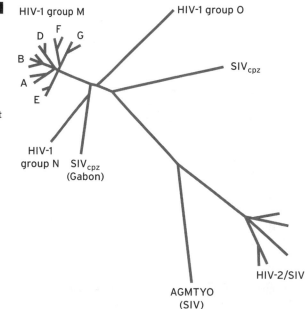

researcher has estimated that at least one mutation takes place in the HIV genome each time a round of replication occurs. Compounding the likelihood of mutation is the presence of two RNA molecules per virus, either of which can mutate. Moreover, mutation is possible in the biochemical step in which RNA is used to synthesize DNA. Most mutations have no effect on the virus, but sometimes the genetic flexibility leads to drug resistance or encourages the virus to escape the body's immune response; it also increases the difficulty of producing an effective vaccine.

Mutations in the HIV genome have probably led to the three groups of HIV-1 now known to exist. The first group is the "main" or **M group**. Viruses in this group are responsible for 99% of all AIDS cases in the world. The M group contains ten subgroups (also called *clades*) based on variations in the *gag* and *env* genes. The subgroups, lettered A to J, are present on all continents. Their presence makes developing a universal vaccine difficult because the vaccine must take all subgroups into account. Subgroup B predominates in North America.

The second group of HIV-1 is the "outlier" or **O group**. Viruses in this group have been found in West African countries such as Gabon and Cameroon. They cause less than 1% of AIDS cases, and their RNA shows less than a 50% watch with the RNA of the M group. Group O viruses were not detected in the United States until 1996, and fewer than five cases of AIDS have been related to viruses in the group. For this reason, tests for group O viruses are not routinely performed in the United States. **Figure 2.18** shows how the O group is related to the M group.

The third group, the **N group**, was discovered in 1998 by French researchers led by François Simon. Appropriately named the "new" (or "non-M-non-O") N group, the viruses are genetically distinct from those in the M and O groups. They were isolated from a small population of individuals in Cameroon. Like the O group viruses, the N group viruses are not routinely sought in the HIV tests. Their similarity to SIV

Mapping HIV

Western and Central Europe

- 820,000 adults and children living with HIV in 2009.

- 31,000 adults and children newly infected with HIV in 2009.

- 8,500 AIDS-related deaths in 2009.

- Unprotected sex between men continues to dominate patterns of HIV transmission.

- In the United Kingdom, 44% of the people newly infected with HIV acquired HIV abroad, mainly in sub-Saharan Africa.

- The rate of HIV infections in Greece has increased sharply since the beginning of the Greek financial crisis (5.4% in 2010 to 8.4% in 2011).

- The government of Greece canceled the needle exchange program in 2011 because of budget cuts.

- Injecting drug users' rate of HIV infection in Greece has risen 1,500% which correlates with the cancellation of the needle exchange program.

Data from Hellenic Center for Disease Control and Prevention (Athens, Greece) and *UNAIDS Report on the Global AIDS Epidemic*, 2010.

found in chimpanzees suggests that the N group's ancestors might have been transmitted to humans from nonhuman primates.

Inhibiting Viruses

Viruses have a structure and replication pattern not found elsewhere in the natural world; however, viruses are composed of chemical compounds, and many of the principles of destruction that apply to all the microorganisms apply to viruses. Viruses, for example, may be inactivated by many of the physical and chemical agents routinely used for other microorganisms. Heat and ultraviolet (UV) radiation are typical of the physical agents. Heat alters the structure of proteins and nucleic acids, and the heat of boiling water destroys most viruses in a matter of seconds. UV radiation affects nucleic acids by binding together adjacent portions of their molecules. Germicidal lamps that emit UV radiation thus eliminate viruses after a few seconds of exposure.

Among the antiviral chemical agents are compounds of chlorine, iodine, and silver as well as phenol, formaldehyde, and detergents. Each has a different effect on viruses, as **Figure 2.19** illustrates, and most are available commercially. It is worth remembering that viruses, including HIV, are no more difficult to destroy than any other microorganisms when they are outside living cells. Indeed, for fragile viruses such as HIV, destruction is rather swift.

In contrast, destroying viruses with drugs is very difficult. Viruses lack complex structures, chemical reactions, or any biochemical activity, and thus, there is little for

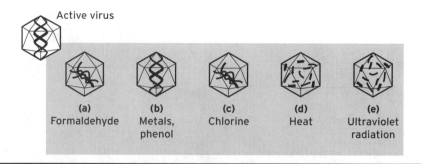

FIGURE 2.19

Six methods of inactivating viruses. (a) Formaldehyde combines with free amino groups on nucleic acid bases. (b) Metals and phenol react with the protein of the viral capsid. (c) Chlorine combines chemically with nucleic acids in viruses. (d) Heat denatures proteins of the capsid. (e) Ultraviolet radiation binds together thymine molecules in the genome and distorts the nucleic acid.

drugs to interfere with. Penicillin destroys the cell wall of a gram-positive bacterium, but viruses don't have cell walls; therefore, penicillin is useless.

Nevertheless, certain antiviral drugs are available for treating people infected with viruses. Acyclovir, for instance, is a drug that interferes with the synthesis of nucleic acids by herpes simplex viruses, and **azidothymidine (AZT)** works similarly with HIV. Useful drugs such as these are rare because drugs often exhibit toxic side effects and can interfere with essential chemical reactions in the body. For HIV, the problem is compounded because the virus remains inside infected T-lymphocytes and brain cells. Still, molecular biologists have identified several steps in the replication cycle of HIV that may yield to interference by antiviral drugs. Among these steps are the binding of HIV to the CD4 receptor site of the host cell, the shedding of the viral coat in the host cell's cytoplasm, the activity of reverse transcriptase (the point at which AZT works), the assembly of viral components into new viruses, and the budding of new HIV particles from the host cell.

The major public health approach to viral diseases is through prevention, rather than treatment, and the major weapon for prevention is **vaccines**. In contemporary medicine, vaccines exist for such viral diseases as polioviruses, measles, mumps, and rubella (the MMR vaccine), chickenpox, hepatitis B, rabies, some strains of influenza, and papillomaviruses. Certain vaccines contain attenuated viruses, that is, viruses that multiply at a low rate. Other vaccines contain inactivated viruses, that is, viruses that cannot multiply any longer but are structurally integrated. Still others contain laboratory-produced viral fragments or subunits. Vaccines stimulate the body's immune system in various ways, one of which is to induce the system to produce protein molecules called **antibodies**. These antibodies then circulate and provide surveillance against the active virus should it enter the body.

It is conceivable that a vaccine may hold the key to interrupting the HIV epidemic. Many problems are associated with producing such a vaccine, however. Among the problems that must be resolved are deciding which form the vaccine should take (attenuated, inactivated, or subunit vaccines), dealing with the various components of the envelope spikes, locating an animal model for testing purposes, and measuring the vaccine's effectiveness in field trials. There is also the possibility that many forms of HIV are causing AIDS in multiple overlapping epidemics. If so, then the usefulness of a single vaccine would be reduced. For the time being, the pessimism of the early 1980s has been replaced with cautious optimism for the new century. In many circles, the question is no longer "If?" but rather "How soon?"

Viruses appear to be transitional forms between inert molecules and living organisms. They are extremely small particles of nucleic acid surrounded by a protein coat and, in some cases, an envelope with spikes. HIV has a genome of RNA, a capsid that is first icosahedral and then bullet shaped and an envelope containing spikes. Molecules of the enzyme reverse transcriptase exist among the strands of RNA in the HIV genome.

Viruses begin their replication process by uniting with host cells. They then penetrate to the cytoplasm of the cells, where the viral nucleic acid is released. Next, the nucleic acid molecule directs the synthesis of new viruses, usually destroying the cell in the process. For HIV, the union between virus and T-lymphocyte involves the glycoproteins gp120 and gp41 of HIV and the CD4 receptor site of the cell as well as several co-receptors. After the RNA has been released in the cell's cytoplasm, reverse transcriptase uses the RNA as a template to synthesize DNA. The resulting DNA then takes up residence in the chromosomal DNA of the cell, where it is called a provirus. From this location, the provirus encodes the synthesis of new HIV particles. Membrane tearing, syncytium formation, and rapid viral replication may lead to the death of T-lymphocytes. Brain cells and macrophages may also be involved.

The genetic information for HIV replication is contained in nine genes, three of which are structural genes. These genes, named *gag, pol,* and *env,* encode the viral capsid proteins, enzymes, and envelope proteins, respectively. The remaining six genes appear to be regulatory genes that oversee such processes as HIV replication, the rate of protein production, and the ability to remain latent in the host cell. Three major groups, the M, N, and O groups, have been identified for HIV-1 based on genetic relatedness. The M group occurs worldwide and is responsible for 99% of AIDS cases in the world.

Like other viruses, HIV can be inactivated with physical and chemical agents such as heat and chlorine; however, drug therapy for viruses is difficult because of the structural and biochemical simplicity of viruses and the toxicity of antiviral drugs to body cells. The fact that HIV resides in cells as a provirus adds to the difficulty. Antiviral drugs, however, are useful in halting the synthesis of DNA and in retarding the formation of the capsid. Vaccines against viruses can contain weakened or inactivated viruses, viral fragments, or subunits. For HIV, however, multiple problems must be resolved before a vaccine becomes available.

Healthline Q & A

Q1 How do viruses multiply?

A Viruses do not multiply by the method of cell division that we associate with most living things. Instead, viruses invade a living cell and release their nucleic acid, and the latter directs the synthesis of new viruses utilizing structures and chemical components of the cell. No other object studied in biology multiplies in this manner. The process results in hundreds or thousands of copies of the virus as well as a poorly functioning or dead cell.

Q2 I've noticed that the incubation period for AIDS is unusually long. Why is that so?

A The HIV that causes AIDS follows the general pattern of viral replication, with an important exception. Once released in a host cell, the viral nucleic acid (RNA) serves as a template for the synthesis of a DNA molecule that incorporates itself into one of the cell's chromosomes. HIV remains in this DNA form in the infected cell, and it may be many months or years before enough viruses are encoded to overwhelm the body's immune system; therefore, the incubation period tends to be very long in some individuals.

Q3 Can viruses be controlled with antiseptics and disinfectants?

A Very definitely. Viruses are particles of nucleic acid, proteins, and other organic substances. As such, they are susceptible to the destructive effects of myriad antiseptics and disinfectants. In broad terms, any agent useful against bacteria will be equally useful against viruses.

Q4 How are vaccines useful for the prevention of viral disease?

A Vaccines contain altered forms of viruses, either weakened viruses that multiply slowly in the body or inactivated viruses that are unable to multiply at all. When introduced into the body, a vaccine stimulates the immune system to produce highly specific antibody molecules. The antibody molecules react with and destroy the active form of the virus if it should appear at a later date in the body. Destruction of the virus prevents the disease.

REVIEW

The major thrust of this chapter has been to survey the structure and mode of replication of the viruses, with particular reference to the HIV. To gauge the extent of your learning, select the letter that corresponds to the best answer for each of the following. Answers are listed in elsewhere in the text.

_____ 1. All viruses consist of two basic components known as
  a. cytoplasm and a cell membrane.
  b. an envelope and a protoplast.
  c. a genome and an envelope.
  d. a cell membrane and a protoplast.
  e. a capsid and a genome.

_____ 2. Which of the following is true regarding the genome of a virus?
  a. The genome contains both DNA and RNA.
  b. An envelope is found within the genome.
  c. The genome contains complex carbohydrates.
  d. Only DNA viruses have a genome.
  e. None of the above is true.

_____ 3. Different forms of HIV are possible because
  a. the capsomeres vary among strains of HIV.
  b. chemical components of the spikes vary among HIV strains.
  c. the envelope is present in some strains of HIV but absent in others.
  d. not all strains of HIV have reverse transcriptase.
  e. capsid proteins vary among strains of HIV.

_____ 4. HIV is able to replicate
  a. only within a living cell.
  b. only if the enzyme amylase functions effectively.
  c. only after the envelope has entered the cytoplasm of a cell.
  d. in either brain cells or liver cells.
  e. in media normally used to cultivate bacteria.

_____ 5. The CD4 molecule is a protein that
  a. is located on the surface of T-lymphocytes.
  b. reacts with an envelope glycoprotein of HIV.
  c. serves as a receptor site for HIV.
  d. All the above are true.
  e. None of the above is true.

6. For HIV to replicate within a host cell,
   a. the envelope must enter the host cell's cytoplasm.
   b. the viral capsid must be converted to carbohydrate.
   c. the virus must be engulfed by a white blood cell.
   d. the viral nucleic acid must be released from the capsid.
   e. the cytoplasm of the host cell must contain CD4 receptor sites.

7. Reverse transcriptase, a key enzyme found in HIV, is so named because it
   a. reverses the generally accepted flow of biological information.
   b. converts DNA to RNA, a reversal of the expected biochemistry.
   c. reverses the replication cycle of HIV and yields broken viruses.
   d. changes capsid proteins into nucleic acids for viral synthesis.
   e. reverses the pathway of HIV from inside the host cell to outside.

8. In its proviral form within the host cell, HIV
   a. can remain for an undetermined period of time.
   b. can escape body defenses such as antibodies.
   c. can be passed from one individual to another.
   d. All of the above are true.
   e. None of the above is true.

9. Which of the following pairs is mismatched?
   a. Spikes/envelope
   b. Reverse transcriptase/T-lymphocytes
   c. Provirus/DNA
   d. Protein/capsid
   e. HIV/retrovirus

10. Blood and semen are able to transmit AIDS because
    a. antibodies do not react with HIV in blood or semen.
    b. blood or semen may contain viral capsids.
    c. drugs are too toxic for use in blood or semen.
    d. viruses do not replicate in blood or semen.
    e. blood or semen may contain infected T-lymphocytes.

11. HIV may destroy its host cell in the body by
    a. inducing the cell to multiply without control.
    b. blocking the passage of important nutrients into the cell.
    c. preventing the cell from uniting with other body cells.
    d. tearing holes in the cell membrane as the virus buds through.
    e. inhibiting the cellular production of reverse transcriptase.

12. Syncytium formation takes place
    a. when HIV particles cling together.
    b. after the transactivator gene has been activated.
    c. when infected and uninfected T-lymphocytes unite to form a mass of cells.
    d. only in uninfected T-lymphocytes.
    e. only when a virus contains DNA in its genome and gp41 in its spikes.

_____ **13.** The drug AZT acts against HIV by
  a. interfering with the construction of viral nucleic acid.
  b. precipitating the protein in the HIV capsid.
  c. altering the structure of the HIV envelope so as to induce precipitation.
  d. All of the above are true.
  e. None of the above is true.

_____ **14.** Compared with other viruses, HIV is generally
  a. more resistant to acid and heat.
  b. more resistant to UV radiation and drying.
  c. more susceptible to physical and chemical treatments.
  d. more susceptible to treatment with drugs such as penicillin.
  e. more susceptible to acyclovir.

_____ **15.** The genetic information for HIV replication is contained in
  a. nine genes.
  b. the capsid of the virus.
  c. the reverse transcriptase of the virus.
  d. the envelope protein molecules in the genome.
  e. 9,000 protein molecules in the genome.

FOR ADDITIONAL READING

Abram, M.E., 2010. "Nature, position, and frequency of mutations made in a single cycle of HIV-replication." *Journal of Virology* 84:9864–9878.

Bronshtein, T., et al., 2011. "Cell derived liposomes expressing CCR5 as a new targeted drug-delivery system for HIV-infected cells." *Journal of Controlled Release* 151:139–148.

Cohen, J., 1994. "The Duesberg Phenomenon: A Berkeley virologist and his supporters continue to argue that HIV is not the cause of AIDS. A three-month investigation by *Science* evaluates their claims." *Science* 266:1643–1644.

Corbyn, Z., 2012. "Paper denying HIV-AIDS link secures publication." *Nature* doi:10.1038/nature.2012.9737.

Duesberg, P. H., et al., 2011. "AIDS since 1984: No evidence for a new, viral epidemic—not even in Africa." *Italian Journal of Anatomy and Embryology* 2:73–92.

Margolis, D. M., 2007. "Confronting proviral HIV infection." *Current HIV/AIDS Report* 4:60–64.

Muoscadet, J-F., et al., 2010. "Resistance to HIV-1 integrase inhibitors: A structural perspective." *Drug Resistance Updates* 13:139–150.

Roux, K. H., et al., 2007. "AIDS virus envelope spike structure." *Current Opinion Structural Biology* 17:244–252.

Smyth, R. P., et al., 2012. "The origin of genetic diversity in HIV-1." *Virus Research* http://dx.doi.org/10.1016/j.virusres.2012.06.015.

Tilton, J.C., Doms, R. W., 2010. "Entry inhibitors in the treatment of HIV-1 infection." *Antiviral Research* 85:91–100.

CHAPTER 3

The Immune System and HIV-1

LOOKING AHEAD

Because it provides a specific defense against infectious disease, the immune system is one of the key systems of the body. This chapter explores how the immune system functions and how the human immunodeficiency virus (HIV) interacts with it, resulting in AIDS. On completing the chapter, you should be able to . . .

- Describe the fetal development of the immune system and understand which cells are central to the system's formation and operation.
- Explain how T-lymphocytes provide specific defense against certain microorganisms, such as fungi and protozoa.
- Summarize the process by which stimulation of B-lymphocytes results in antibodies against viruses, bacteria, and other microorganisms, and describe how antibodies act in body defense.
- Conceptualize how HIV depresses the functions of the immune system by depressing CD4+ T-cell activity.
- Describe the genetics of HIV resistance with reference to the immune system.
- Summarize the features of AIDS that make it a unique disease, accompanied by phenomena that do not occur in other infectious diseases.
- Discuss new research developments toward a "cure" for HIV infection.

INTRODUCTION

The human body is shaped like a doughnut. Just as the hole passes through the doughnut, the gastrointestinal tract passes through the body. Being in the hole does not mean being in the doughnut, nor does being in the gastrointestinal tract mean that an object is in the body. The respiratory tract and urinary tract technically lie outside of the body as well. One is led to conclude that the body is a closed container (**Figure 3.1**).

The construction of the body is of more than passing interest. Indeed, it is essential to the body's resistance to disease. If a microorganism is to invade the body tissues, it must pass the cellular barrier separating the interior of the body from the exterior. Resistance like that provided by the cellular barrier is said to be **nonspecific resistance** because it operates against all forms of microorganisms, not just one or more specific forms. Other forms of nonspecific resistance include powerful enzymes in the tears and saliva, which help resist infection in the eyes and mouth, respectively; the high acid content of the stomach, which presents a chemical barrier to the intestine beyond; the sticky mucus that traps microorganisms in the

FIGURE 3.1

The body as a closed container. (a) An object is clearly outside the body. (b) The object remains outside the body as the outer surface bulges inward. (c) Despite additional inward bulging, the object remains outside. (d and e) The object continues to be outside the body, even though many surface alterations have occurred. Being within a body cavity such as the gastrointestinal tract or lung does not necessarily imply that an object is in the body's internal environment.

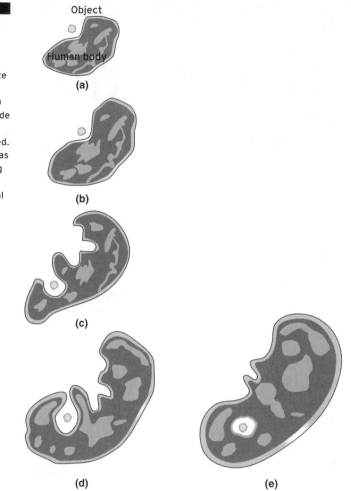

respiratory tract; and phagocytosis, the complex process by which the body's white blood cells engulf and destroy microorganisms.

Sometimes, however, nonspecific resistance needs to be supplemented. The cells of the skin, for example, may be penetrated during an arthropod bite (the microorganisms causing Lyme disease, West Nile encephalitis, and malaria are introduced this way); microorganisms may be able to resist the body's protective enzymes, or microbial toxins may injure or destroy the white blood cells dispatched to eliminate the microorganisms. In cases such as these, the body responds with a second major type of resistance called **specific resistance.**

Specific resistance is resistance mounted against a single species of microorganism or a single type of chemical compound. The resistance is stimulated by that microorganism or compound and is directed solely at the stimulant. The specific resistance that arises during an attack of measles, for instance, helps rid the body of the measles virus only. The resistance then remains in the body for many years (often for life) so that should the measles virus return it cannot gain a foothold in the body. Ultimately, specific resistance prevents serious injury and death (**Figure 3.2**).

The body system that provides specific resistance is the **immune system.** The term "immune" is derived from the Latin *immuno,* meaning "to be free from"; the immune system frees the body of foreign microorganisms and chemicals and keeps it free of

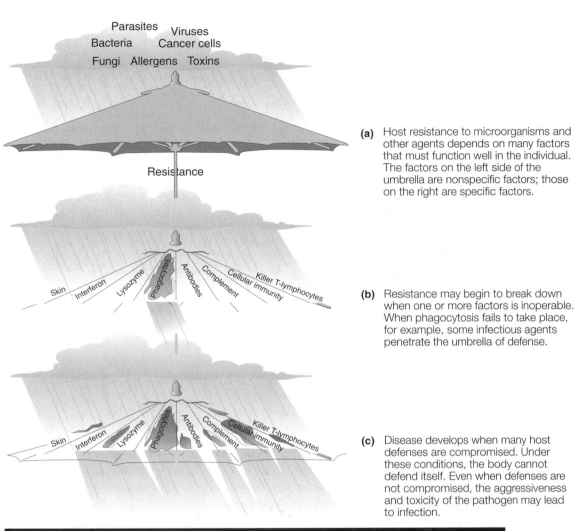

(a) Host resistance to microorganisms and other agents depends on many factors that must function well in the individual. The factors on the left side of the umbrella are nonspecific factors; those on the right are specific factors.

(b) Resistance may begin to break down when one or more factors is inoperable. When phagocytosis fails to take place, for example, some infectious agents penetrate the umbrella of defense.

(c) Disease develops when many host defenses are compromised. Under these conditions, the body cannot defend itself. Even when defenses are not compromised, the aggressiveness and toxicity of the pathogen may lead to infection.

FIGURE 3.2

The relationship between host resistance and disease.

them. The system is a complex series of cells, chemical factors, and processes in which white blood cells, called **lymphocytes**, respond to and eliminate foreign agents. The agents that elicit the response are called **antigens**. Elimination of the antigens can occur through direct destruction by lymphocytes or through indirect destruction by specialized protein molecules called **antibodies**. As we discuss here, the antibodies destroy the antigen directly and target it for destruction by other cells.

The immune system bears the brunt of attack in most individuals infected by HIV. The consequence of HIV infection and destruction of the immune system is that the individual is susceptible to infection from a wide range of organisms with no way of establishing a protective immunity. The result of these **opportunistic infections** is serious illness and death. Thus, to comprehend fully the nature of HIV infection and AIDS, it is important to have a clear idea of the workings of the immune system. Indeed, the very name **"acquired *immune deficiency* syndrome"** (AIDS) refers to a poorly functioning immune system.

We study the immune system by first exploring its origin in the fetus. We then investigate the substances that stimulate its operation and outline the two major branches of the system. Once a firm foundation on the mechanics of the system has been set, we study the effects of HIV on the system's cells and explore how AIDS comes about.

Overview of the Immune System

The human immune system is composed of specialized cells known as **leukocytes** that develop in the **bone marrow** and **thymus** and traffic throughout the body in a system of vessels known as the **lymphatic system** (**Figure 3.3a**). Leukocytes must traffic throughout the system of lymphatic vesicles in surveillance of foreign infectious agents such as bacteria, fungi, parasites, and viruses. These lymphatic vesicles occasionally meet at intersections, known as **lymph nodes** (Figure 3.3a), which are meeting locations, or hubs, for the leukocytes to communicate the presence of an infection. Frequently, lymph nodes then become swollen as the cells grow in preparation to destroy the infection. Often the swelling node, for this reason, is near to the site of infection. In addition to lymph nodes, other leukocyte hubs include the adenoids, tonsils, spleen, and Peyer's patches (Figure 3.3a).

Once in the lymphatic system of vessels, the leukocytes can also enter the circulatory system, which broadens the ability of these cells to migrate throughout the body and recognize and target infectious agents. This is made possible at capillary beds, where the leukocytes can traverse the walls of tangles of narrow lymphatic vessels that have come into close contact with tangles of narrow blood vessels (**Figure 3.3b**). At these capillary beds, leukocytes can traffic between the lymphatic and circulatory systems in both directions and multiple times; therefore, any leukocyte that encounters an infection can use the circulatory system to find its way back to the lymph node to report the infection and alert additional cells.

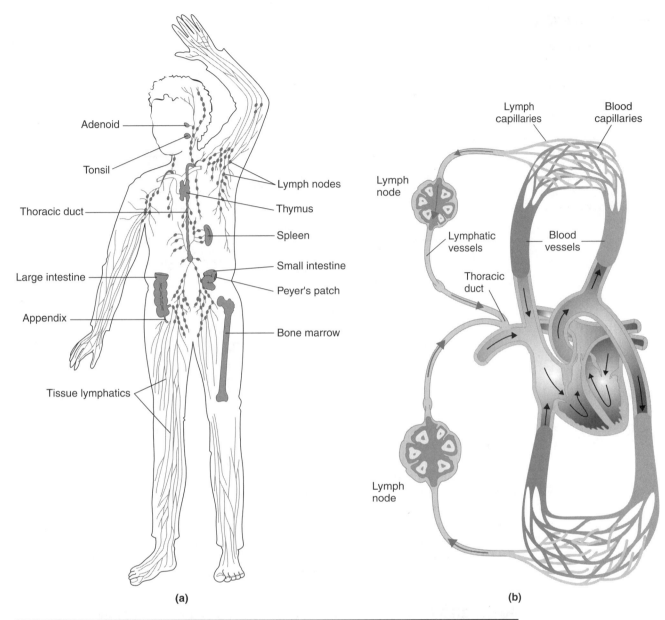

FIGURE 3.3

Overview of the immune system. The interrelationship of the circulatory system and the lymphatic system.
(a) The human lymphatic system consists of lymphocytes, lymphatic organs, lymph vessels, and lymph nodes
located along the vessels. The lymphatic organs are illustrated, and the preponderance of lymph nodes in the
neck, axilla, and groin is apparent. (b) Fluid passes out of the blood from the arteries in the upper and lower parts
of the body. It enters a system of lymphatic ducts that arise in the tissues. The fluid, called lymph, passes through
lymph nodes and on the right side makes its way back to the general circulation via the thoracic duct. The thoracic
duct enters a main vein just before the vein enters the heart. A similar system exists on the left side.

In order to coordinate the presentation of infections to the immune system with the
appropriate response, there are many different types of leukocytes, each of which take
on different functions. All of the different types of leukocytes are generated by the same
originating cell known as the **hematopoietic cell**. The hematopoietic stem cell is found in
the bone marrow (**Figure 3.4),** where many leukocytes, including B-cells, reach maturity

FIGURE 3.4

Trafficking of cells in the immune system. The bone marrow and thymus are the primary tissues in which the cells of the immune system develop. The cells that fully mature in the bone marrow are B-cells, while those that leave the bone marrow to finish their development in the thymus are T-cells. Mature cells move from the bone marrow and thymus into the lymphatic system, from which point they can traffic to the circulatory system and back again into the lymphatic system.

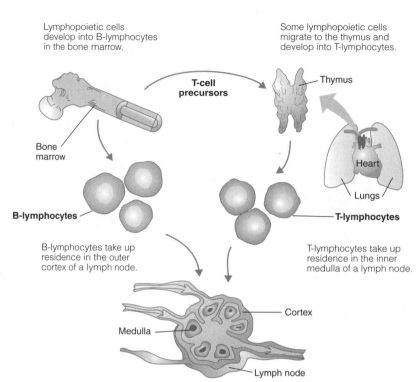

and are prepared to fight infections; however, some cells, known as T-cells leave the bone marrow and reach their final state of preparedness in the thymus (Figure 3.4). Although there are many more types of cells of the immune system, the **B-cells** (or **B-lymphocytes**) and **T-cells** (or **T-lymphocytes**) are of particular importance in beginning to understand the immune system and how HIV attacks and deteriorates the immune response.

T-cells are central to the immune response because they have receptors on their cell surface that can bind directly to bacteria or viral components and subsequently secrete enzymes that destroy the bacteria or virally infected cells. B-cells accomplish a similar task by secreting proteins, known as *antibodies*, into the circulatory system and areas covered by mucous membranes. These antibodies can also bind directly to bacterial and viral components and tag them for destruction. By interfering with the activation of these T- and B-cells, HIV is able to ultimately compromise the entire immune system.

Antigens

When infected with viruses or bacteria, the immune system generally targets specific proteins or other molecules of the infectious agent. The immune system generally does not specifically recognize the entire infectious agent as a single entity. The individual molecules of infectious agent that stimulate an immune response are known as immunogens. **Immunogens** stimulate an immune response in the host that is exposed to them. The immunized host can then recognize these immunogens. Anything that B-lymphocytes or cytotoxic T-lymphocytes can recognize is known as an *antigen*. This recognition is mediated by cell surface receptor proteins on the T-cells and by the secreted antibody proteins from B-cells, and again involves direct physical binding to the antigen; therefore, antigens are the targets bound by the proteins of the immune system, which are involved in specific recognition. Antigens are usually proteins but can be polysaccharides, lipids, or nucleic acids and, under normal conditions, are components from infectious agents including parasites, bacteria, viruses, and fungi, or other foreign molecules such as those from food, blood components, tissues, and venoms.

Normally, a person's own proteins and polysaccharides do not stimulate an immune response because they are interpreted as "self." Research evidence suggests that before birth the proteins and polysaccharides of body cells contact and paralyze immune system cells that might later respond to them. Thus, the individual becomes tolerant of "self" and remains able to respond only to antigens interpreted as "non-self," or foreign. The paralysis of responsive cells must continue throughout life for tolerance to self to persist. If an individual's immune system reacts with **self-antigens**, the reaction is known as an **autoimmune condition** and can lead to serious disease.

The Immune Process in Humans

The immune system reaches maturity several months after a person's birth and continues to function until a person's death. As antigens enter the body, either as free molecules or in association with microorganisms or viruses, the immune process begins in earnest. There are special cells whose job it is to provide surveillance for these infectious agents. These surveillance cells are **phagocytic cells**, which mean that they are able to surround the entire infectious agent with the cell membrane and engulf and internalize the infectious agent in a process known as **phagocytosis**. Once inside the phagocyte, the infectious agent can be destroyed (**Figure 3.5**). Chief among the phagocytes are large, amoeboid cells called **macrophages.**

These macrophages set the immune process into motion by taking microorganisms into their cytoplasm through phagocytosis and digesting them into their component

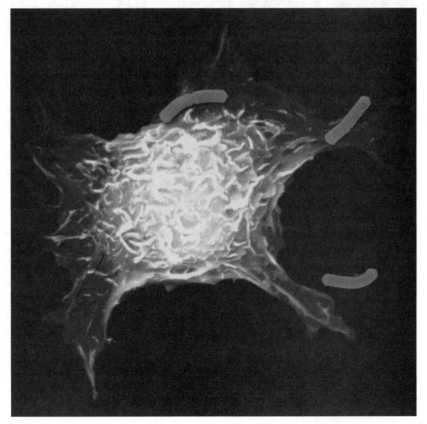

FIGURE 3.5

Phagocytosis. In this scanning electron micrograph a phagocytic cell called a macrophage is engulfing old, misshapen red blood cells. Macrophages also engulf bacteria and other microorganisms and set the immune process in motion. © Phototake Inc./ Alamy Images.

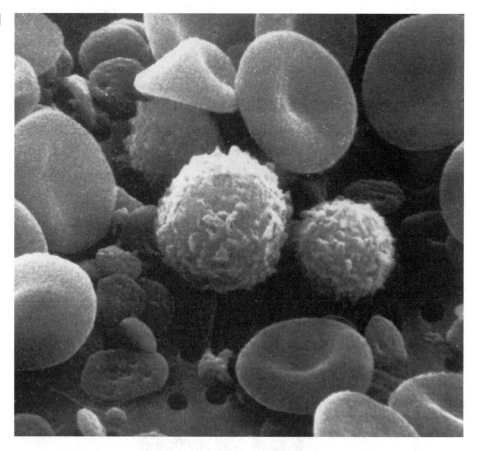

parts. Fragments of the destroyed foreign organism are then brought back to the macrophage membrane and displayed on the surfaces of the macrophages and serve as antigens. These antigens are held on the macrophage surface by special host proteins known **human leukocyte antigens (HLAs)**, which are encoded by the **major histocompatibility complex (MHC) genes.**

Macrophages then proceed to the lymph node in order to present (show) the antigen held on its surface to the immune system. The first cell that the macrophage presents the antigen to is of central importance because it is this cell that will alarm the immune system and call all of its arms into action in order to fight the infection. This central cell of the immune system is known as a **helper T-cell** and is also a CD4+ T-cell. The helper T-cell uses the **CD4 molecule** and an additional T-cell receptor to recognize the antigen-MHC complex on the surface of the presenting macrophage (**Figure 3.6**). After being presented with the antigen, the CD4+ helper T-cell activates and coordinates the immune response through the release of a series of highly reactive proteins called **lymphokines.** Lymphokines then stimulate either the B-lymphocytes and/or the **cytotoxic T-lymphocytes,** depending on the nature of the antigen that began the process. The immune system, therefore, diverges at this point into two major functional branches. One branch is dominated by the cytotoxic T-lymphocytes, and the immunity that results is called **cell-mediated immunity (CMI);** the other branch is dominated by the B-lymphocytes, and the resulting immunity is called **antibody-mediated immunity (AMI)**. Both of these two branches of the immune system are set into action by the antigen-stimulated CD4+ helper T-cell. In the following section, we discuss each type of immunity in turn, beginning with CMI.

Cell-Mediated Immunity

CMI is so named because the defense imparted by this branch of the immune system involves a direct assault by cells on microorganisms or foreign molecules. The *lymphokines* released by the helper T-lymphocytes begin the process of CMI by stimulating other T-lymphocytes, called *cytotoxic T-lymphocytes*, to multiply rapidly (**Figure 3.7**). Cytotoxic T-lymphocytes are also sensitized by antigen presenting macrophages, and these cytotoxic T-lymphocytes then enter the circulatory system and search for body cells displaying the antigens that sensitized the lymphocytes. Fungi, protozoa, and certain virus-infected or bacteria-infected cells can also be the targets of the cytotoxic cells. A cytotoxic T-lymphocyte can target a virally infected cell by recognizing viral antigen on the surface of infected cells. This "lethal hit," as it is called, is assisted by an enzyme released from the cytotoxic cell, which becomes inserted into the surface membrane of the infected cell and opens holes in the target cell membrane, causing cytoplasmic leakage and death. In this regard, the cytotoxic T-lymphocytes can kill "self" cells, but only when those cells express foreign antigens, as in the case of a virally infected cell.

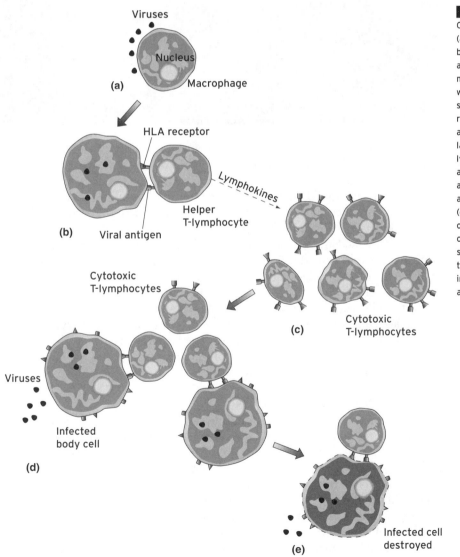

FIGURE 3.7

Cell-mediated immunity. (a) Microorganisms are engulfed by a macrophage, and microbial antigens are displayed on the macrophage surface together with MHC antigens. (b) Receptor sites on a helper T-lymphocyte react with the microbial and MHC antigens. This reaction stimulates the helper cell to release lymphokines. (c) The lymphokines activate cytotoxic T-lymphocytes and induce them to react with antigen-marked infected cells. (d) The lethal hit of cytotoxic cell on infected cell leads to infected cell death (e). Lymphokines also stimulate cytotoxic T-lymphocytes to multiply and attack other infected cells, thereby continuing and expanding the immune process.

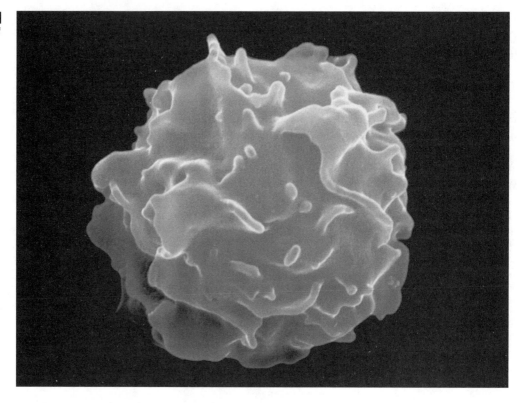

Cytotoxic T-lymphocytes are also the source of additional lymphokines. Secreted at the site where the antigens were first detected by macrophages, the lymphokines draw a host of fresh macrophages to the infection site and stimulate them to destroy the microorganisms.

Another kind of T-lymphocyte is the **natural killer cell**. This cell also attacks microorganisms and infected cells, but it is less specialized than the cytotoxic T-lymphocyte. Natural killer cells appear to be a primary mechanism of defense by the body against tumor cells, which are regarded as foreign. Together with cytotoxic T-lymphocytes, the natural killer cells provide an important line of defense in the body (**Figure 3.8**). The immunity they offer is called cell-mediated because cells are the actual modes of defense.

To prevent the immune process from becoming too exaggerated and destroying normal body cells, another T-lymphocyte comes into play. This is the **suppressor T-lymphocyte**. The suppressor T-lymphocyte dampens the activity of cytotoxic. T-lymphocytes and natural killer cells slow the immune process as the antigen stimulus lessens. With the gradual elimination of microorganisms and the antigens they carry, the process of CMI comes to a halt.

Antibody-Mediated Immunity

The second major branch of the immune system depends on the activity of B-lymphocytes and results in AMI. To begin this process, B-lymphocytes are stimulated by helper T-lymphocytes after the latter have interacted with the human leukocyte antigen-bearing macrophages. As before, the helper T-lymphocytes produce lymphokines to activate the B-lymphocytes and stimulate their division, but there are also additional stimulants required. These second stimulants are, again, the human leukocyte antigens on the macrophage surfaces. As a macrophage moves among the hundreds of thousands of different types of B-lymphocytes, it eventually encounters one type that has surface antibody molecules that can bind to the antigen. The binding

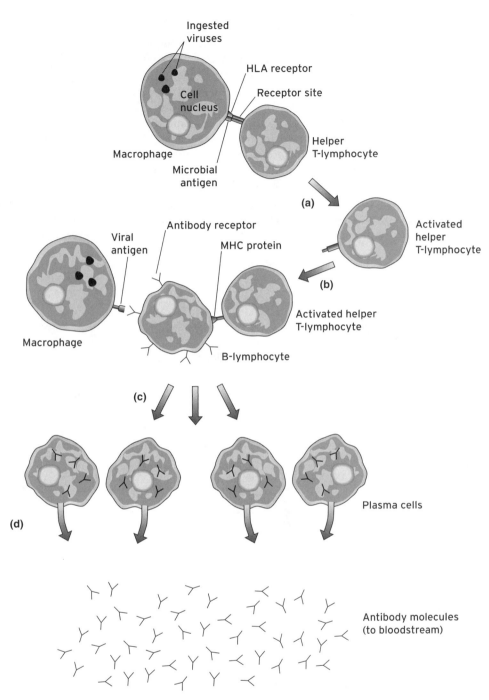

FIGURE 3.9

Antibody-mediated immunity.
(a) The process of antibody
mediated immunity begins with
the reaction between the receptor site of a helper T-lymphocyte
and a macrophage bearing MHC
proteins and microbial antigens.
This process activates the helper
cells. (b) The activated helper
T-lymphocytes produce lymphokines and stimulate uncommitted
B-lymphocytes. At the same time,
the macrophage "searches" among
uncommitted B-lymphocytes until
it locates a B-lymphocyte with surface antibody receptors that match
both the microbial antigen and the
MHC protein. (c) The binding of
macrophage and helper T-lymphocyte commits the B-lymphocyte,
which now undergoes proliferation
and differentiation to plasma cells.
Lymphokines assist the process.
(d) Plasma cells produce highly
specialized antibody molecules to
lend immunity.

of the antigen and antibody molecules, along with the helper T-lymphocytes' intervention, activates or "commits" the B-lymphocytes.

Once committed, B-lymphocytes undergo cell division and give rise to a colony
(or clone) of cells programmed to produce antibodies. Antibodies designed to react
with the antigen pour forth from the B-lymphocytes at a rate of more than 2,000
molecules per second. Within hours, other biochemical signals convert many of the
B-lymphocytes into plasma cells, a group of highly active antibody-producing cells
(**Figure 3.9**). The antigens initiating the process of AMI are found primarily in viruses

and bacteria. Other biological molecules such as milk protein, bee or snake venom, food components, and ragweed proteins can also stimulate the process.

Antibody molecules are proteins. There are five types of antibodies, the most common of which is composed of four chains of amino acids arranged as two long chains and two small chains. Such an antibody molecule has a hinge point where the chains diverge, and the molecule is, therefore, depicted as a Y. One end of the molecule is highly specific for the antigen that elicited its production. This implies that the antibody will interact with that antigen only. The immune system has the capacity to produce perhaps a million different kinds of antibodies, one for each different antigen. Thus, a measles antibody will react only with proteins in a measles virus, a chickenpox antibody only with a chickenpox virus, and so forth.

Antibodies circulate through the body and soon encounter the microorganism or virus whose antigens stimulated their production. They then chemically combine with the antigen molecules and neutralize the microorganism by any of several mechanisms.

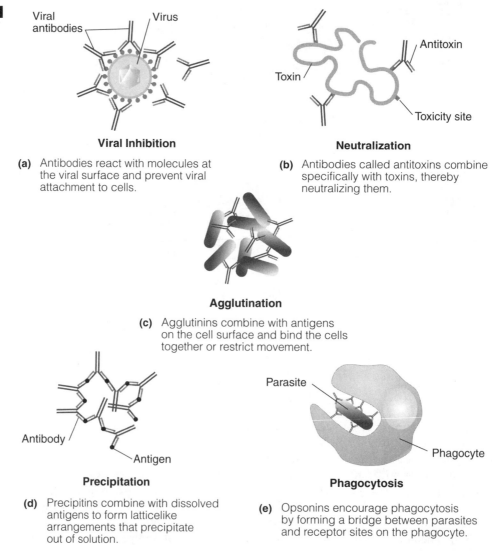

FIGURE 3.10

Five mechanisms by which antibodies interact with antigens.

Viral Inhibition

(a) Antibodies react with molecules at the viral surface and prevent viral attachment to cells.

Neutralization

(b) Antibodies called antitoxins combine specifically with toxins, thereby neutralizing them.

Agglutination

(c) Agglutinins combine with antigens on the cell surface and bind the cells together or restrict movement.

Precipitation

(d) Precipitins combine with dissolved antigens to form latticelike arrangements that precipitate out of solution.

Phagocytosis

(e) Opsonins encourage phagocytosis by forming a bridge between parasites and receptor sites on the phagocyte.

For example, some antibodies bind to viral capsids and prevent the viruses from entering cells by covering their receptor sites. Other antibodies combine with antigens on the surface of bacteria and bind the bacteria together in a mesh-like pattern that can easily be phagocytized. Still others form a bridge between microorganisms and macrophages to encourage *phagocytosis*, and others set off a cascading series of reactions that tear apart microbial membranes (**Figure 3.10**). The antigen–antibody reaction usually results in destruction of the microorganism.

From the foregoing, we can see that the immune system is the source of specific resistance originating with T-lymphocytes or B-lymphocytes. The common thread between the two branches of the system is the helper T-lymphocyte. This cell can stimulate either branch to function, depending upon which antigens are delivered by macrophages. As we have noted, antigens primarily from fungi, protozoa, certain types of bacteria, and viruses stimulate the T-lymphocytes, whereas antigens primarily from bloodborne bacteria and viruses set the B-lymphocytes into action. Once activated, the T- or B-lymphocytes serve as the underlying process for the immune response that provides specific defense for the body. Should that defense be compromised, the results can be extraordinarily detrimental. Herein lies the basis for AIDS.

HIV-1 and the Immune System

In its most simplified form, the general pattern of infectious disease is one in which the body and a population of microorganisms compete for supremacy. It is a battle of sorts, with each side trying to outdo the other. Some microorganisms or viruses attempt to overwhelm the body with sheer numbers, while consuming nutrients or processes needed by the cells of the body. Some microorganisms produce toxins that interfere with critical cellular processes such as nerve transmission. Some synthesize enzymes that interfere with body defenses (e.g., some bacteria produce enzymes that destroy white blood cells). The body, in turn, responds via its immune system. Although some diseases, such as malaria, plague, and typhoid fever, often terminate in death, most diseases have a favorable prognosis because the specific resistance provided by the immune system is substantial. For example, we do not fear death from noroviruses (e.g., noroviruses sometimes cause diarrhea outbreaks on cruise ships), chickenpox, or a common cold.

It is now generally accepted that HIV is the causative agent of AIDS. This virus was first isolated in 1984 by Luc Montagnier and his group at the Pasteur Institute. It is a retrovirus containing two RNA molecules as its genome as well as the enzyme reverse transcriptase, which synthesizes the DNA provirus using RNA as its template.

The Focus of HIV-1

A key focus of HIV-1 is the collection of helper T-lymphocytes. These T-lymphocytes have on their surface protein molecules called *CD4*. The T-lymphocytes are, therefore, called CD4 cells, CD4+ cells, or CD4 T-lymphocytes. These CD4 molecules provide the receptor sites for HIV. The CD4 T-lymphocytes also contains **CCR-5** or **CXCR-4** chemokine receptors at their surface; CCR-5 and CXCR-4 act as receptors for chemokines. **Chemokines** are a family of small proteins that act as messengers attracting T-lymphocytes to locations where problems are occurring (e.g., chronic inflammation, tumors, and an infection).

When these cells are experimentally infected with HIV-1, they are observed to be distorted, with numerous cytoplasmic projections and a tendency toward disorganization. When the gp120 surface protein of HIV-1 binds to CD4 present on a T-lymphocyte, it causes a conformational change of gp120 and gp41 such that it can access the nearby CCR-5 or CXCR-4 co-receptors. Essentially, gp120 acts like a key that begins to open a lock with the help of gp41 and CCR-5. The CCR-5 or CXCR-4 interact with HIV-1 and the host cell fuses, ultimately like a cellular door opening as HIV-1 enters and viral uncoating occurs in the cytoplasm of the host cell. The discovery of the CCR-5 and CXCR-4 HIV-1 co-receptors prompted research efforts to develop "**antagonists**" or drugs that would block or plug the CCR-5 or CXCR-4 co-receptors so that HIV-1 could not attack or enter host cells. Clinical trials in 2009 showed that oral CXCR-4 *antagonists* reduced viral loads in patients. Further research into the "HIV doorway" led to curing "the Berlin patient" of his HIV infection (**Box 3.1**)!

As HIV takes over the cellular machinery, an "early warning system" of the cell is activated to warn other cells of the HIV invasion. During the early hours of HIV-1 infection, the infected cells force viral peptides or fragments of proteins to the surface of the infected cell. These bits and pieces of foreign viral material are displayed by the **HLA receptors** of the cell (acting like a warning sign to neighboring cells of the body). The HLA receptors attract T-helper and cytotoxic T-cells to destroy HIV-1 infected cells. The warned T-helper cells trigger the production of B-cells to produce antibody molecules to bind to HIV released from infected cells. Unfortunately, antibodies are not effective in combatting the virus, which is why the infection persists.

The competition between HIV and helper T-lymphocytes is intense. The virus accumulates in the lymph nodes of the infected individual and multiplies at an extraordinarily high rate, yielding more than 100 billion new viral particles per day. At the same time, the helper T-lymphocytes multiply at a high rate, yielding 1 to 2 billion new cells per day, and a race is on between these cells and the viruses. To retain its edge, the body inactivates an estimated billion viral particles per day but also loses about a billion helper T-lymphocytes per day.

Normally, every one of the body's helper T-lymphocytes is replaced within 14 days, but that HIV-1 counteracts this by replenishing its numbers every three days. The research conducted by David Ho and his associates apparently contradicts the notion that HIV kills by gradually undermining the immune system's ability to produce helper T-lymphocytes. Rather, the virus outlasts the immune system (the so-called **empty sink model**) and overwhelms it because its regenerative capacity is finite. Insights into how the virus exhausts the immune system was offered by mathematical biologists Robert May and Martin Nowak of Oxford University. These investigators postulated that AIDS is triggered when the infecting virus mutates and diversifies into so many different strains that the immune system is suddenly overwhelmed. The diversification of HIV introduces new strains different enough from the original to elude the immune system. This imaginative theory was based on observations that the number of viral strains in HIV-infected men kept escalating until the men developed AIDS. Thus, viral diversity is a possible cause of degenerating immune systems. HIV is not a subtle intruder, as many believed, but a raw aggressor.

At odds with the theory of destroying established cells (the *empty sink model*) is an alternative theory that HIV interferes with the production of new cells (the disease "turns down the tap" rather than empties the sink). This theory is based on the study of **telomeres**, the extreme ends of chromosomes that shorten slightly each time a cell divides. Dutch researchers found that the *telomeres* from helper T-lymphocytes of infected patients were not shorter than telomeres from cells of uninfected patients, indicating that the turnover rate of the cells was the same. Thus, it did not appear likely that established cells were being lost; rather, destruction of new cells was the key to understanding the pathology of HIV. Another group, also working with telomeres,

BOX
3.1

Shutting the Cellular Door to HIV-1: Research Toward a Cure

The identification of CCR-5 as a major HIV-1 co-receptor provided an explanation for why certain groups of people were resistant to HIV-1 infection. Turns out that people with two copies of a defective *CCR-5* gene known as *CCR-5-Delta32* were naturally resistant to HIV-1 even after having been exposed to HIV-1 multiple times and did not become infected. The *CCR-5-Delta32* gene is missing 32 nucleotides compared to the normal *CCR-5* gene. This deletion or defect results in a shortened gene product or truncated CCR-5 protein that doesn't present itself on the surface of host cells, making the cells highly resistant to HIV-1 infection because HIV-1 can't latch onto it in order to enter its host cell (**Figure 3.B1**). The mutation is rare in Native Americans, Africans, and Asians. Individuals with this genetic anomaly are healthy, although they may be more vulnerable to West Nile virus infection. About 1% of the Caucasian population has inherited two copies of the defective gene. Individuals who inherit one copy of *CCR-5* and one copy of *CCR-5-Delta32* are susceptible to HIV-1 infection, but it takes longer on average for them to progress to the later stages of disease (full blown AIDS).

Over a period of years, scientists focused efforts on developing inhibitors of CCR-5 and CXCR-4. In 2007, the U.S. Food and Drug Administration (FDA) approved Selzentry (maraviroc) as a CCR-5 co-receptor antagonist for the treatment of HIV infection. Another CCR-5 inhibitor, vicriviroc, had disappointing results in Phase III clinical trials and its manufacturer did not apply for FDA licensure.

In 2007, Timothy Ray Brown was *cured* of his HIV infection by receiving a bone marrow stem cell transplant from a donor match that had two gene copies of *CCR-5-Delta32*. The aim of the bone marrow transplant was to treat an unrelated health condition that was life-threatening (leukemia). Mr. Brown's leukemia went into remission and he had undetectable HIV loads. The leukemia returned, requiring a second transplant. More than five years have passed since the first transplant. Brown is in remission from cancer and does not take HIV medication. His viral loads are undetectable. In 2012, two more men were reported to be HIV-free following bone marrow transplants. Both men were also being treated for cancers.

A different approach to antagonize CCR-5 was to create HIV-resistant cells using a "customized DNA scissors" to cleave the normal *CCR-5* gene present in cells. A designer "zinc finger nuclease" was engineered to bind to the *CCR-5* gene and cut or chew a section of the *CCR-5* gene sequence out. The altered gene would then be repaired by the host cell machinery. Its gene product would lose function similar to those individuals with the *CCR-5-Delta32* gene. Results *in vitro* were encouraging. In other words, T-cell populations treated with the zinc finger nuclease were resistant to HIV infection. A safety trial in people is in progress. These different approaches bolster hope that there are several possible paths to an AIDS cure. It is the closest any researcher has come to shutting the door on HIV in 30 years.

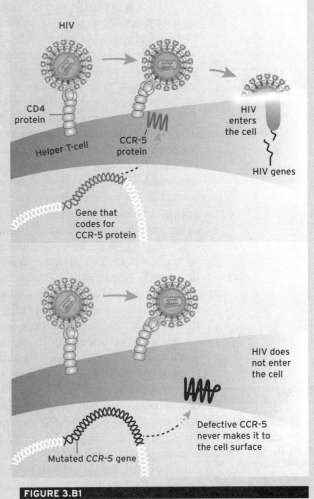

FIGURE 3.B1

HIV enters T-helper cells by attaching to CD4 receptors located near a chemokine receptor such as CCR-5. HIV cannot enter cells that express a defective CCR-5 chemokine receptor. Therapies are targeting CCR-5 or CXCR-4 as a way to block HIV's entry into cells. Adapted from June, C., Levine, B. 2012. "Blocking HIV's attack." *Scientific American.* March, 2012, pp. 54–59. Massie, Bob, *A Song in the Night: A Memoir of Resilience.* Nan A. Telese Books, 2012.

found that the lengths of *telomeres* in lymphocytes from HIV-infected patients are equivalent to those in 100-year-old individuals, a sign that immune system cells were replicating so often that they were becoming worn out. Research on both of these approaches is continuing.

HIV-1 affects the production of new T-lymphocytes more than it induces destruction of mature T-lymphocytes. Using innovative methods in which patients were infused with isotope-containing nucleic acid precursors, the biochemists found that the rate of T-lymphocyte production was no higher in HIV-infected patients than in uninfected individuals. Moreover, the average T-lymphocyte lifespan in untreated patients was a third of that in uninfected individuals, a factor consistent with a certain level of T-lymphocyte death caused by HIV-1; antiretroviral therapy encourages the immune system to boost its production of new T-lymphocytes above normal levels. Taken together, these data appear to support the "turn down the tap" model of HIV pathogenesis.

HIV-1 is probably carried from the initial infection site to the brain and lymph nodes by macrophages, the cells that function in normal immune processes discussed earlier. In the lymph nodes, cells called **follicular dendritic cells** then come into play. These cells are found within the germinal centers of the lymph nodes (the follicles) and have tree-like branches (they are dendritic). The cells remove the HIV from the macrophages and "trap" the viruses. Virtually every other cell in a lymph node is touched by the branches of the follicular dendritic cells, and it is only a matter of time before T-lymphocytes come in contact with these cells and are activated by receiving the HIV particles. Molecules of a protein extending like fingers on the branches of the follicular dendritic cells act like an adhesive to bind the cells with T-lymphocytes and pass HIV into the lymphocytes. Later in the infection, the follicular dendritic cells dissolve, die, and become part of the general destruction taking place in the lymph nodes. The activated T-lymphocytes appear to be ones not previously stimulated by any type of antigen. These so-called **naïve T-lymphocytes** are different from those previously committed to destroying a particular microorganism.

What happens, however, when the activated T-lymphocytes have been destroyed by replicating HIV particles? Will the virus run out of suitable host cells and itself be destroyed? Apparently not. The *tat* gene of HIV encodes a protein (the **tat** protein) that activates naïve T-lymphocytes and transforms them into host cells suitable for HIV infection. Their evidence came in part from observations that naïve cells exposed to tat protein display a set of properties consistent with those of activated cells. Also, *tat*-activated cells were more supportive of HIV replication than untreated cells. It did not escape the researchers that a drug directed at the tat protein could conceivably slow down HIV replication in cells.

In addition to transporting HIV, macrophages represent another important focus of infection, especially in late stages of infection, where they probably contribute to the high blood level of HIV seen in patients at this point. Indeed, while helper T-lymphocytes die within a few days of HIV infection, macrophages appear to persist for months, while continuing to release HIV. Of significance is the finding that the brain's macrophages, called **microglia,** are related to HIV encephalopathy, the brain infection that characterizes AIDS. Perhaps, scientists say, the infected macrophages produce and release a number of neurotoxic substances that induce inflammation. Thus, the macrophage that normally helps rid the body of deadly foes has turned into a foe itself, an HIV-harboring "Trojan horse."

Turns out, there are a group of people infected with HIV known as elite HIV controllers. These individuals are infected with HIV but have extremely low or even undetectable viral loads. This group does not possess the *CCR-5-Delta32* gene anomaly and has puzzled researchers for 20 years. After a long and grueling process that involved the analysis of DNA samples from elite controllers, the missing puzzle piece was found (**Box 3.2**).

Effects of HIV

HIV has a profound effect on macrophages as well as helper T-lymphocytes. By destroying helper cells, HIV affects the entire immune system because helper cells set into action both the cytotoxic T-lymphocytes and the B-lymphocytes. Because helper T-lymphocytes are also known as T4 cells (from CD4) in the jargon of immunology, it is not uncommon to hear scientists speak of "T4 cell destruction."

The effect of HIV on suppressor T-lymphocytes is much less severe, partly because they contain CD8 receptor sites to which HIV does not bind easily. Although this factor may appear advantageous at first glance, the effects work against the body because suppressor T-lymphocytes dampen the activity of the immune system's cytotoxic T-lymphocytes. Normally, a person has twice as many helper cells (T4 cells) as suppressor cells (also known as T8 cells); however, as the helper cells disappear, the relative number of suppressor cells climbs. Put another way, the ratio of T4 to T8 cells (helpers to suppressors) is normally 2 to 1, but with a reduction in the helper cell population, the ratio of T4 to T8 cells gradually reverses and eventually becomes 1 to 2. As that occurs, the effects of cytotoxic T-lymphocytes are dampened.

Properly functioning suppressor T-lymphocytes may be critical to controlling HIV in the body. Scientists depleted the population of suppressor cells in the blood and lymph tissues of monkeys and found that increased viral replication took place and the disease progressed more rapidly. Furthermore, the level of HIV in the bloodstream increased significantly; however, the suppressor cells reappeared, the level of HIV in the bloodstream declined, and the disease regressed. The results appeared to confirm the importance of CMI in controlling HIV infection, while pointing up the role of suppressor T-lymphocytes in HIV destruction.

HIV also multiplies in and destroys the cytotoxic T-lymphocytes as well as the natural killer cells. With the loss of these cells, an entire branch of the immune system is first depressed and then eliminated. Now the affected individual cannot mount a defense against fungi, protozoa, and various viruses and bacteria. Persons with AIDS often suffer serious bouts of illness from protozoa such as *Pneumocystis jirovecii* and *Toxoplasma gondii* and from fungi such as *Candida albicans* and *Cryptococcus neoformans*. Before the advent of AIDS, the names of these organisms were largely unfamiliar to most people because T-lymphocytes normally kept them and their diseases under control. In the HIV-infected individual, however, the populations of helper and cytotoxic T-lymphocytes diminish, and the microbial populations increase dramatically. Disease and death may follow. In many cases, the multiplication of HIV within the lymphocyte literally "uses up" the lymphocyte's nutrient molecules to construct new viruses. This intense level of parasitism leaves the T-lymphocyte in such a damaged state that it undergoes lysis and disintegrates.

T-lymphocytes can also be destroyed during the release phase of the HIV replication cycle. In this case, the virus moves through the cell membrane (a process called **budding**), thereby altering the membrane and causing cellular leakage. Moreover, T-lymphocytes infected with HIV are marked at their surfaces with virus specified molecules that induce the infected T-lymphocytes to cling to uninfected T-lymphocytes. As molecular cross-bridges form, there develops a giant multinucleated cell mass, called a **syncytium**. Another effect of being marked with virus-specified molecules is that normal cytotoxic T-lymphocytes will attack and destroy the infected T-lymphocytes as if the latter were foreign to the body. Thus, we see the unusual and ironic phenomenon of body cells destroying other body cells.

Focus on HIV

CCR-5 is located on chromosome 3, CXCR-4 is located on chromosome 2, and the HLA gene is located on chromosome 6.

The Massie Puzzle Piece Hiding on Chromosome 6

Surviving AIDS was originally broadcast on the PBS science television series NOVA on February 2, 1999. The documentary featured the cutting edge research on HIV/AIDS. It included a segment on Bob Massie, a hemophiliac who contracted HIV after receiving contaminated blood clotting factor 8 in 1978. At the time, HIV infection was considered a death sentence, but Massie remained healthy without taking any antiviral drugs and his T-helper cell counts were normal. Researchers determined that Massie's DNA does not possess the defective *CCR-5* gene. What was different about Massie's immune system?

His unique genetic make-up offered hope for the prevention and treatment of HIV/AIDS, but it would take two decades to unravel the genetic mystery through many research collaborations led by physician-scientist Bruce D. Walker at the Ragon Institute of MGH, MIT, and Harvard. The application of the latest advances in human genetics technology in the twenty-first century were needed to gain a better understanding of why Massie and others were elite controllers of HIV infection.

MIT professor Eric Lander, a leader in the latest advances in technology, explained to Walker that he would need about 1,000 elite HIV controllers to identify the differences in their responses to control HIV infection compared to individuals susceptible to the effects of HIV infection. Specimens of blood or a swab of saliva were collected from a total of 974 elite controllers and 2,638 HIV-positive individuals who had progressed to AIDS.

Imagine the daunting task of analyzing the variations of the DNA/genetic code of over 3,000 individuals. And the funding needed to do this project? Millions of dollars, time, and massive computing services were needed to make this happen. The human genome has 3 billion nucleotides. Finding the "Massie" puzzle piece was no easy task. The permutations between the elite controllers and noncontrollers were narrowed down to 300 differences or genetic markers referred to as **single nucleotide polymorphisms** (**SNPs**). By 2009, four of the 300 SNPs were correlated with controlling HIV infection.

All four of the SNPs were located on chromosome 6, which houses many genes that affect the immune system. Xiaoming "Sherman" Jia, a medical student at Harvard with an interest in bioinformatics and computational genetics, solved the puzzle with his computer algorithm. The elite HIV controllers had differences in the **HLA receptor gene**, resulting in HLA receptors that display HIV antigens in such a way that they are "seen" better by cytotoxic T-cells, which then destroy HIV-infected cells. The cytotoxic T-cells keep the number of virally infected cells low, protecting the remainder of the T-helper cells from HIV infection (**Figure 3.B2**). This puzzle was solved in 2010. Now scientists need to figure out how to apply this information to develop treatments that would fine-tune the HLA warning system in those individuals that are not elite HIV controllers.

There is also the effect of **apoptosis**. Apoptosis is programmed cell death, that is, the natural mechanism whereby a cell dies at the end of its normal life cycle. Apoptosis is believed to be an altruistic phenomenon, one in which the body rids itself of "old" cells as new ones are produced through mitosis and cell division. In the case of the normal immune system, between 10 billion and 20 billion T-lymphocytes are produced daily, and an equal number undergo apoptosis to make room for the new cells.

When a virus infects a cell, the body seeks to get rid of that cell as quickly as possible in order to eliminate the viral infection. Examples of this phenomenon are seen in cells infected with cowpox virus, cytomegalovirus, and Epstein-Barr virus. To get rid of the infected cell, one mechanism the body apparently uses is speeding up the process of apoptosis; however, when the infected cells are T-lymphocytes, their elimination brings about a loss of immune function at a time when the body most needs it. Thus, the body must confront a dilemma: Encouraging apoptosis eliminates HIV but also lowers the body's resistance to HIV.

There are some complex applications of the apoptosis theory. For example, some researchers believe that the binding of gp120 molecules to the CD4 receptor sites

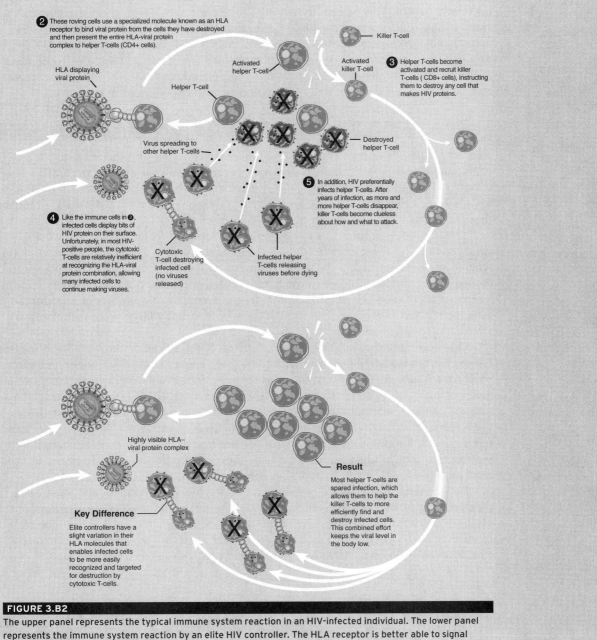

FIGURE 3.B2

The upper panel represents the typical immune system reaction in an HIV-infected individual. The lower panel represents the immune system reaction by an elite HIV controller. The HLA receptor is better able to signal cytotoxic T-cells to destroy infected T-helper cells. Modified from Walker, B. D. "Secrets of the HIV Controllers." *Scientific American* July, 2012 pp. 44-51.

The following labels appear within the figure:

2. These roving cells use a specialized molecule known as an HLA receptor to bind viral protein from the cells they have destroyed and then present the entire HLA-viral protein complex to helper T-cells (CD4+ cells).

HLA displaying viral protein

Helper T-cell

Activated helper T-cell

Killer T-cell

Activated killer T-cell

3. Helper T-cells become activated and recruit killer T-cells (CD8+ cells), instructing them to destroy any cell that makes HIV proteins.

Virus spreading to other helper T-cells

Destroyed helper T-cell

5. In addition, HIV preferentially infects helper T-cells. After years of infection, as more and more helper T-cells disappear, killer T-cells become clueless about how and what to attack.

4. Like the immune cells in 2, infected cells display bits of HIV protein on their surface. Unfortunately, in most HIV-positive people, the cytotoxic T-cells are relatively inefficient at recognizing the HLA-viral protein combination, allowing many infected cells to continue making viruses.

Cytotoxic T-cell destroying infected cell (no viruses released)

Infected helper T-cells releasing viruses before dying

Highly visible HLA–viral protein complex

Key Difference

Elite controllers have a slight variation in their HLA molecules that enables infected cells to be more easily recognized and targeted for destruction by cytotoxic T-cells.

Result

Most helper T-cells are spared infection, which allows them to help the killer T-cells to more efficiently find and destroy infected cells. This combined effort keeps the viral level in the body low.

"mis-activates" the helper T-lymphocytes, rendering them unable to respond to further stimulations and speeding up their programmed death. Another possibility is that gp120 molecules bind to uninfected T-lymphocytes and trigger their apoptosis, whereas the combination of gp120 molecules with host cells actually inhibits apoptosis via a gene-encoded protein of HIV. Another intriguing possibility is that HIV "highjacks" lymphocytes by activating the molecules that cause lymphocytes to migrate into body tissues. This possibility would also explain the depletion of lymphocytes seen in the circulation and the observation that HIV also enters the central nervous system and can cause apoptosis in neurons. Thus, the devastating effects of HIV on the immune system are often compounded by impairment of neural function. To be sure, AIDS is a very complex disease, with many facets and implications. Many researchers and physicians feel that nothing like it has ever been experienced in medicine.

LOOKING BACK

The body depends on two types of resistance—nonspecific and specific—to protect itself from disease. Nonspecific resistance operates against all foreign organisms and is centered in the body's enzymes, its covering structure, and the process of phagocytosis, among other mechanisms. Specific resistance operates only against the organism that elicited it; it involves the immune system.

The immune system arises during fetal development from bone marrow cells called stem cells. These cells become T-lymphocytes or B-lymphocytes, depending on where they are modified in the body. Both T- and B-lymphocytes carry highly specific receptor sites and accumulate in the lymph nodes, spleen, and similar tissues. The lymphocytes respond via their receptor sites to chemical components of microorganisms interpreted as foreign. The foreign chemicals, usually large proteins or polysaccharides, are called antigens.

One branch of the immune system, CMI, is based in the cytotoxic T-lymphocytes. Antigens from macrophages stimulate these cells, using helper T-lymphocytes as intermediaries. The cytotoxic T-lymphocytes travel to the antigen site, where they promote phagocytosis via the activity of lymphokines. Moreover, they attack microorganisms directly and kill them. Natural killer cells provide additional defense by destroying foreign cells. B-lymphocytes are also stimulated by antigens. The B-lymphocytes remain in the lymphoid tissues and produce antibodies, and some are converted to plasma cells that produce additional antibodies. Antibodies are protein molecules that neutralize microorganisms by reacting with their antigens.

HIV attacks both helper T-lymphocytes and cytotoxic T-lymphocytes, destroying them in the process. As CMI is depressed and the entire immune system is compromised, minor organisms that are normally of minor consequence can cause life-threatening illnesses. The destruction of T-lymphocytes by HIV is a unique phenomenon in medicine. So far as is known, AIDS has no parallel in any other infectious disease.

Healthline Q & A

Q1 How does the AIDS virus weaken a person's immunity to disease?

A The AIDS virus targets cells of the immune system for destruction and, in so doing, diminishes a person's ability to react to disease. Organisms normally held under control by the body then invade the tissues and bring on the symptoms of AIDS. These organisms are called "opportunistic organisms" because they seize the opportunity to invade. Death often results from their effects.

Q2 What does it mean to be immune to a disease?

A "Immune" is derived from Latin stems meaning "to be free from." The term implies that once a person has been subjected to an infectious disease, his or her body will develop highly specialized mechanisms to prevent future recurrences of that disease.

Q3 Are antibodies the same as antibiotics?

A No, antibodies and antibiotics are totally different. Antibodies are protein molecules manufactured by the human immune system when it is stimulated by fragments of microorganisms or by chemical molecules. The antibodies enter the body tissues and neutralize the microorganisms or molecules by carefully specified means. Antibiotics, in contrast, are drug molecules that are toxic to microorganisms, usually without producing any serious side effects in the body. Antibiotics can be synthetically produced by pharmaceutical companies or naturally produced by microorganisms. They are administered by physicians during times of illness.

Q4 Why does HIV attack cells of the immune system and brain but not other cells?

A HIV and its host cell bear a resemblance to a key and a lock. The virus is the key, and whichever cell contains the lock is the cell that will be attacked. In humans, certain cells of the immune system and certain cells of the brain have the necessary "lock" in the form of receptor protein molecules called CD4 and co-receptors. Other body cells lack CD4 molecules, and these cells do not attach to HIV; therefore, they do not become infected by the virus.

REVIEW

Having completed this chapter, you should be familiar with the structures and activities of the human immune system and how HIV affects it. To review your knowledge, match the term on the right side with the description or characteristic listed on the left side by placing the appropriate letter in the space. (A letter may be used more than once.) The correct answers are listed elsewhere in the text.

_____	1. Forerunners of the immune system	**A.** Thymus
_____	2. Large, amoeboid cells that function as phagocytes to initiate the immune process	**B.** MHC proteins
_____	3. Type of immunity resulting from activity of B-lymphocytes	**C.** Helper T-lymphocyte
_____	4. Prevents the immune process from becoming too exaggerated	**D.** Spleen
_____	5. Also known as T8 cell	**E.** Antigen
_____	6. Products of B-lymphocytes and plasma cells	**F.** Lymphokines
_____	7. Organ in the embryonic chick where B-lymphocytes are modified	**G.** Cytotoxic T-lymphocyte
_____	8. Type of immunity involving a direct assault on microorganisms by body cells	**H.** Antibodies
_____	9. Site where T-lymphocytes and B-lymphocytes are found	**I.** Phagocytes
_____	10. General class of body cells to which lymphocytes belong	**J.** Macrophages
_____	11. Attacks microorganisms but is less specialized than cytotoxic T-lymphocyte	**K.** Lymphopoietic cells
		L. Bursa of Fabricius
		M. Cell-mediated immunity

_____ **12.** Antigens present on all body cells that define an individual's uniqueness

_____ **13.** Also called a CD4 cell

_____ **14.** Type of immunity resulting from activity of cytotoxic T-lymphocytes

_____ **15.** Exerts a "lethal hit" on fungus-infected and protozoa-infected cells

_____ **16.** Composed of four chains of amino acids

_____ **17.** Organ in which cells are modified to form T-lymphocytes

_____ **18.** Type of T-lymphocyte unaffected by HIV

_____ **19.** A substance, usually a protein or polysaccharide, which stimulates the immune system

_____ **20.** Located on the surface of unstimulated B-lymphocytes

_____ **21.** Normally twice as common as suppressor T-lymphocyte

_____ **22.** White blood cells that specialize in engulfing and destroying foreign materials

_____ **23.** First lymphocyte encountered by an antigen-bearing macrophage

_____ **24.** Highly reactive proteins from helper T-lymphocytes that stimulate other lymphocytes

_____ **25.** Type of immunity involving activity of antibodies

N. Antibody-mediated immunity

O. Suppressor T-lymphocyte

P. White blood cells

Q. Natural killer cells

FOR ADDITIONAL READING

Allers, K., et al., 2011. "Evidence for the cure of HIV infection by CCR5Δ32/Δ32 stem cell transplantation." *Blood* 10:2791–2799.

Cannon, P., June, C., 2011. "Chemokine receptor 5 knockout strategies." *Current Opinions in HIV and AIDS* 6:74–79.

Carrington, M, Walker, B. D., 2012. "Immunogenetics of spontaneous control of HIV." *Annual Review of Medicine* 63:131–145.

Hutter, G. et al., 2009. "Long-term control of HIV by CCR5 Delta 32/Delta 32 stem-cell transplantation." *New England Journal of Medicine* 360:692–698.

June, C., Levine, B., 2012. "Blocking HIV's attack." *Scientific American* March, 54–59.

Massie, B., 2012, *A Song in the Night: A Memoir of Resilience*. Nan A. Telese Books/ Doubleday.

Munawwar, A., Singh, S., 2012. "AIDS associated tuberculosis: A catastrophic collision to evade the host immune system." *Tuberculosis (Edinb)* 92:384–387.

Rosenberg, T., 2011. "The man who had HIV and now does not." *New York Magazine* May 29.

Stevenson, M., 2012. "Review of Basic Science Advances in HIV." *Topics in Antiviral Medicine* 20:26–29.

The International HIV Controllers Study, et al., 2010. The major genetic determinants of HIV-1 control affect HLA class I peptide presentation." *Science* 10:1551–1557.

Walker, B. D., 2012. "Secrets of the HIV controllers." *Scientific American* July:44–51.

4

Defining and Recognizing AIDS

LOOKING AHEAD

AIDS is a very complex disease, with multiple phases and symptoms. This chapter discusses how public health officials define AIDS and points out the symptoms associated with various categories or phases of HIV infection and AIDS. On completing the chapter, you should be able to . . .

- Understand the case definition for AIDS and appreciate why it is important to spell out the definition clearly.
- Describe how AIDS develops through multiple phases.
- Define the key symptoms that accompany HIV infection and AIDS.
- Identify the different regions of the body affected by HIV.
- Describe the important elements of HIV wasting syndrome and AIDS dementia complex.
- List several opportunistic diseases that accompany AIDS, identify the agents that cause these diseases, and discuss the nature of each disease.
- Discuss the impact of HAART and prophylaxis on opportunistic infections.
- Compare the subtle changes in opportunistic infections associated with AIDS from the 1980s to present day.
- Discuss efforts in progress to reduce mother-to-child-transmission resulting in pediatric AIDS.

INTRODUCTION

Physicians and researchers are often asked to provide a clear and precise definition for **acquired immune deficiency (AIDS)**. After a few tries, they realize that defining AIDS is a difficult task because AIDS is a very complex disease. They must ask themselves: Does a person infected with **human immunodeficiency virus (HIV)** have AIDS? What about the HIV-positive individual who has swollen lymph nodes and vague flulike symptoms but is in generally good health? Does that person have AIDS? If so, then what do we call the condition in which patients are wracked with pneumonia and are reduced almost to skeletons?

One might suggest that these distinctions are merely semantics and that "AIDS" is a label we can apply to any stage of the disease and still be correct. Perhaps so, but in public health matters, it is important to define terms carefully. One reason is that AIDS is an extremely volatile issue in our society, and we must be very cautious before stigmatizing an individual. For instance, a person who has tested positive for HIV antibodies does not have AIDS; rather, he or she has HIV infection.

Defining AIDS is also important from the viewpoints of record-keeping and public health. Since 1981, the CDC has been tracking the AIDS epidemic in the United States by asking physicians to report cases they have treated. The statistics generated from such reports are only as reliable as the physicians' ability to recognize AIDS when they see it. Physicians must therefore have a clear definition at their disposal. Finally, it is important to define AIDS so that we can understand the disease more fully. As we see later here, there are many phases to the disease, and only in the final stages does the patient truly have AIDS. In the earlier stages, the virus may be present, but the disease is not.

The Case Definition

A **case definition** was needed for public health surveillance of the HIV/AIDS epidemic. A case definition includes individual criteria (e.g., age, gender, ethnicity) and clinical characteristics including symptoms of those infected as well as information on time and location of an outbreak (e.g., school, a restaurant, town, state, etc.). In 1981, officials at the CDC spelled out a relatively simple case definition for AIDS: (1) A person with AIDS was suffering from Kaposi's sarcoma (a rare cancer) or an infectious disease that occurs only when the immune system is impaired, and (2) there was no other reason except AIDS for the impaired immune system (if a person were taking an immune-suppressing drug after organ transplant surgery, for example, that would be a reason for an impaired immune system not related to AIDS).

In the early years of the epidemic, this case definition for AIDS worked well (**Table 4.1**), and the CDC was able to follow the epidemic's development based on reports from physicians. Then, in 1984, HIV was identified, and soon thereafter, scientists developed a blood test for antibodies produced by the body when infected with this virus. Physicians could now test their patients to see whether they were harboring HIV. In doing so, they discovered that a number of conditions accompany HIV infection. It was clear that a spectrum of AIDS-related conditions exists. The CDC therefore revised its case definition for AIDS, and in conjunction with physicians promulgated an updated version in 1987 (**Table 4.2**). The new definition took into account the presence of HIV and 23 conditions that can accompany infection with HIV.

Then, on January 1, 1993, the CDC revised the case definition once again. The latest definition (**Table 4.3**) took into account the number of helper T-lymphocytes (CD4+ cells) present in the patient and added three new conditions to the previous 23. According to this case definition, a person infected with HIV had AIDS if the count of

TABLE 4.1	1983 Case Definition for Acquired Immune Deficiency Syndrome
1.	Presence of a reliably diagnosed disease at least moderately indicative of cellular immunodeficiency (e.g., *Pneumocystis* pneumonia).
2.	Absence of known causes of underlying reduced resistance to the disease other than due to HTLV-III/LAV infection (e.g., immunosuppressive therapy, Hodgkin's disease).*

*HTLV-III/LAV: human T-cell lymphotropic virus type III/lymphadenopathy-associated virus.

Source: CDC, 1983. Current Trends Update: Acquired Immunodeficiency Syndrome (AIDS)—United States. *MMWR* 32:389–391.

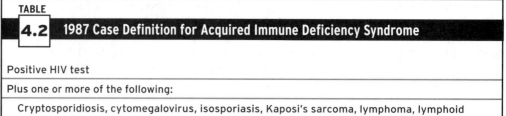

TABLE 4.2	1987 Case Definition for Acquired Immune Deficiency Syndrome

Positive HIV test
Plus one or more of the following:
Cryptosporidiosis, cytomegalovirus, isosporiasis, Kaposi's sarcoma, lymphoma, lymphoid pneumonia (hyperplasia), *Pneumocystis carinii* pneumonia, progressive multifocal leukoencephalopathy, toxoplasmosis, candidiasis, coccidioidomycosis, cryptococcosis, herpes simplex virus, histoplasmosis, extrapulmonary tuberculosis, other mycobacteriosis, salmonellosis, other bacterial infections, HIV encephalopathy (dementia), and HIV wasting syndrome

Source: CDC, 1987. Human Immunodeficiency Virus in the United States. *MMWR* 36: Supplement S-6.

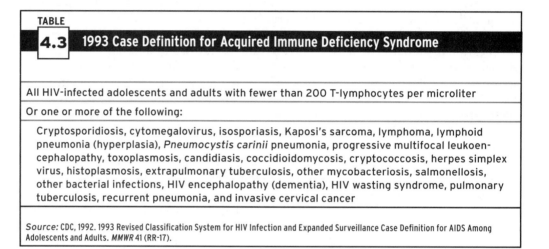

TABLE 4.3	1993 Case Definition for Acquired Immune Deficiency Syndrome

All HIV-infected adolescents and adults with fewer than 200 T-lymphocytes per microliter
Or one or more of the following:
Cryptosporidiosis, cytomegalovirus, isosporiasis, Kaposi's sarcoma, lymphoma, lymphoid pneumonia (hyperplasia), *Pneumocystis carinii* pneumonia, progressive multifocal leukoencephalopathy, toxoplasmosis, candidiasis, coccidioidomycosis, cryptococcosis, herpes simplex virus, histoplasmosis, extrapulmonary tuberculosis, other mycobacteriosis, salmonellosis, other bacterial infections, HIV encephalopathy (dementia), HIV wasting syndrome, pulmonary tuberculosis, recurrent pneumonia, and invasive cervical cancer

Source: CDC, 1992. 1993 Revised Classification System for HIV Infection and Expanded Surveillance Case Definition for AIDS Among Adolescents and Adults. *MMWR* 41 (RR-17).

helper T-lymphocytes fell below 200 per microliter of blood or if the person had one or more of 26 specified conditions, including the new conditions of invasive cervical carcinoma, pulmonary tuberculosis, and recurrent bacterial pneumonia. The new definition reflected T-lymphocyte testing that had become increasingly common, and consequently, the number of AIDS cases reported in 1993 took a significant turn upward, as **Figure 4.1** illustrates. It also included more HIV-infected women because early cervical cancer often occurs concurrently with HIV infection. Moreover, the helper T-lymphocyte (CD4+) count per microliter was used to classify a person's HIV infection into three categories: A count above 500 represents the least severe level of disease, between 200 and 499 reflects intermediate severity, and below 200 is the severest form.

Another version of the case definition took effect in January 2000 when the CDC asked all states and territories to implement a program to detect persons having HIV infection (as well as those with AIDS). This expansion of the national surveillance effort to seek out those with HIV infection arose because of the impact of new therapies introduced in 1996 and the increased need for data on persons having all stages of the disease. Officials at the CDC hoped that the new data would enhance efforts to prevent HIV transmission, improve allocation of resources for treatment services, and assist evaluation of public health interventions.

The seminal element of the revised case definition is the inclusion of surveillance for HIV infection in adults and children. In issuing the guidelines, the CDC

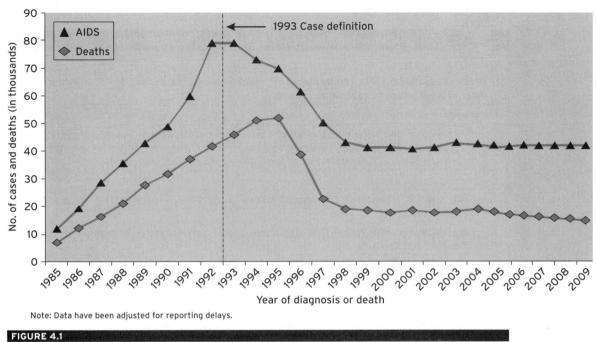

Note: Data have been adjusted for reporting delays.

FIGURE 4.1

The estimated incidence of AIDS cases and deaths associated with AIDS in adults by quarter-years through 2009. The graph shows the increase in AIDS cases resulting from the 1993 expansion of the case definition for AIDS. Note that the revised case definition had no effect on the incidence of death, which continued to rise until 1996 when new therapies were introduced to control HIV in the body.

Data from: CDC, *HIV/AIDS Surveillance Report*, vol. 22, 2010.

acknowledged that AIDS is the end stage of the natural history of HIV infection and that although monitoring AIDS had been valuable in the past, recent advances in therapy have slowed the progression to AIDS and reduced the ability of AIDS surveillance data to reflect trends—thus, the need to survey HIV infection as well as AIDS.

The revision is shown in **Table 4.4**. It applies to infections by both HIV-1 and HIV-2, and it incorporates the previous 1993 reporting criteria for HIV infection and AIDS into a single case definition. Among the new criteria are the use of RNA or DNA tests (the "**viral load**" tests) to detect HIV and the isolation of HIV in viral culture. The ensuing years will show how well the new definition works.

The most recent change in the case definition took effect in January 2009. The revisions were aimed to improve standardization and comparability of surveillance data and not as a guide for clinical diagnosis. The case definition for HIV infection now requires laboratory-confirmed evidence of HIV infection among adults, adolescents, and children aged 18 months to less than thirteen years of age. The CDC and the Council of State and Territorial Epidemiologists recommend that all states and territories conduct surveillance. This case definition has been simplified to focus on CD4+ T-lymphocyte counts and percentages that are objective measures of immune suppression that are routinely used to care for HIV-infected patients (**Table 4.5**). In February 2012, several groups convened to consider another revision of the HIV surveillance case definition. Some of the recommendations at this meeting were to change the case definition to include criteria to differentiate between HIV-1 and HIV-2. New revisions were published in an *MMWR* in 2012.

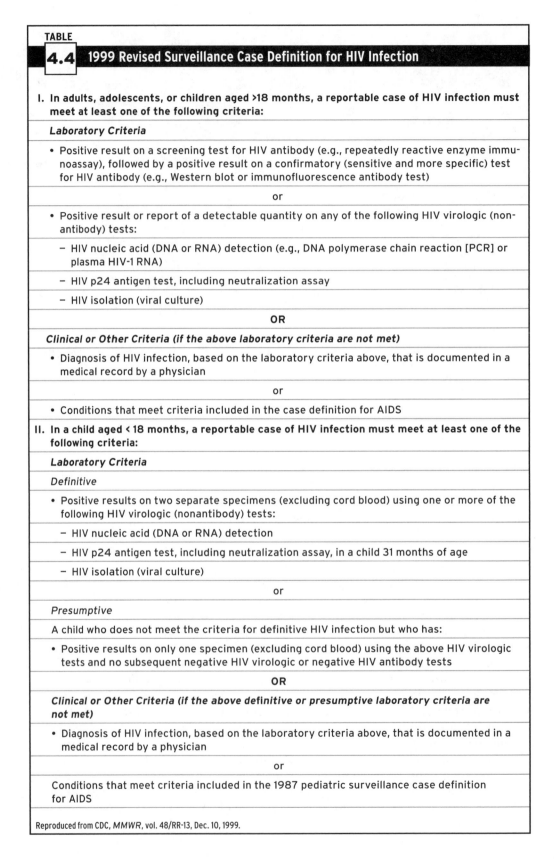

TABLE

4.4   1999 Revised Surveillance Case Definition for HIV Infection

I. In adults, adolescents, or children aged >18 months, a reportable case of HIV infection must meet at least one of the following criteria:

Laboratory Criteria

- Positive result on a screening test for HIV antibody (e.g., repeatedly reactive enzyme immunoassay), followed by a positive result on a confirmatory (sensitive and more specific) test for HIV antibody (e.g., Western blot or immunofluorescence antibody test)

or

- Positive result or report of a detectable quantity on any of the following HIV virologic (non-antibody) tests:
  - HIV nucleic acid (DNA or RNA) detection (e.g., DNA polymerase chain reaction [PCR] or plasma HIV-1 RNA)
  - HIV p24 antigen test, including neutralization assay
  - HIV isolation (viral culture)

OR

Clinical or Other Criteria (if the above laboratory criteria are not met)

- Diagnosis of HIV infection, based on the laboratory criteria above, that is documented in a medical record by a physician

or

- Conditions that meet criteria included in the case definition for AIDS

II. In a child aged < 18 months, a reportable case of HIV infection must meet at least one of the following criteria:

Laboratory Criteria

Definitive

- Positive results on two separate specimens (excluding cord blood) using one or more of the following HIV virologic (nonantibody) tests:
  - HIV nucleic acid (DNA or RNA) detection
  - HIV p24 antigen test, including neutralization assay, in a child 31 months of age
  - HIV isolation (viral culture)

or

Presumptive

A child who does not meet the criteria for definitive HIV infection but who has:

- Positive results on only one specimen (excluding cord blood) using the above HIV virologic tests and no subsequent negative HIV virologic or negative HIV antibody tests

OR

Clinical or Other Criteria (if the above definitive or presumptive laboratory criteria are not met)

- Diagnosis of HIV infection, based on the laboratory criteria above, that is documented in a medical record by a physician

or

Conditions that meet criteria included in the 1987 pediatric surveillance case definition for AIDS

Reproduced from CDC, *MMWR*, vol. 48/RR-13, Dec. 10, 1999.

TABLE 4.5	2008 CDC Revised Case Definition for HIV Infection: for HIV Infection in Adults, Adolescents and Children > 13 Years of Age—United States
Stage I	
Laboratory Evidence	
Laboratory confirmation of HIV infection AND	
CD4+ T-lymphocyte count of ≥ 500 cells/µl or CD4+ T-lymphocyte ≥ 29%	
Clinical Evidence	
None required (but no AIDS-defining condition)	
Stage II	
Laboratory Evidence	
Laboratory confirmation of HIV infection AND	
CD4+ T-lymphocyte count of 200–499 cells/µl or CD4+ T-lymphocyte of 14%–28%	
Clinical Evidence	
None required (but no AIDS-defining condition)	
Stage III (AIDS)	
Laboratory Evidence	
Laboratory confirmation of HIV infection AND	
CD4+ T-lymphocyte count of < 200 cells/µl or CD4+ T-lymphocyte of < 14%	
Clinical Evidence	
Or documentation of an AIDS-defining condition (with laboratory confirmation of HIV infection)	
Stage UNKNOWN	
Laboratory Evidence	
Laboratory confirmation of HIV infection AND	
No information on CD4+ T-lymphocyte count or %	
Clinical Evidence	
And no information on presence of AIDS-defining conditions	

Reproduced from CDC, *Morbidity and Mortality Weekly Report*, vol. 57/RR-10, Dec. 5, 2008.

HIV Infection

One of the primary targets of HIV is the T-lymphocyte of the immune system. More specifically, the host cells for HIV are the helper T-lymphocytes and the cytotoxic T-lymphocytes. Destruction of these cells, as determined by a T-lymphocyte count, points to impending symptoms. Determining the **viral load** is another way of knowing that HIV infection is present. Some of the first signs of the presence of HIV occur within weeks or months after the virus has entered the body. This is the period of acute primary HIV infection. Patients report fatigue, mild fever, sore muscles, occasional diarrhea, and the onset of swollen lymph nodes ("swollen glands"). The symptom of swollen lymph nodes is known as **lymphadenopathy**, and it usually reflects the fact that HIV has activated B-lymphocytes to become plasma cells and secrete HIV antibodies. Apparently, HIV continues such activation for a long period of time, which accounts for the persistent swollen lymph nodes, but because HIV penetrates and resides in T-lymphocytes in the provirus form, the antibodies are ineffective against the virus.

During this stage of the disease, a test for HIV antibodies will normally give a positive result. A viral load test will also be positive; however, the count of helper T-lymphocytes does not drop to a level that compromises body defense. The count, normally about 800 cells per microliter of blood, will remain in that range. Infections caused by opportunistic organisms are not typically experienced. Some persons may experience extended headaches or brain inflammation that could indicate infection of the brain cells and AIDS dementia, as we discuss later.

An individual in the earliest stages of disease is said to be suffering from HIV infection. The CDC has created a classification system for HIV infection. There are three clinical categories and three CD4 cell count categories (**Table 4.6**). Examples of symptomatic conditions of HIV infection of categories B and C are summarized in **Table 4.7**. Some symptoms often remain for several weeks and then disappear;

TABLE 4.6	CDC Classification System for HIV-Infected Adults and Adolescents		
CD4 Cell Count Categories	**Clinical Category A: Asymptomatic, Acute HIV, or Persistent Lymphadenopathy**	**Clinical Category B: Symptomatic Conditions, not A or C**	**Clinical Category C: AIDS-Indicator Conditions**
≥ 500 cells/μl	A1	B1	C1
200–499 cell/μl	A2	B2	C2
< 200 cells/μl	A3	B3	C3

Reproduced from CDC.

TABLE 4.7	Examples of Symptomatic Conditions of HIV Infection	
Category B Symptomatic Conditions	**AIDS-Indicator Conditions**

Candidiasis
Angiomatosis
Pelvic inflammatory disease
Hairy leukoplakia
Herpes zoster (shingles)
Diarrhea lasting longer than a month
Constant low-grade fever
  (about 37.8°C, or 100°F)
	Recurrent bacterial pneumonia (2 or more
	  bouts in 12 months)
	Candidiasis of the bronchi, trachea, lungs,
	  esophagus
	Invasive cervical carcinoma
	Systemic coccidioidomycosis
	Cryptococcosis
	Cryptosporidiosis
	  (lasting longer than a month)
	Cytomegalovirus
	Chronic herpes simplex ulcers
	Kaposi's sarcoma
	Lymphoma
	Systemic *Mycobacterium avium* or
	  Mycobacterium kansasii infection
	Tuberculosis
	Pneumocystis jirovecii pneumonia
	Recurrent *Salmonella* septicemia
	Toxoplasmosis of the brain
	Wasting syndrome caused by HIV

Data from HIV Classification: CDC and WHO Staging Systems: Guide for HIV/AIDS Clinical Care, HRSA HIV/AIDS Bureau June 2012.

Mapping HIV

The Middle East and North Africa
Reliable data on the epidemics in this area of the world are in short supply.

- 460,000 adults and children were living with HIV in 2009.

- 75,000 adults and children were newly infected with HIV in 2009.

- 0.2% adult (ages 15-49 years) prevalence of HIV.

- 24,000 AIDS-related deaths among adults and children in 2009.

- The Islamic Republic of Iran has the largest number of people who inject drugs; an estimated 14% of the people who inject drugs were living with HIV in 2007.

- It is estimated that nearly half (45%) of the Iranian prison population are incarcerated for drug-related offenses.

- Approximately 80% of people in prison in Tehran inject drugs and have hepatitis C and the spread of HIV has high potential.

- Sex between men is heavily stigmatized in this region and a punishable offense in many countries .

- HIV treatment for men who have sex with men is limited.

- In surveys in Sudan, 8%-9% of men who have sex with men were living with HIV.

- Paid sex networks (prostitution) exist, but surveys in 2006 indicate that 1% of female sex workers in Egypt were living with HIV, suggesting a low HIV prevalence in sex networks to date.

Source: Data are from *UNAIDS Report on the Global AIDS Epidemic*, 2010.

however, this ensuing period of quiet should not be misinterpreted. It is only a period of "clinical latency" because HIV is multiplying rapidly in the lymph nodes and attempting to outrace the body's replacement of infected and destroyed T-lymphocytes.

At this point, it is conceivable that the body could successfully clear HIV from its tissues and restore itself to a normal state of health. This possibility was pointed out in 1995 when physicians reported on a number of babies who initially tested positive for HIV and then showed up free of infection weeks or months later. The observations, described in respected scientific journals, included one case where pediatric HIV infection was confirmed by three different test methods and on three separate occasions up to 51 days of age. By the age of one year, however, the child was negative by every possible HIV test, and no medical intervention had occurred. Four years later, the child was still HIV negative.

It is even possible that a person may not progress beyond HIV infection for a long period of time, perhaps a decade or more. Individuals in this status are known as **long-term nonprogressors**. Several definitions have been used for long-term nonprogressors resulting in 1%–25% of HIV-infected individuals who are classified as a long-term nonprogressor after 8 to 10 years of infection. The definitions are based mainly on immunological criteria (e.g., normal T-cell counts) in patients that have never received antiviral therapy. Some studies distinguish a subgroup of long-term nonprogressors who have persistently low HIV-1 viral loads.

AIDS-Related Complex

After the period of primary acute HIV infection, there is a quiet period that may be prolonged. For many individuals, the next consequence of infection with HIV is a set of constitutional symptoms known as **AIDS-related complex (ARC)**. Health agencies

A photograph of nurse bathing an AIDS patient at Edendale Hospital, South Africa. The patient is displaying the severe weight loss that commonly accompanies AIDS. Also visible on his back are opportunistic skin infections associated with AIDS. © Gideon Mendel/Corbis.

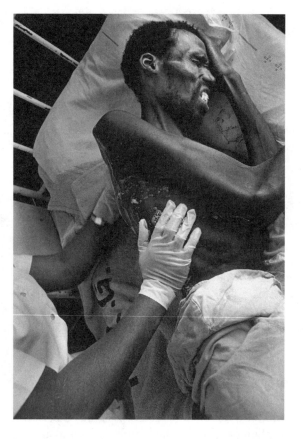

have begun to delete "ARC" from their official vocabulary (preferring to refer to all pre-AIDS conditions as "HIV infection"), but this book uses the term because it is ingrained in the public lexicon and is part of the vernacular of AIDS. Many symptoms of ARC are similar to those of HIV infection, but usually the symptoms are exaggerated in persons with ARC.

When a person experiences ARC, lymphadenopathy once again develops in lymph nodes of the neck, armpits, and groin. Swelling can remain for 3 months or more. Patients also experience weight loss of up to 10% of baseline body weight or more than 10 pounds (**Figure 4.2**). Also, there is constant low-grade fever at about 37.8°C (100°F) and diarrhea extending over several weeks. In addition, the fatigue may be so overwhelming that patients cannot lift their heads from the pillow on waking in the morning.

One of the most troublesome aspects of ARC is the night sweats. Individuals perspire so heavily at night that the bed linens and nightwear become drenched with sweat. Saturation can be extensive enough to necessitate linen changes, and sleep is fitful at best. Few other microbial diseases are accompanied by such heavy sweating.

During this stage of the disease, the number of helper T-lymphocytes has generally dropped to fewer than 400 cells per microliter, an indication of viral replication. Evidence for the declining cell numbers comes from a number of skin tests that depend on T-lymphocyte activity. These tests, similar to the tuberculin test, elicit a response from a normal individual but give negative results in a person with ARC because of the reduced numbers of T-lymphocytes.

Some persons with ARC also suffer an infection of the mouth known as **thrush**. This is a fungal disease caused by a yeast-like microorganism called *Candida albicans*. Commonly found on the skin surface and in the intestine, *C. albicans* is normally held under control by T-lymphocytes. With destruction of T-lymphocytes, however, the fungus grows on the tongue, gums, and lining of the cheeks, producing ulcers and milky, white flakes of fungal growth (**Figure 4.3**). Infected individuals are not inclined to eat, especially when the disease reaches the esophagus, erodes the surface tissue there, and exposes the nerve endings. A loss of appetite increases the weight loss.

For some individuals, ARC is a mild illness with limited bouts of symptoms. For others, it may be severe and may lead to physical deterioration. Moreover, the person with ARC suffers psychological stress from fear that AIDS lies ahead. No widely accepted estimate as to how many persons with ARC will develop AIDS is available, but public health officials generally agree that a substantial number of ARC cases will progress to AIDS.

Acquired Immune Deficiency Syndrome

As HIV continues to replicate itself, the destruction of T-lymphocytes goes on methodically and inexorably. Soon the count of helper T-lymphocytes has dropped below 100 cells per microliter of blood, and crippling of the immune system is virtually complete. At any time, the symptoms of ARC can return in severe form, together with brain disease, a **wasting phenomenon** (Table 4.8), and/or **opportunistic infection**. Then the person has AIDS.

TABLE 4.8	Diagnostic Criteria for HIV Wasting Syndrome
Findings of involuntary loss of more than 10% of baseline body weight plus either chronic diarrhea (two or more loose stools per day for 30 or more days) or chronic weakness and documented fever (for 30 days or longer, intermittent or constant) in the absence of a concurrent illness or condition other than HIV infection that could explain the findings (e.g., cancer, tuberculosis, cryptosporidiosis, or other specific enteritis)	

Source: Kaplan, J.E. et al., 2009. Guidelines and Treatment of Opportunistic Infections in HIV-Infected Adults and Adolescents. *MMWR* 58 (RR-04):1–198.

The average time that passes from HIV infection until AIDS (i.e., the **incubation period**) has been a subject of controversy for many years. At one time, the incubation period was believed to be seven years, but according to a more recent study performed at the University of California at San Francisco, the incubation period for AIDS averages about 10 years. The San Francisco study was based on blood samples obtained from homosexual men in the 1970s and stored for later analysis. It also included an elaborate system of confidential blood testing in the San Francisco homosexual community. Researchers were able to track men whose blood tested positive for a period of 10 years and found that it took that long, on average, for the disease to develop.

Antiretroviral therapy, however, can lengthen the incubation period significantly. Before 1996, few antiretroviral treatment options were available. **Monotherapy** (treatment with one anti-HIV drug) began as treatment in the early 1990s. By the mid 1990s, the standard care of HIV-1 patients evolved to cocktail or combination-therapy known as **HAART** (highly active antiretroviral therapy). HAART dramatically suppresses replication of HIV-1 and reduces viral loads. Many cohort studies have been done to compare life expectancies of recently diagnosed HIV-infected patients and uninfected individuals from the general population.

A 2010 report was published involving the study of 4,612 patients in the Netherlands diagnosed between 1998 and 2007 and still on antiretroviral therapy. During the study, 118 deaths occurred (6.7% mortality rate per 1,000 persons). The report concluded that an individual diagnosed around the age of 25 would live approximately 27 years (males) or 32 years (females) if the individual complied to antiretroviral therapy. Comparing these results to individuals in the general population (uninfected) of the same age and gender, the HIV-1 infected patients lived 0.4 years less (if diagnosed at 25) and 1.4 years less if diagnosed at age 55 than uninfected individuals. HIV antiviral drugs have changed the perception of HIV/AIDS from a fatal to a chronic and potentially manageable disease.

Indeed, studies show that about 1% to 25% of individuals infected with HIV live for 12 years or more without developing AIDS. These individuals are the long-term nonprogressors. At the other end of the spectrum, AIDS is known to develop in as little as six months, depending largely on the type of lifestyle the individual leads. Clinical illness cofactors such as co-infection with hepatitis B viruses or other microbes may be another determining factor.

As full-blown AIDS develops in the patient, the lymph nodes become enlarged, tender, and stiffer than normal lymph nodes. This lymphadenopathy persists for months. Also, a low-grade fever at about 37.8°C (100°F) returns. It is commonly present in the late afternoon and early evening. When another infection is present, the fever may be higher and more difficult to control. Chills and periods of sweating often accompany the fever.

Persons with AIDS suffer from substantial nausea and fatigue. The nausea is often accompanied by vomiting, and the fatigue does not seem to end. Both sap the individual's strength, making normal routines extremely difficult. Job efficiency declines and the will to care for oneself diminishes. A headache may come and go in some individuals; in others, it remains for days. The night sweats resemble those in ARC, only they are worse.

The diarrhea associated with AIDS can also be considerable. In its most severe form, a patient may lose a quart of water in a 24-hour period (an opportunistic disease of the intestine may make the diarrhea worse). Exhausting fatigue and considerable weight loss usually are observed. In milder cases of AIDS, two to four loose stools are passed per day, often containing blood or mucus.

Tissue-Specific HIV-Associated Diseases

Although the predominant complications of HIV and AIDS are the opportunistic infections, there is a set of tissue-specific HIV-associated diseases that do not result from opportunistic infections; nevertheless, they do contribute to the morbidity of AIDS patients and can lead to death. These syndromes include the HIV wasting syndrome, AIDS-dementia complex, HIV-nephropathy, and possibly Kaposi's sarcoma. Although Kaposi's sarcoma may be related to a second viral infection, these tissue-specific manifestations of HIV infection are thought to be the result of local tissue infection with HIV, or the infiltration of tissues by HIV-infected lymphocytes and macrophages. Recent research has shown that HIV activates some of the inflammatory mechanisms in cells of the immune system, making it likely that infected cells will traverse blood vessel barriers and localize in tissues. Once localized in tissues, these infected cells may release, or leak, HIV proteins that cause damage to the nearby tissue. Indeed, individual HIV proteins have been found to be neurotoxic and cause apoptosis in the central nervous system.

HIV Wasting Syndrome

For some patients, the diarrhea associated with AIDS can be so profound that a condition called **HIV wasting syndrome** ensues. This condition is accompanied by a dramatic weight loss of more than 10% of a person's baseline body weight plus either chronic diarrhea lasting at least 30 days or chronic weakness and fever lasting 30 days. For the definition of AIDS to apply, no other illness, such as opportunistic disease, or any condition other than HIV infection need be present. In such an instance, the diagnostic criteria for *HIV wasting syndrome*, shown in Table 4.8, have been fulfilled.

HIV wasting syndrome is a major cause of morbidity and mortality in patients with AIDS. In addition to decreased quality of life, the syndrome has been associated with a lower survival rate independent of the count of T-lymphocytes. Although the precise cause remains unknown, various researchers have related the syndrome to altered energy intake, secondary infections, altered body metabolism, and abnormalities in the body's hormone balance. The appetite loss is due to such things as mouth soreness, vomiting, lethargy, and depression.

The most common hormone abnormality associated with the wasting syndrome is **hypogonadism**, or low level of the sex hormone testosterone. Low testosterone levels may manifest themselves as poor tissue preservation, and researchers are investigating whether treating patients with anabolic steroids is beneficial. Over 50% of men with advanced HIV disease display low testosterone levels.

Another approach to forestalling HIV wasting syndrome is to strengthen the muscles with exercise. Many AIDS sufferers are reluctant to work with weights for fear of stimulating further muscle loss, but research reported in 1998 indicates that HIV levels do not rise as a result of strenuous workouts.

AIDS-Dementia Complex

There is much accumulated evidence that HIV itself causes **AIDS-dementia complex (ADC)**. ADC typically occurs in HIV-1 infected individuals whose CD4+ T-lymphocyte counts fall below 200 cells/μl and may be the first sign of AIDS. HAART has reduced the frequency of ADC, which has declined from 30%–60% of people infected with HIV-1 to 20%. HAART may not only prevent or delay the onset of ADC in people with HIV infection, but also improve mental function in people who already have ADC. Early and severe symptoms of ADC are listed in **Table 4.9**.

ADC may account for periods of depression. Persons with AIDS also display signs of confusion and have trouble coordinating their muscular activities. Apathy, along

TABLE 4.9	Symptoms of AIDS-Dementia Complex (ADC)

Early Symptoms	Late, Severe Symptoms
Mental slowness	Sleep disturbances
Apathy	Psychosis (severe mental and behavioral disorder)
Confusion	Extreme agitation
Word-finding difficulty	Loss of contact with reality
Forgetfulness	Inability to respond appropriately to
Difficulty learning new things	the environment
Changes in behavior	Delusions
Withdrawal from hobbies or social activities	Hallucinations
Depression	Mania (extreme restlessness, hyperactivity, very
Poor concentration	rapid speech, poor judgment)
Decreased sex drive	Seizures

Source: Kaplan, J.E. et al., 2009. Guidelines and Treatment of Opportunistic Infections in HIV-Infected Adults and Adolescents. *MMWR* 58 (RR-04):1–198.

with fatigue, nausea, and loss of concentration, is another signal that the nervous system is involved. AIDS patients often experience a sudden, strong emotion unprecipitated by anything they can recall. They suffer memory loss, concentration lapses, minor disturbances in thought processes, and difficulty in sleeping. Headache, disorientation, and a general feeling of the "blues" are often experienced. Unsteady gait, slurred speech, and tremors may be other signs that the AIDS-dementia complex has developed. The patient often experiences absence of feeling in the arms and legs, leading to loss of such basic functions as walking (Table 4.9).

Mounting evidence indicated that ADC was a true viral infection of the brain. For example, HIV has been isolated from brain cells called astrocytes, shown in **Figure 4.4**, and from their surrounding fluid. Moreover, infection of the brain can be induced by inoculating animals with brain tissue obtained at autopsy from deceased individuals. The actual mechanism of HIV involvement with the brain cells is unclear,

FIGURE 4.4

An electron micrograph of human brain tissue showing astrocytes, the star-shaped cells with multiple long extensions. Astrocytes may be the targets of HIV, and the result of the infection may be AIDS-related dementia. Image courtesy of National Research Council Canada.

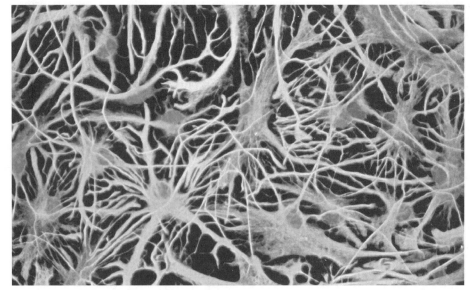

but one possibility is that the gp120 protein of HIV may be released in brain tissue and act as a neurotoxin that interferes with normal maintenance processes in the brain. Most researchers agree that the virus does not destroy brain cells simply by multiplying within them, but by mechanisms not yet understood. One possibility is apoptosis, a form of cell suicide.

For HIV to infect human brain tissue, the virus would have to interact with brain cells much as it interacts with T-lymphocytes. Such an interaction would require receptor molecules on brain cells similar to the CD4 molecules on T-lymphocytes. As early as 1986, scientists from New York's Columbia University located protein molecules at the surface of brain cells and provided the clue to how HIV and brain cells interact.

How HIV enters the brain tissue was another problem requiring resolution. Normally the brain is shielded from the blood and its constituents by the so-called blood–brain barrier. This barrier of membranes and substances allows only selected materials and cells to pass from blood to brain tissue. In 1987, scientists learned how HIV passes this barrier. Apparently, the virus remains sequestered inside a macrophage, whereas the latter squeezes through the barrier (macrophages are one of the few types of cells that can pass the barrier). Thus, the macrophage appears to be the key to HIV delivery into the brain. Indeed, the emerging consensus is that the central nervous system is a reservoir for HIV, probably in the macrophages of the brain tissue.

The ability of HIV to attack brain cells as well as T-lymphocytes has changed the outlook on AIDS significantly. No longer is AIDS considered solely a disease of the immune system. Rather, the substantial effects of HIV on the nervous system indicate that AIDS affects at least a second system of the body and pushes concerns about the disease another notch higher. Moreover, the infection of brain tissue implies further demands on already overburdened public health facilities. People with mild dementia, for example, cannot perform routine day-to-day activities, and those with severe dementia resemble senile patients—they cannot get around on their own or find their way in their own homes. The irrational and unpredictable behavior they sometimes display adds a mental component to the loss of immune function.

Another disquieting note was sounded in 2000 when researchers observed the inability of protease inhibitor drugs to cross the blood–brain barrier and penetrate the brain tissue. This observation raised concerns that although infected patients may live longer as a result of the drug therapy, they may face an increased risk of developing AIDS-dementia complex. Opponents of this conclusion point out that the incidence of the complex has declined since the drugs were introduced.

AIDS-dementia complex necessitated modification of the definition for AIDS. As previously mentioned, the original definition of AIDS specified that some other infectious disease involving immunodeficiency must be present. As we discuss here, that requirement still holds for the majority of cases; however, the case definition was expanded in 1987 to include neurologic disease (technically called HIV encephalopathy), even though some other microbial disease is not present. Thus, a person may fulfill the case definition for AIDS if they have AIDS-dementia complex. The new term, **HIV-associated neurocognitive disorder (HAND)** is used to describe the full spectrum of neurological disease observed in HIV patients.

HIV Nephropathy

HIV can also cause glomerular disease in the kidney in the absence of any secondary infections. Before developing full-blown AIDS, roughly 5% to 10% of HIV individuals experience kidney failure. **HIV-associated nephropathy** usually begins with loss of blood proteins in the urine (proteinuria), and within 1 year, the patient usually develop what is known as end-stage renal failure. With end-stage renal failure, the patient's kidneys are not functioning. Dialysis can prolong life; however, the best treatment is a kidney transplant. Kidney transplantation often requires the use of **immunosuppressant drugs**, which is risky in the case of HIV patients because of their already compromised state of the immune system. Also, because we do not understand why some people develop HIV nephropathy and others do not, the chances that the new kidney will have the same fate as the original must be considered.

Kaposi's Sarcoma

One of the conditions associated with AIDS is a type of cancer known as **Kaposi's sarcoma**, named for Moritz Kaposi, who first described it in 1924. The original case definition of AIDS included Kaposi's sarcoma because it was frequently present in homosexual men who displayed immune deficiency. As of 1990, almost one quarter of all persons with AIDS suffered from Kaposi's sarcoma, and the condition remains one of the most common diseases associated with AIDS. The case definition for AIDS, set down in 1987 and revised in 1993, continues to include the presence of Kaposi's sarcoma as a basis for AIDS.

Before the AIDS epidemic, Kaposi's sarcoma was a rare form of cancer found primarily in older men of Mediterranean descent. In this population, Kaposi's sarcoma is characterized by slow-growing tumors in the blood-vessel linings, with patches of red to violet-red on the lower extremities (**Figure 4.5**). Eventually, the patches become darkened purplish brown nodules. Kaposi's sarcoma by itself is rarely fatal; patients live for many years and often die of an unrelated disease.

When associated with AIDS, Kaposi's sarcoma is a much more serious disease. Nodules appear on the lower extremities as well as in scattered patches all over the

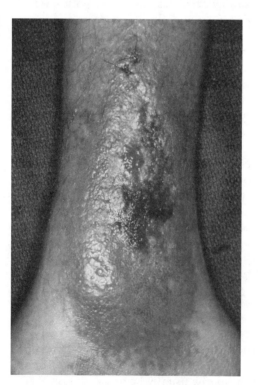

FIGURE 4.5

The foot of a patient with AIDS, displaying the patches of Kaposi's sarcoma that commonly develop on the lower extremities. Courtesy of Dr. Steve Kraus/CDC.

body. The upper body and face are often the sites of the purplish brown patches, and the oral cavity may be involved. Internal organs such as the lymph nodes, gastrointestinal tract, liver, and spleen are possible nodule sites, and any soft tissue appears to be susceptible. Kaposi's sarcoma is often accompanied by diarrhea, weight loss, fever, and intense sweating at night. The disease is so aggressive that affected individuals survive an average of less than two years. This span may be even shorter if a microbial infection occurs concurrently.

Evidence is lacking on why Kaposi's sarcoma is so prevalent in persons with AIDS and why it is so aggressive; however, research results indicate that HIV itself may promote the disease. Scientists at the National Cancer Institute transplanted genes from HIV into unborn mice. They then noted that a significant percentage of mice born with certain of the transplanted genes had abnormal skin growths similar to those in Kaposi's sarcoma. The HIV genes were located in cells distant from the growths, suggesting that a protein encoded by the genes was responsible for the cancer-promoting effect. Scientists have reported that the *tat* gene of HIV was the critical protein-encoding gene.

Observations of a different sort indicate that Kaposi's sarcoma may be caused by a separate viral agent transmitted coincidentally with HIV. In 1990, a prominent New York physician reported 12 cases of Kaposi's sarcoma in homosexual men testing negative for HIV. The suggestion was made that a Kaposi's sarcoma agent may have entered the human population at approximately the same time as HIV but was becoming uncoupled from AIDS. This speculation was fueled by a 1998 report that the number of cases of Kaposi's sarcoma in homosexual men with AIDS had dropped to 21% (compared with nearly 50% in the early years of the AIDS epidemic).

Further insight was provided in 1995 when Patrick S. Moore and Yuan Chang of Columbia University identified human herpesvirus type 8 (HHV-8) as the cause of Kaposi's sarcoma. The investigators determined the DNA sequence of the virus and identified the sequence in lesions associated with Kaposi's sarcoma, including lesions from patients living in widely separated areas of the United States. Many physicians believe that HIV infection promotes HHV-8 replication by impairing the immune system of the host; however, in 1997, Robert Gallo and his coworkers postulated that the tat protein encoded by HIV acts as a toxin and attracts HHV-8 into the cellular area that becomes a Kaposi's sarcoma lesion. This work suggests that HHV-8 may be a mere "visitor" to the site, a virus that takes advantage of the conditions altered by HIV. Thus, HHV-8 may be necessary to establish Kaposi's sarcoma, although it does not act alone. In some reports, the virus is called KS-associated herpesvirus. Its isolation from salivary secretions in 1997 may give a glimpse of how it is transmitted. The virus lives in the back of the mouth and throat, and kissing is thought to be one route of transmission.

Opportunistic Diseases

Another hallmark of AIDS is the presence of disease caused by an **opportunistic organism**. Opportunistic organisms are organisms that often exist in the body but cause no harm because the body's immune system and other natural defenses keep them under control. When the natural defenses are compromised, however, the organisms seize the "opportunity" to invade the tissues and cause disease. The occurrence of diseases caused by opportunistic organisms illustrates the delicate balance that exists between infectious organisms and natural defenses. After the defenses break down, the opportunists invade.

The majority of persons with AIDS suffer from one or several diseases caused by opportunistic organisms. These diseases have euphemistically been called **opportunistic diseases**. Although technically incorrect (because the organisms are opportunistic, not the diseases), the term opportunistic disease has become part of AIDS terminology, and so we shall use it here.

When the original definition of AIDS was promulgated in 1981, the CDC specified that an opportunistic disease or Kaposi's sarcoma had to be present before doctors could report a case as AIDS. With the realization that dementia and wasting could be key factors, the 1987 definition of AIDS incorporated neurological disease (HIV encephalopathy) and HIV wasting syndrome as other bases for identifying a case of AIDS. Then the 1993 case definition specified a count of helper T-lymphocytes (CD4 cells). Nevertheless, the presence of an opportunistic disease remains a prime criterion in the case definition of AIDS, and a majority of persons with AIDS suffer from at least one opportunistic disease.

The opportunistic disease most prevalent among persons with AIDS is ***Pneumocystis* pneumonia**, sometimes referred to as **PCP**. The disease is well studied in rodents and was first described as a protozoan by John Carini in 1910 and subsequently named *Pneumocystis carinii* (**Figure 4.6a**). However, we now know that the organism is a yeast-like fungus and the prevailing species in humans is *Pneumocystis jirovecii* and not *Pneumocystis carinii*. *Pneumocystis jirovecii* is a benign inhabitant of the lungs in many people. In fact, the majority of children in the world have detectable antibodies (meaning they have been exposed and their bodies have produced an immune response against *Pneumocytis jirovecii*) by 2–4 years of age.

Disease occurs when the *Pneumocystis* organisms multiply furiously and take up all the lung's air spaces (**Figure 4.6b**), which usually occurs only in immunodeficient individuals, such as the elderly, or those with diseases of the immune system, such as AIDS. The patient experiences rapid, labored breathing, a nonproductive (no mucus) cough, and extreme anxiety because enough oxygen cannot be drawn from the air into the bloodstream. Symptoms may be relieved by administering high concentrations of oxygen through a facemask or, when symptoms are extreme, by using a respirator machine connected to a tube placed in the windpipe. The respirator reduces the effort to breathe, but the procedure can be frightening and cause some physical discomfort.

Patients with *Pneumocystis* pneumonia can be treated with a drug called **pentamidine isethionate**. Although this drug kills the fungal pathogen, the damage to the lungs may be irreversible, and the drug may impair kidney and liver function and cause painful abscesses at the injection site. As of 1996, more than 50% of individuals who died from complications of AIDS died from *Pneumocystis* pneumonia. Because the disease is rarely seen elsewhere, a diagnosis of *Pneumocystis* pneumonia is a reliable indicator that HIV is present. At its peak in the United States, PCP was the leading AIDS-defining diagnosis and was responsible for more than 20,000 new AIDS-defining cases per year from 1990 to 1993. In Europe, PCP was the leading AIDs indicator in the World Health Organization *2008 HIV/AIDS Surveillance in Europe Report's* accounting for 16% of the AIDS cases diagnosed that year. PCP continues to be an important cause of HIV-associated pneumonia but rates have decreased. For example, at San Francisco General Hospital, nearly 1,000 cases of HIV-associated PCP were diagnosed from 1990 to 1993. This number has decreased to 20 or 30 cases per year. Most of the cases occurred in individuals who were not receiving antiretroviral therapy or PCP prophylaxis and many were unaware that they were infected with HIV (**Figure 4.7**).

Another fungus that causes an opportunistic disease is *Cryptococcus neoformans* (**Figure 4.8a**). This organism, a common resident of the lungs, is usually inhaled from the air. It grows actively in the droppings of pigeons and enters the air in windborne particles. The fungus is generally noninfectious, but in persons with AIDS, it multiplies in the lungs, spreads to the blood, and localizes on the brain and its coverings.

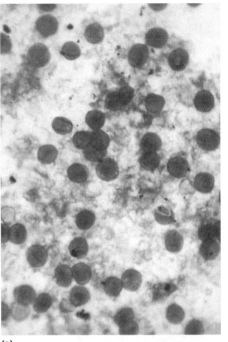

(a)

FIGURE 4.6

Pneumocystis jirovecii (formerly *carnii*), an opportunistic fungal pathogen associated with AIDS. (a) The fungi is seen in the throat washings of an AIDS patient. It appears as dark spheres in this specimen as it might be viewed by a pathologist (×250). Courtesy of Lois Norman/CDC. (b) An x-ray of the lungs of an AIDS patient infected with *P. jirovecii*. The dark areas reflect regions of fungi involvement and demonstrate the extensive nature of the infection. The organisms occupy the spaces normally used for the transfer of gases and make breathing very difficult. Courtesy of National Cancer Institute.

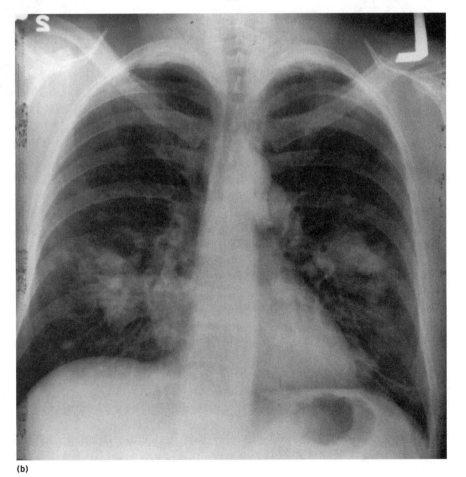

(b)

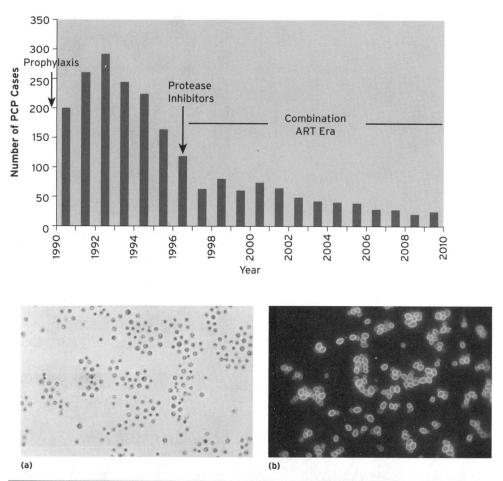

FIGURE 4.7

Confirmed cases of PCP microscopically diagnosed at San Francisco General Hospital, 1990–2009. Data from Huang, L. et al., 2011. "HIV-associated Pneumocystis pneumonia." *Proceedings of the American Thoracic Society.* 8:294–300.

FIGURE 4.8

Two opportunistic fungi associated with AIDS. (a) A light microscope view of *Cryptococcus neoformans*, the agent of cryptococcosis. Each cell is round to oval and is surrounded by a distinctive capsule. The capsule provides resistance to phagocytosis and enhances the pathogenicity of the fungus. Courtesy of Dr. Lucille K. George/CDC. (b) *Candida albicans*, the cause of thrush. In this light micrograph, the fungus appears as oval cells with a tendency to cling together in long filament-like strands. Courtesy of Maxine Jalbert and Dr. Leo Kaufman/CDC.

Piercing headaches, stiff neck, paralysis, and mental dysfunction accompany the infection. Changes in behavioral patterns and mental confusion may also be observed. The disease is termed **cryptococcosis**.

Candida albicans, the yeast-like microorganism mentioned earlier in the discussion on ARC, is of substantial consequence for AIDS patients. In a person with AIDS, *C. albicans* grows in the mouth as patches of white, curd-like material, producing a condition called **thrush** or **candidiasis**. When scraped off, the patches reveal a painful, red, inflamed base. As the fungus spreads to the esophagus, tissue erosion occurs, and the exposed nerve endings make eating an excruciating experience. The appetite is quickly lost, and body deterioration ensues, sometimes necessitating liquid feeding through a tube inserted through the nose into the stomach. Though several drugs are available to control candidiasis, repairing the damage to the esophagus takes considerable time.

Another opportunistic disease often associated with AIDS is toxoplasmosis. The agent of this disease is *Toxoplasma gondii*, a protozoan shown in **Figure 4.9a**.

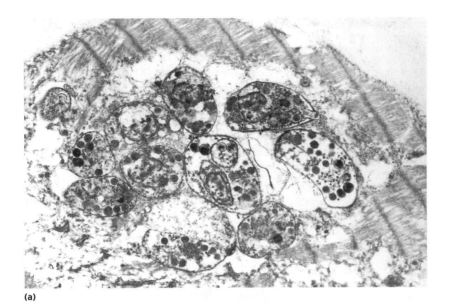

(a)

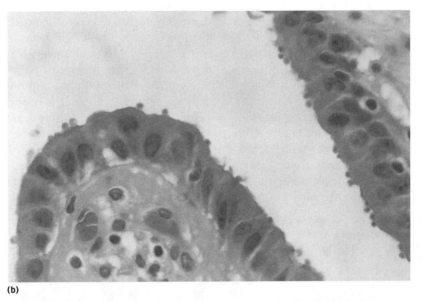

(b)

FIGURE 4.9

Two opportunistic protozoa associated with AIDS. (a) Pseuodocysts of *Toxoplasma gondii*, the cause of toxoplasmosis visible in myocytes of an AIDS patient's heart. Courtesy of Dr. Edwin P. Ewing, Jr./CDC. (b) Histopathology of gallbladder epithelial cells containing *Cryptosporidium*. The 4–5 micron protozoans are seen here lining the lumen of the gallbladder of this patient with AIDS. Courtesy of Dr. Edwin P. Ewing, Jr./CDC.

Normally, toxoplasmosis is a mild, mononucleosis-like disease, but in persons with AIDS, the protozoan attacks the brain tissue, causing lesions, cerebral swelling, and seizures. Severe headaches, sensitivity to light, neck stiffness, and loss of some motor or sensory functions may also occur. Several drugs are available to control toxoplasmosis, but because the person is usually quite weak, these are of limited value. Domestic house cats often harbor *T. gondii*, and thus, persons with HIV infection are advised to keep their cat indoors and to avoid cleaning the litter box.

Another protozoan opportunistic pathogen is *Cryptosporidium coccidia* (**Figure 4.9b**). This organism infects the intestine in AIDS patients and induces unrelenting, voluminous, watery diarrhea, progressing to dehydration, loss of important body salts, and malnutrition. The results are extreme weight loss, wasting of muscle tissue, and loss of skin tone as well as poor tissue healing and repair. In its severest form, **cryptosporidiosis** may result in loss of up to a gallon of fluid per day. Dehydration and emaciation accompany this disease, and intravenous fluid replacement is required to maintain the body's water balance.

One virus that can cause an opportunistic disease is the **cytomegalovirus (CMV)**. This virus is so named because in the laboratory it causes cells to assume an enlarged size (*cyto* means "cell," and *megalo* refers to "giant" size). Normally, CMV is present without consequence in many body tissues, but when the immune system is compromised, the virus multiplies aggressively in lung tissues, where it induces pneumonia. The virus also multiplies in other tissues: in the liver and kidney tissues, where it causes tissue death; in salivary gland tissues, where it brings on swollen glands; and on the retina (CMV retinitis), where it leads to partial or complete blindness. Indeed, over 30% of AIDS patients develop eye disease because of CMV and experience "floaters," flashes of light, blind spots, or blurred vision.

In addition, the CMV appears to invade helper T-lymphocytes, thus accelerating the destruction of those cells begun by HIV. An interesting corollary to this observation was made in 1997 when French investigators reported that cells invaded by CMV produce proteins that act as receptors identical to CCR-5. The implication is that cells infected with CMV become unusually receptive to HIV. Although the evidence is controversial and somewhat contradictory, if proved true, it would imply a tight relationship between HIV and its accomplice CMV. Skeptics of the findings point out, however, that although CMV is known to infect brain cells (as well as cells of the retina), there is little evidence that co-infection of helper T-lymphocytes with both HIV and CMV is a common event. **Figure 4.10a** shows the CMV.

Opportunistic diseases of bacterial origin have also been observed. One example, tuberculosis, has always been a significant cause of death worldwide. In the United States, for example, tuberculosis was responsible for one death in seven (from all causes) in the early 1900s. Today's statistics, although improved, are still

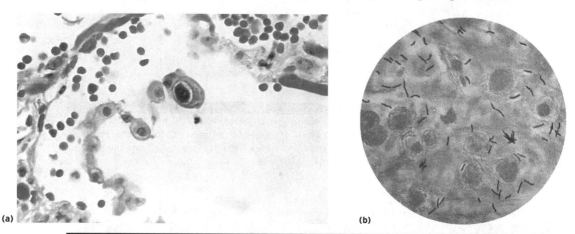

(a)   (b)

FIGURE 4.10

Viral and bacterial opportunists associated with AIDS. (a) An electron micrograph of the cytomegalovirus cultivated in tissue cells. Some viral particles are within the nucleus of the cell, but others have budded through the nuclear membrane and are now in the cellular cytoplasm. The thicker covering of these viruses reflects the envelope they have acquired during budding. Note the very small size of the viruses relative to the host cell (×45,000). Courtesy of Dr. Edwin P. Ewing, Jr./CDC. (b) A photomicrograph of *Mycobacterium tuberculosis*, the agent of tuberculosis. This rod-shaped bacillus has been cultivated outside of living tissue in laboratory medium (×1,000). Courtesy of CDC.

alarming: Over 14,500 cases of the disease are reported each year in the United States. The cause of tuberculosis is a small bacterial rod called *Mycobacterium tuberculosis* (**Figure 4.10b**). A generally good quality of life together with natural controls centered in T-lymphocytes prevent the disease from proliferating in most individuals. For those with AIDS, however, the controls break down and tuberculosis develops. Progressive deterioration of the lung tissue leads to labored breathing and a cough that brings up a mucousy pus and sometimes blood. Chest pain is substantial. Shortness of breath is obvious, and the disease may spread to other organs. Without treatment, the patient often becomes emaciated and dies. It is not coincidental that the spread of HIV in recent years has paralleled an increase in the incidence of tuberculosis.

Two other similar species of *Mycobacterium* can also be a serious threat to those with HIV. These are *Mycobacterium avium* and *Mycobacterium intracellulare*, also referred to as ***Mycobacterium avium* complex, (MAC)**. Like *M. tuberculosis*, the organisms invade lung tissue as the T-lymphocyte count drops and causes a progressive destruction of tissue that can be fatal. Drugs are available for treating MAC and tuberculosis, but drug therapy must be aggressive and must extend over a period of many months.

With the advent of HAART therapy and more effective **prophylaxis** against these "classic" opportunistic infections, the incidence of these infections has decreased. The kind of opportunistic infections encountered has shifted subtly. PCP and candidiasis remain a significant cause of opportunistic infections. MAC, cryptosporidiosis, and **aspergillosis** (a fungal infection) are rare.

In addition to these "classic" opportunistic infections, healthcare workers are encountering a range of infections that would not normally be called opportunistic infections but behave as opportunistic infections. These infections can occur in HIV-negative populations as well. **Table 4.10** contains a list of these infections associated with changing patterns in people living with HIV.

Focus on HIV

One in four AIDS-related causes of death is tuberculosis, a preventable and curable disease.

TABLE 4.10	Changing Patterns of Infections in People Living with HIV
Infection Pattern	**Infections/Diseases**
Opportunistic infections still common but overall incidence decreasing	*Pneumocystis jirovecii* pneumonia (PCP) Esophageal candidiasis Cryptococcal meningitis
Less commonly opportunistic infections observed	*Mycobacterium avium* complex Cytomegalovirus infections of the eye Cryptosporidiosis Aspergillosis
Community-acquired infections increasing, not classically considered opportunistic infections but behaving as opportunistic infections	*Streptococcus pneumoniae* infection *Haemophilus influenzae* infection *Clostridium difficile* infection
Community-acquired infections causing complications	Influenza A infection
Geographically restricted infections that are increasing in people living with HIV because of immigration or travel	*Falciparum* malaria Leishmaniasis Chagas disease Dimorphic fungi infection

Modified from Dockrell, D. H. et al., 2011. "Evolving controversies and challenges in the management of opportunistic infections in HIV-seropositive individuals." *Journal of Infection* 63:177-186.

Pediatric AIDS

In 1982, when a prominent researcher wrote that babies were contracting AIDS, few colleagues believed him. Today, there is little doubt that HIV can be passed from a pregnant woman to her offspring. New HIV infections among children have declined since 2003. Approximately 430,000 children were infected in 2008 and 330,000 were infected in 2011. This represents a 43% decline since 2003. More than 90% of the children who acquired HIV in 2011 live in sub-Saharan Africa.

HIV in children most often occurs during pregnancy and labor or postnatally during breastfeeding. Mother-to-child transmission has virtually been eliminated in industrialized countries, but remains common in Africa. Children with AIDS fail to thrive (**Figure 4.11**). They do not demonstrate the expected growth patterns after birth, nor do they respond to aggressive nutritional therapy. Many children achieve their normal height for a particular age, but they experience weight loss that does not reverse itself. Persistent diarrhea may contribute to the weight loss.

Another symptom of pediatric AIDS is decreased cognitive skills, most likely because of HIV infection in the brain. Researchers have discovered that brain disease in the fetus can begin during the first stages of pregnancy and can lead to dementia. Expected intelligence milestones are not reached, and on tests of cognitive skills, children with AIDS consistently score in the lowest 2% of their age group. In some cases, children lose the use of their arms and legs and become blind and deaf by the age of two years.

Children with AIDS suffer from opportunistic diseases, but the sources are generally bacteria; in contrast, in adults, protozoa, fungi, and viruses are important. In newborns and children with AIDS, the common opportunistic diseases are due to *Salmonella* species (intestinal infections), streptococci (respiratory infections), and staphylococci (skin and blood infections). It is conceivable that susceptibility to these bacterial diseases may reflect the inability of infected children to develop the antibody-producing B-lymphocytes necessary for protection against them. Candidiasis is also a significant symptom in children with AIDS, as are lymphadenopathy and swollen salivary glands. The latter symptom usually does not occur in adults.

Focus on HIV
48% of HIV-positive women received treatment to prevent HIV transmission to their child in 2010.

FIGURE 4.11

A nurse holding an infant boy suffering from pediatric AIDS at the Victor Babes Hospital in Bucharest, Hungary. © Peter Turnley/Corbis.

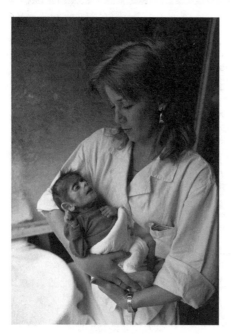

Probably the only common denominator among children with AIDS is that their immune systems have been compromised. No two cases are identical. Indeed, no two cases in adults are alike because the symptoms of AIDS vary within very broad parameters. In the final analysis, there is no one single entity called AIDS but, instead, myriad conditions and diseases. For this reason, defining and recognizing AIDS can be a formidable task.

In 2009, the Joint United Nations Programme on HIV/AIDS called for the virtual elimination of **mother-to-child-transmission (MTCT)** of HIV. This call has been embraced by several other agencies, regional coordinating bodies and national governments. In 2010, the World Health Organization (WHO) published a set of guidelines for the best scientific, programmatic tools, and new advice for safer infant feeding to eliminate MTCT. UNAIDS and the Earth Institute developed a partnership to establish MTCT-free zones in the Millennium Villages. Clinics, agricultural enhancement, electrification, mobile telephone connections, road construction, water and sanitation improvement, and improving school are occurring in these **Millenium Villages Project sites (Figure 4.12.)** Ante- and post-natal care, infant diagnosis services, antiretroviral

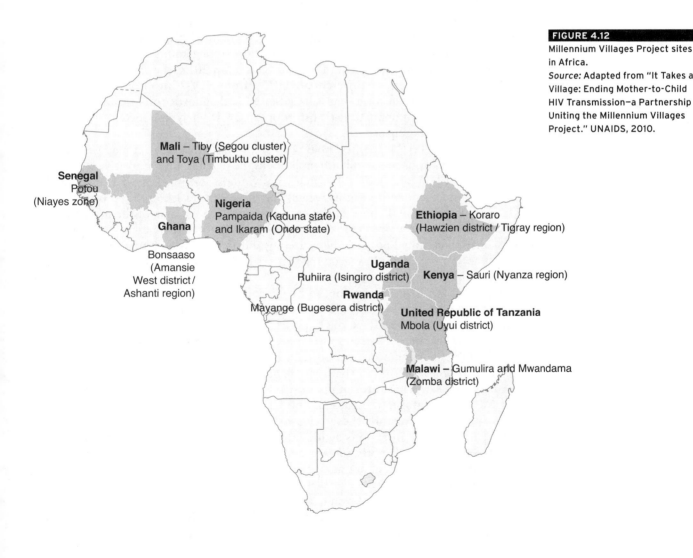

FIGURE 4.12
Millennium Villages Project sites in Africa.
Source: Adapted from "It Takes a Village: Ending Mother-to-Child HIV Transmission—a Partnership Uniting the Millennium Villages Project." UNAIDS, 2010.

treatment, and family planning services are available in the MTCT-free villages or zones. The project is aimed to also foster community conversations to reduce gender inequality and HIV-related stigma that in turn can help to create a family-centered approach to HIV prevention and health promotion.

Defining AIDS is a complex matter because there are many phases to the disease. It is important that the disease be defined, however, because diagnosis, treatment, and public health record-keeping are based on a clear understanding of it. The earliest definition, promulgated by the CDC in 1981, spelled out AIDS as a condition in which certain diseases exist in the patient and reflect an impaired immune system. The revised definition, issued in 1987, included in the case definition evidence of the presence of HIV and three distinctive conditions that can accompany HIV infection: HIV wasting syndrome, AIDS-dementia complex, and opportunistic diseases. The 1993 case definition includes a helper T-lymphocyte (or CD4 cell) count below 200 cells per microliter of blood or pulmonary tuberculosis, recurrent pneumonia, or invasive cervical cancer, and the 2000 expansion of the case definition includes reporting of individuals with HIV infection. In 2008 the case definition was revised requiring laboratory-confirmed evidence of HIV infection among adults, adolescents, and children aged 18 months to less than 13 years of age. In 2012, the case definition was updated to include criteria to differentiate HIV-2 from HIV-2 infection, the addition of an acute or stage 0 infection that is based on HIV diagnostic tests regardless of T-cell counts. The role of opportunistic infections and CD4-T lymphocyte counts were further defined in staging criteria.

Infection with HIV is the first phase of a multiphase disease that may progress to AIDS. Among its signs are swollen lymph nodes, headaches, and a positive viral load test and test for HIV antibodies. For many individuals, ARC follows. Extensive and persistent lymphadenopathy, night sweats, diarrhea, and thrush are signals of ARC. AIDS is actually the final phase of the progression, when wasting of the body is substantial, when brain involvement is severe, or when opportunistic diseases develop. Kaposi's sarcoma, now known to be caused by a different virus (human herpesvirus-8), is accompanied by slow-growing tumors in the blood vessels and splotches on the skin.

Opportunistic diseases are caused by microorganisms normally present in the body and controlled by a healthy immune system. When the system is weakened, the microorganisms seize the "opportunity" to infect. The most prevalent opportunistic disease is *Pneumocystis* pneumonia, a severe lung disease accompanied by lung consolidation and difficult breathing. Toxoplasmosis of the brain, *Cryptosporidium* infection of the intestine, and *Cryptococcus* infection of the nervous system are other rarer opportunistic diseases. In many cases, an opportunistic disease is fatal to the patient who has AIDS. HAART therapy and more effective **prophylaxis** against these "classic" opportunistic infections has decreased the incidence of these infections significantly. The kind of opportunistic infections encountered has shifted subtly toward community-acquired and specific geographically-defined infections that act as opportunistic infections but can also infect HIV-negative populations.

Pediatric AIDS, as acquired by a newborn from its mother (MTCC), primarily occurs in countries in Africa and India. Children with AIDS fail to thrive. They display reduced cognitive skills, loss of use of their motor organs, and opportunistic diseases, often due to bacteria. Although antiretroviral therapy in pregnant women has successfully reduced the number of cases of pediatric AIDS, the number of cases remains substantial. Millennium Villages Project sites have been set up in Africa by partnerships to reduce MTCT, the HIV-related stigma, as well as a positive impact on health.

Healthline Q & A

Q1 Are there any special facilities for the care of AIDS patients?

A Persons with AIDS are treated in hospitals, physicians' offices, clinics, and other healthcare facilities along with other patients. As long as normal precautions are observed, a person with AIDS poses no more risk to healthy individuals than any other ill person.

Q2 Is any one opportunistic disease more common than the others?

A Very definitely. Among patients with AIDS, more than 50% have died from *Pneumocystis* pneumonia. Although this fungal disease can be treated with drugs, the therapy is less successful in AIDS patients than in patients who do not have AIDS or HIV infection.

R E V I E W

Having studied the pathology of AIDS, you should be familiar with the signs, symptoms, and progress of the stages of the disease. To test your knowledge, enter at the left the word or words that best complete each thought. The correct answers are listed elsewhere in the text.

1. _____ is the name of the brain disorder occurring in roughly one third of persons with AIDS.

2. _____ is the hospital whose physicians devised a six-stage classification system of the stages leading to AIDS.

3. _____ is the cancer of the skin often observed in persons with AIDS.

4. _____ is the virus that forms giant cells in the laboratory and that invades many cells of many organs as an opportunist.

5. _____ is the name of the fungus that causes lung disease as an opportunistic disease accompanying AIDS.

6. _____ is the approximate incubation period for AIDS.

7. _____ is the name assigned to the condition where a patient harbors HIV but shows no symptoms of disease.

8. _____ is the term given to the swollen lymph nodes commonly associated with HIV infection, ARC, and AIDS.

9. _____ is the approximate number of cases of pediatric AIDS reported through 2011.

10. _____ is the normal count of T-lymphocytes per microliter of blood.

11. _____ is the yeast-like fungus that causes the infection of the mouth called thrush, an early sign of impending AIDS.

12. _____ is the system of the body attacked by the opportunistic microorganism *Cryptosporidium.*

13. _____ is the name given to the chronic diarrhea and dramatic weight loss that can fulfill the definition for AIDS.

14. _____ is the drug used to treat patients who experience *Pneumocystis* pneumonia.

15. _____ are the animals that harbor *Toxoplasma* and that should be avoided by a person infected by HIV.

FOR ADDITIONAL READING

Arts, E. J., Hazuda, D. J., 2012. "HIV-1 antiretroviral drug therapy." *Cold Spring Harbor Perspective Medicine* 2:a007161.

Dockrell, D. H., 2011. "Evolving controversies and challenges in the management of opportunistic infections in HIV-seropositive individuals." *Journal of Infection* 63:177–186.

Gallo, R. C., 1998. "The enigmas of Kaposi's sarcoma." *Science* 282:1837–1839.

Gartner, S., 2000. "HIV infection and dementia." *Science* 287:602–605.

Huang, L., et al., 2011. "HIV-associated *Pneumocystis* pneumonia." *Proceedings of the American Thoracic Society* 8:294–300.

Hughes, W., 1987. "*Pneumocystis carinii* pneumonitis." *New England Journal of Medicine* 317:1021–1023.

Jose, J. S., 2012. "Opportunistic and fungal infections of the lung." *Medicine* 40:335–339.

Miceli, M. H., et al., 2011. "Emerging opportunistic yeast infections." *Lancet Infectious Diseases* 11:142–151.

Schneider, E., et al., 2008. "Revised surveillance case definitions for HIV infection among adults, adolescents, and children aged < 18 months and for HIV infection and AIDS among children aged 18 months to < 13 years—United States, 2008." *MMWR* 57/RR-10:1–12.

Singh, D., 2012. "What's in a name? AIDS dementia complex, HIV-associated dementia, HIV-associated neurocognitive disorder or HIV encephalopathy." *African Journal of Psychiatry* 15:172–175.

Tan, I. L., et al., 2012. "HIV-associated opportunistic infections of the CNS." *Lancet Neurology* 11:605–617.

Van Sighem, A., et al., 2010. "Life expectancy of recently diagnosed asymptomatic HIV-infected patients approaches that of uninfected individuals." *AIDS* 24:1527–1535.

5 The Epidemiology of AIDS

LOOKING AHEAD

Scientists have identified several methods by which AIDS spreads among individuals. This chapter explores those methods and discusses the principles that underlie them. On completing the chapter, you should be able to . . .

- Describe the work of epidemiologists during the AIDS epidemic and appreciate their contributions to public health care.
- Summarize current statistics on reported cases and groups affected by the AIDS epidemic.
- Understand the methods by which the human immunodeficiency virus (HIV) spreads among sexually active homosexual men, transgender people, injection drug users, and heterosexual couples including seniors, and understand how HIV can pass to newborns.
- Compare the patterns by which the AIDS epidemic is spreading in the United States, Eastern Europe, Africa, and other parts of the world.
- Make some generalizations about the spread of AIDS in society, correctional facilities, schools and colleges, and healthcare settings.
- Discuss prevention challenges that different affected groups face.

INTRODUCTION

During July 1976, the Bellevue-Stratford Hotel in Philadelphia was the site of the 58th annual convention of the Pennsylvania contingent of the American Legion. Toward the end of the convention, 149 Legionnaires and 72 others in or near the hotel experienced fever, coughing, and pneumonia. Within days, 34 had died. Microbiologists began an intensive search for a causative agent, but no pathogenic microorganisms could be found within the tissues of victims.

As the weeks continued, the illness came to be known as **Legionnaires' disease**, and its enigma deepened. There were hints that a cause might never be identified. There were even some allegations of foul play, including a charge that the disease was caused by genetically engineered microorganisms stolen from U.S. Army laboratories.

Then, on January 6, 1977, less than six months after the outbreak, the mystery was resolved. Scientists from the **Centers for Disease Control and Prevention (CDC)** identified the responsible bacterium and named it *Legionella pneumophila*. Two things about *L. pneumophila* were significant: First, the organism was a species of bacterium never before encountered; second, the organism existed where water collected and was blown into the air by wind currents. Public health officials now had two important facts to help them interrupt the epidemic: They knew what was causing it, and they understood the source of the causative agent. They soon brought the outbreak of Legionnaires' disease under control by identifying the bacterium in contaminated water supplies and treating the water with disinfectants.

A somewhat similar situation has existed during the AIDS epidemic. Since the spring of 1984, scientists have known that HIV is the responsible agent, and as early as 1982, they were fairly certain how the agent is transmitted. Armed with this information, public health officials could predict which groups the epidemic would affect the most and how it was likely to spread in the future. Their predictions were largely correct. Then came the task of educating those at risk and showing individuals how to protect themselves from infection. As we see in this chapter, that goal has been difficult to accomplish because protection often involves changes in lifestyle and behavior. Interrupting the epidemic of Legionnaires' disease, by comparison, was relatively easy because it involved disinfecting the water where bacteria collected. In addition, Legionnaires' disease can be successfully treated with antibiotics, and the patient will return to good health. No such option is yet available for AIDS.

AIDS and the Epidemiologist

Epidemiology is the study of relationships that influence the frequency and distribution of diseases in a community. Discovering that an epidemic is in progress and defining the circumstances under which it spreads are two tasks of the epidemiologist. In the public health system, the epidemiologist also studies a disease's pattern of distribution and makes recommendations on controlling it in a population. Controlling an epidemic can be a substantial challenge because the epidemiologist must often work in a situation where no drugs or vaccines are available.

The best epidemiologists were put to the test in June 1981 when doctors at the CDC described five cases of *Pneumocystis* pneumonia in homosexual men living in the Los Angeles area (**Figure 5.1** shows the original report). At approximately the same time, the CDC received several requests for **pentamidine isethionate**, a drug used to treat *Pneumocystis* pneumonia (only two requests had been received in the previous 10 years). They also received reports of an unusual number of cases of Kaposi's sarcoma and suppressed immune systems in patients.

By the end of 1982, epidemiologists had given the name **acquired immune deficiency syndrome (AIDS)** to the disease and were collecting data on its symptoms. Because the cities involved were far apart (New York, San Francisco, Miami, Los Angeles), epidemiologists surmised that there was a common denominator within the separate populations. As data on the long incubation period became apparent, they guessed that a larger pool of individuals was infected and would manifest symptoms of the disease in the future.

To gather additional data, epidemiologists used **case control studies**. In such a study, people having the disease ("cases") are compared with people from a population whose members are not ill ("controls"). One of the first case-control studies was performed in 1981. This study led scientists to believe that among homosexual men, the number and frequency of a person's sexual contacts were factors affecting the incidence of AIDS. Results of another study conducted in 1982 indicated that the sexual partners of homosexual and bisexual men were at risk and that more than 20%

CENTERS FOR DISEASE CONTROL

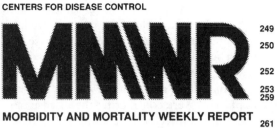

MORBIDITY AND MORTALITY WEEKLY REPORT

June 5, 1981 / Vol. 30 / No. 21

Pneumocystis Pneumonia — Los Angeles

In the period October 1980-May 1981, 5 young men, all active homosexuals, were treated for biopsy-confirmed *Pneumocystis carinii* pneumonia at 3 different hospitals in Los Angeles, California. Two of the patients died. All 5 patients had laboratory-confirmed previous or current cytomegalovirus (CMV) infection and candidal mucosal infection. Case reports of these patients follow.

Patient 1: A previously healthy 33-year-old man developed *P. carinii* pneumonia and oral mucosal candidiasis in March 1981 after a 2-month history of fever associated with elevated liver enzymes, leukopenia, and CMV viruria. The serum complement-fixation CMV titer in October 1980 was 256; in May 1981 it was 32.* The patient's condition deteriorated despite courses of treatment with trimethoprim-sulfamethoxazole (TMP/SMX), pentamidine, and acyclovir. He died May 3, and postmortem examination showed residual *P. carinii* and CMV pneumonia, but no evidence of neoplasia.

Patient 2: A previously healthy 30-year-old man developed *P. carinii* pneumonia in April 1981 after a 5-month history of fever each day and of elevated liver-function tests, CMV viruria, and documented seroconversion to CMV, i.e., an acute-phase titer of 16 and a convalescent-phase titer of 28* in anticomplement immunofluorescence tests. Other features of his illness included leukopenia and mucosal candidiasis. His pneumonia responded to a course of intravenous TMP/SMX, but, as of the latest reports, he continues to have a fever each day.

Patient 3: A 30-year-old man was well until January 1981 when he developed esophageal and oral candidiasis that responded to Amphotericin B treatment. He was hospitalized in February 1981 for *P. carinii* pneumonia that responded to oral TMP/SMX. His esophageal candidiasis recurred after the pneumonia was diagnosed, and he was again given Amphotericin B. The CMV complement-fixation titer in March 1981 was 8. Material from an esophageal biopsy was positive for CMV.

Patient 4: A 29-year-old man developed *P. carinii* pneumonia in February 1981. He had had Hodgkins disease 3 years earlier, but had been successfully treated with radiation therapy alone. He did not improve after being given intravenous TMP/SMX and corticosteroids and died in March. Postmortem examination showed no evidence of Hodgkins disease, but *P. carinii* and CMV were found in lung tissue.

Patient 5: A previously healthy 36-year-old man with a clinically diagnosed CMV infection in September 1980 was seen in April 1981 because of a 4-month history of fever, dyspnea, and cough. On admission he was found to have *P. carinii* pneumonia, oral candidiasis, and CMV retinitis. A complement-fixation CMV titer in April 1981 was 128. The patient has been treated with 2 short courses of TMP/SMX that have been limited because of a sulfa-induced neutropenia. He is being treated for candidiasis with topical nystatin.

---

*Paired specimens not run in parallel.

FIGURE 5.1

The issue of *Morbidity and Mortality Weekly Report* in which Centers for Disease Control epidemiologists summarized five cases of *Pneumocystis carinii* pneumonia. This issue of June 5, 1981, is considered to be one of the beginning points of the AIDS epidemic.
Source: CDC, *Morbidity and Mortality Weekly Report*, 30(21).

of all known cases were occurring in men who had had a sexual relationship with an infected person. It was becoming clear that a transmissible agent was involved.

By 1983, epidemiologists had accumulated enough data to develop a case definition for AIDS. Then they developed a set of recommendations for avoiding the disease (even though they did not know what was causing it). As physicians continued to report cases, the data made it apparent that blood and semen were the two principal sources of the disease agent. Hemophiliacs were contracting the disease from blood products; transfusion recipients were being infected by contaminated blood. The number of cases was rising among sexually active homosexual men, and heterosexuals were contracting the disease.

The identification of the **human immunodeficiency virus (HIV)** in the spring of 1984 and the development of the AIDS antibody test in 1985 gave public health officials the ability to track the epidemic with more confidence. It was assumed that if a person's blood had **HIV antibodies**, exposure to HIV had taken place. Now epidemiologists could confirm a diagnosis, measure the extent of the epidemic, and screen the blood supply. They could also make more specific recommendations on halting the spread of AIDS. They targeted certain groups for prevention and control campaigns and devised a more complete case definition. Some of the major goals of epidemiology were thereby fulfilled.

The Fragility of HIV

AIDS is a bloodborne disease caused by a fragile virus, HIV (**Figure 5.2**). Outside the body, HIV quickly disintegrates because its molecular structure, especially the outside envelope, cannot resist environmental pressures such as drying, heat, alcohols,

Some people think you can catch AIDS from a glass.

You can't.

The California Medical Association and public health officials agree: AIDS is not spread through the air. AIDS is not spread by touching someone. AIDS is not spread by hot tubs. AIDS is not spread through the preparation or serving of food or beverages in restaurants or homes.

The virus that causes AIDS is spread by unprotected sex with an infected person, or by contaminated blood entering the blood stream—such as by sharing drug needles.

Fight the fear with the facts:
800-367-AIDS/800-922-AIDS
(Toll-free in No. Calif.)   (Toll-free in So. Calif.)

detergents, and various other chemical substances. In a loose sense, the virus "dies." This is why soap and water can disrupt the virus. Certain other viruses, in contrast, are more resistant to environmental agents. The virus of hepatitis A, for example, can remain "infectious" outside the body and, therefore, can be transmitted by contaminated food such as raw shellfish or by water contaminated with intestinal matter.

Because of its fragile nature, HIV must pass directly from person to person in such fluids as blood and semen. In addition, a large number of the viral particles must be passed during transmission because many of them will be unable to survive the transfer to the recipient host and remain active in the new environment. Although HIV has been found in saliva, tears, and sweat, the number of viral particles in these fluids is extraordinarily low—too low, public health officials maintain, to affect HIV transmission. Infected semen and blood remain the major transferring substances, and for this reason, individuals coming in contact with infected semen and/or blood are at risk. A report in the July 27th, 2012 *MMWR* from the CDC indicates that progress in reducing HIV-related risk behaviors among high school students did not change significantly in from 1991–2011. In 2011, 47.4% of high school students surveyed were sexually active. Of these, 15.3% had sexual intercourse with four or more partners in their life. About 40% did not use condoms and 2.3% injected illegal drugs. More educational efforts and other risk reduction interventions are warranted.

Other factors that contribute to risk of infection are the volume of fluid introduced into the recipient (higher volume, higher risk), the general state of health of the recipient (better health, lower risk), and the inoculation site of the fluid (blood or semen must enter the recipient's blood system). As we see here, maternal–fetal transfer is also possible because the virus can cross the placenta. Still other factors shall become apparent as we proceed.

The Spreading Epidemic in the United States

In the early 1980s, public health officials hoped that the number of AIDS cases would be small and that researchers would quickly develop an effective drug or vaccine. Their hopes, unfortunately, would not be fulfilled. At the end of 2006, nearly 1.3 million Americans had been diagnosed with AIDS or found to be HIV positive, and over 550,000 had died. There was some good news with indication that AIDS deaths were on the decline in 1996 (primarily as a result of new anti-HIV therapies); however, unfortunately the rates of new HIV infections and deaths caused by AIDS have not significantly changed in the United States since 1998. Over the past decade, roughly 40,000 new HIV infections and 15,000 AIDS-associated deaths have been seen each year. Approximately 48,100 new HIV infections occurred in the United States in 2009. Furthermore, CDC officials estimate 1.3 million Americans continue to live with HIV infection and that at least 20% do not know that they are infected. Moreover, 1 in 3 newly diagnosed HIV infections in the United States are transmitted by heterosexual activity. It is clear that AIDS will not fade from memory until well into the twenty-first century.

According to the CDC, since the inception of the AIDS epidemic in 1981, a cumulative total of nearly 1.3 Americans were diagnosed with either HIV infection or active AIDS. Of this total number, roughly 80% occurred in men and nearly 20% in women (**Table 5.1**). When the cumulative cases of HIV and AIDS (HIV/AIDS) is examined by race, as of 2010, 38% of the cases occurred in whites, 42% in blacks, 17% in Hispanics, and less than 1.5% in Asians, Pacific Islanders, American Indians, Alaskan Natives, and individuals of multiple races. The 1990s and 2000s has seen a gradual shift of the epidemic away from sexually active homosexual men and toward blacks, Hispanics,

Focus on HIV

The Constella Group and NIOSH conducted an national survey of paramedics between 2002 and 2003 to measure their incidence of blood exposure on the job. Twenty-two percent of all paramedics indicated they had at least one blood exposure in their previous year on the job.

Source: CDC. *Workplace Solutions, Preventing Exposures to Bloodborne Pathogens among Paramedics.* DHHS (NIOSH) Publication No. 2010-139. April 2010.

TABLE 5.1 Reported HIV/AIDS Cases in the United States by Race and Gender, 2010 and Cumulative, 1981–2010

Gender	2010	Cumulative (since the beginning of the epidemic through 2010)
Male	24,749 (75%)	875,116 (80%)
Female	8,242 (25%)	220,955 (20%)

Race/Ethnicity	2010	Cumulative (since the beginning of the epidemic through 2010)
American Indian/ Alaska Native	170 (0.5%)	3,639 (0.3%)
Asian	480 (1.5%)	8,422 (0.8%)
Black/African American	16,188 (49%)	462,647 (42%)
Hispanic/Latino	6,636 (20%)	192,586 (17%)
Native Hawaiian/Other Pacific Islander	44 (0.1%)	843 (0.08%)
White	8,875 (27%)	423,517 (38%)
Multiple races	622 (1.9%)	14,676 (1.3%)
Totals	33,015	1,106,499

Figures represent reported cases among adults and adolescents.

Data from: CDC, *HIV/AIDS Surveillance Report*, Vol. 22, 2010.

and women. Of note is the fact that the proportion of women among AIDS patients in the United States—according to annual statistics—increased steadily until 2000, when it reached 30%. This number has dropped slightly to 25% in 2010 (Table 5.1).

The CDC has tracked the AIDS epidemic in various groups based on mode of transmission of the virus. While in 1990, 61% of all reported cases of AIDS occurred in homosexual or bisexual men who had no history of injection drug abuse, this number was reduced as a proportion (not absolute values) to 48% by June 2003 and held somewhat steady at 50% in 2006 but climbed to 77% in 2010 (**Figure 5.3**). So which groups have increased their proportional representation of diagnosed HIV cases? In 1990, injection drug users accounted for 20% of HIV diagnosed infections, and this increased to 27% by 2003; however, injection drug transmission dropped to 13% (2006) and 7% in 2010 of diagnosed HIV infections (Figure 5.3). Thus, although the injection drug users are of significant concern, this group has not been on a dramatic increase.

HIV and AIDS by Geographic Distribution

The HIV and AIDS epidemic is not evenly distributed across states and regions (**Figure 5.4**). At the end of 2010, the South accounted for 45% of the estimated new AIDS diagnoses in the 50 states and Washington D.C., followed by the Northeast (24%), the West (19%), and the Midwest (13%). The rate of HIV infection is disproportionally distributed among the various ethnic groups in the United States.

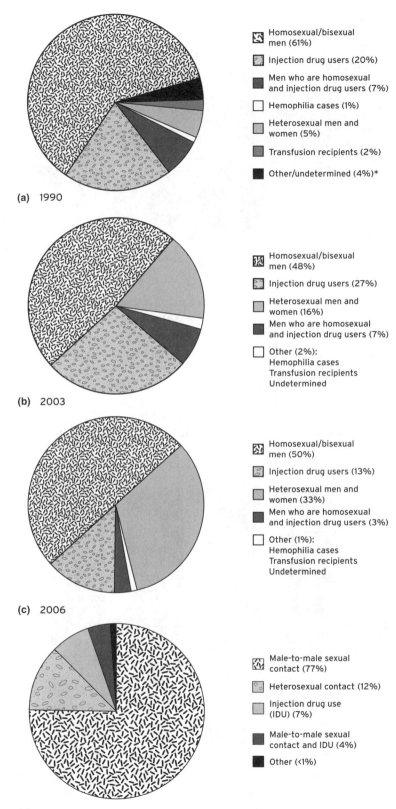

(a) 1990

☒ Homosexual/bisexual men (61%)

▨ Injection drug users (20%)

◼ Men who are homosexual and injection drug users (7%)

☐ Hemophilia cases (1%)

◻ Heterosexual men and women (5%)

◼ Transfusion recipients (2%)

◼ Other/undetermined (4%)*

(b) 2003

☒ Homosexual/bisexual men (48%)

▨ Injection drug users (27%)

◻ Heterosexual men and women (16%)

◼ Men who are homosexual and injection drug users (7%)

☐ Other (2%):
Hemophilia cases
Transfusion recipients
Undetermined

(c) 2006

☒ Homosexual/bisexual men (50%)

▨ Injection drug users (13%)

◻ Heterosexual men and women (33%)

◼ Men who are homosexual and injection drug users (3%)

☐ Other (1%):
Hemophilia cases
Transfusion recipients
Undetermined

(d) 2010

☒ Male-to-male sexual contact (77%)

▨ Heterosexual contact (12%)

◻ Injection drug use (IDU) (7%)

◼ Male-to-male sexual contact and IDU (4%)

◼ Other (<1%)

FIGURE 5.3

An analysis of cases of AIDS (not asymptomatic HIV infection) among different groups as reported to the CDC (a) through September 1990, (b) through June 2003, (c) through 2006, and (d) through 2010. The greatest percentage of cases has occurred among homosexual men, and the second greatest percentage among intravenous drug users. In recent years, the percentage among homosexual/bisexual men and high risk heterosexual activity has been increasing, Percentages do not add to 100% because of rounding. Reproduced from CDC, *HIV/ AIDS Surveillance Report*, vol. 22, 2010.

FIGURE 5.4

Rates of individuals (18–64 years of age) living with a diagnosis of HIV infection, year-end 2008 by geographic location. U.S. States in the South and Northeast report the highest rates. Reproduced from CDC HIV and AIDS in the United States by Geographic Distribution Fact Sheet, June, 2012.

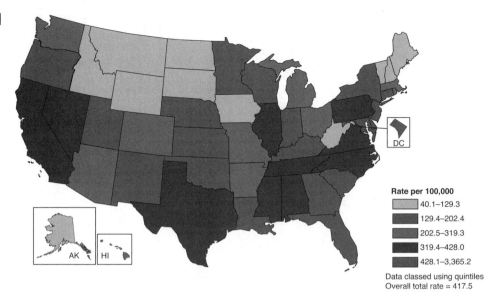

Rate per 100,000

	40.1–129.3
	129.4–202.4
	202.5–319.3
	319.4–428.0
	428.1–3,365.2

Data classed using quintiles
Overall total rate = 417.5

The greatest number of diagnosed AIDS cases is the black population (**Figure 5.5**). Although this figure shows that the increase in number of people diagnosed with AIDS increased in parallel for all ethnic groups, a more careful analysis of proportion reveals that the black community still suffers as a proportion of the population to a much greater extent than other ethnic groups in the United States. For example, in 2010, blacks accounted for the largest proportion of AIDS diagnoses in all regions except the West, where whites accounted for the highest proportion of diagnoses (Figure 5.5). It should be noted that some smaller ethic groups can be

FIGURE 5.5

AIDS in the United States by race/ethnicity in 2010 in the 50 states and Washington, D.C. Reproduced from CDC HIV and AIDS in the United States by Geographic Distribution Fact Sheet, June, 2012.

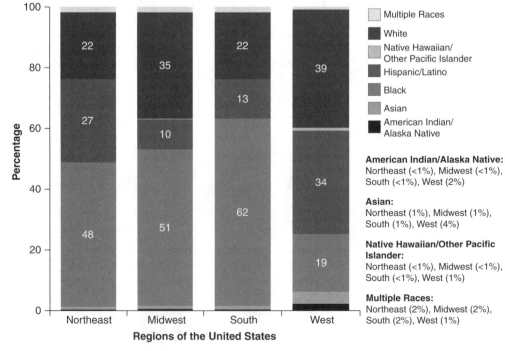

Multiple Races
White
Native Hawaiian/Other Pacific Islander
Hispanic/Latino
Black
Asian
American Indian/Alaska Native

American Indian/Alaska Native:
Northeast (<1%), Midwest (<1%), South (<1%), West (2%)

Asian:
Northeast (1%), Midwest (1%), South (1%), West (4%)

Native Hawaiian/Other Pacific Islander:
Northeast (<1%), Midwest (<1%), South (<1%), West (1%)

Multiple Races:
Northeast (2%), Midwest (2%), South (2%), West (1%)

The Epidemiology of AIDS

TABLE 5.2	HIV Diagnoses of Male Adults or Adolescents by Race/Ethnicity High-Risk Heterosexual Transmission*, 2010—46 States	
Race/Ethnicity	Cases	Percentage
American Indian/Alaska Native	14	0.6%
Asian	29	1.2%
Black/African American	1,466	62%
Hispanic/Latino	479	20%
Native Hawaiian/Other Pacific Islander	1	0.04%
White	356	15%
Multiple races	30	1.3%
Subtotal (all race/ethnicities)	2,375	

*Data include persons with a diagnosis of HIV infection regardless of their AIDS status at diagnosis. Data are from 46 states with confidential name-based HIV infection reporting.

Data from: CDC, *HIV/AIDS Surveillance Report*, Vol. 22, 2010.

strongly affected but are not mentioned because of the relatively small number of cases among the groups. For example, by the end of 2010, male American Indians/Alaska Natives had a higher rate of persons diagnosed (18.1/100,000 people) with HIV infection than whites (15.3/100,000).

This disproportionate representation of blacks/African Americans in the HIV positive population is even further highlighted when heterosexual transmission is considered. As noted previously here, heterosexual spread is the fastest growing mode of transmission of HIV in the United States (Figure 5.3). The proportion of blacks/African Americans that contributes to this group presents a significant problem for this community and must be addressed if one is to understand and change the picture of the HIV/AIDS epidemic in America. Indeed, 66% (2006) and 62% (2010) of the diagnosed HIV infections that resulted from heterosexual contact were black/African American (**Table 5.2**). The CDC has targeted large-scale HIV testing initiatives and other resources into the 12 metropolitan areas with the highest AIDS prevalence in the United States. The CDC is allocating additional funding for state and local health departments, which is the most significant HIV investment to help those geographic locations most affected by the epidemic.

HIV and AIDS in Women

According the CDC's *HIV/AIDS 2012 Surveillance Report*, women represent 25% of all diagnoses of HIV infection in the United States. As noted, the spread of HIV through high-risk heterosexual contact is on the rise. High-risk sexual contact is defined as having sex with a person known to have, or to be at high risk for acquiring HIV infection. If the heterosexual population is examined by gender, race/ethnicity again disproportionate spread of HIV is observed (**Figure 5.6**). Of the total number of new HIV infections among women, 57% occurred in blacks, 21% were in whites, and 16% were in Hispanics/Latinos. In 2009, the rate of HIV infections in black women was 15 times that of white women and over 3 times that of Hispanic/Latino women.

FIGURE 5.6

Estimates of new HIV infections by race/ethnicity, risk group, and gender in the United States, 2009. Note that infections of women are on the rise. MSM= men who have sex with men; IDU = intravenous drug users. Data from: Prejean J, et al. Estimated HIV incidence in the United States, 2006-2009. *PLoS One* 2001;6(8):1-13. Reproduced from CDC.

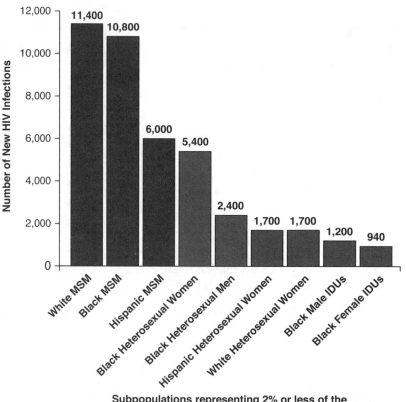

Subpopulations representing 2% or less of the overall US epidemic are not reflected in this chart.

Focus on HIV

Black and Hispanic/Latino women are at increased risk of being diagnosed with HIV infection (1 in 32 black women and 1 in 106 Hispanic/Latino women will be diagnosed with HIV, compared with 1 in 182 Native Hawaiian/other Pacific Islander women; 1 in 217 American Indian/Alaska Native women; and 1 in 526 for both white and Asian women).

In 1985, only 7% of all AIDS cases to that time had occurred in women. By 1994, 18% of all U.S. cases of AIDS since the beginning of the epidemic were reported in women, but by December 2003, this number rose 25.5%. By 2006, 26% of all HIV/ AIDS cases were in women; however, if this number is analyzed by mode of transmission, we see that for women in 2010, 86% of transmission is through heterosexual sex (**Figure 5.7**). This rate of heterosexual transmission highlights two important facts. First, women are more susceptible than men to the spread of HIV through heterosexual activity, as we discuss later, and second, most sex events in the human population are heterosexual. This forecasts an eventual HIV distribution in the United States that is predominantly heterosexual and female. Indeed, as we shall see later in this chapter, this is already the case in Africa where HIV/AIDS affects more women than men in part because it has had a longer time to spread among the population.

There are many prevention challenges women face. Some women become infected because they may be unaware of a male partner's risk factors for HIV infection. Relationship dynamics may also play a role. For example, some women may not insist on condom use because they fear that their partner will physically abuse or leave them. Women who have experienced sexual abuse may be more likely than women with no abuse history to use drugs as a coping mechanism. They may exchange sex for drugs or engage in high-risk sexual activities.

To deal with their special needs, women's groups have started preventive counseling programs and care centers for the children of infected women to supplement publicly sponsored programs. Educational campaigns have also been directed at women, including suggestions that women carry condoms (**Figure 5.8**), and researchers have

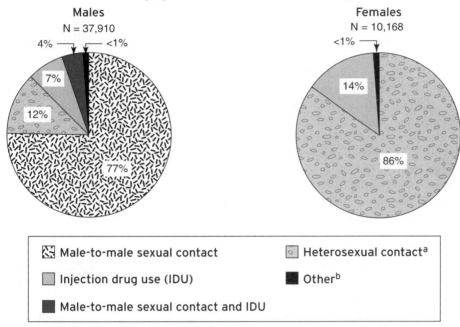

Diagnoses of HIV infection among adults and adolescents, by sex and transmission category, 2010–46 states and 5 U.S. dependent areas

Males
N = 37,910

4% → <1%
7%
12%
77%

Females
N = 10,168

<1%
14%
86%

Legend:
- Male-to-male sexual contact
- Injection drug use (IDU)
- Male-to-male sexual contact and IDU
- Heterosexual contact[a]
- Other[b]

Note. All displayed data have been statistically adjusted to account for reporting delays and missing risk-factor information, but not for incomplete reporting.

[a]Heterosexual contact with a person known to have, or to be at high risk for, HIV infection.
[b]Includes hemophilia, blood transfusion, perinatal exposure, and risk factor not reported or not identified.

FIGURE 5.7

Proportion of HIV/AIDS cases among adults and adolescents, by sex and transmission category, 2010 (46 states).
Note: Data include persons with a diagnosis of HIV infection regardless of their AIDS status at diagnosis.
Source: Data are from 33 states with confidential name-based HIV infection reporting. Reproduced from CDC, HIV Surveillance Report: Diagnoses of the HIV Infection and AIDS in the United States and Dependent Areas, 2010. Vol. 22.

Talk about AIDS before it hits home.

"I really don't have to tell Linda about AIDS. They're teaching about it in school."
"Why discuss AIDS with my Johnny? He isn't gay."
"If I talk to them about AIDS, they'll think it's okay to have sex."

If you're thinking any of these thoughts, you're not doing all you should to protect your teenager from AIDS.

So put your embarrassment and your fear of encouraging sex aside.

Just sit down and tell them the facts.

Tell them that you just can't be sure who's infected with the AIDS virus. Sometimes it can be carried for years without any symptoms.

Tell them that since they can't possibly know who's infected they must use precautions to protect themselves.

Tell them if they're having sex, they must always use a condom. And not having sex is still the best protection.

Tell them that AIDS is incurable, there's no vaccine, and once you get it you'll likely die.

Then tell them it's preventable.

Tell them everything you can about AIDS. But make sure you tell them now.

AIDS Because by the time you think they're old enough to know, it might be too late.

If you think you can't get it, you're dead wrong.

NEW YORK CITY DEPARTMENT OF HEALTH. FOR MORE INFORMATION CALL: 1 (718) 485-8111

FIGURE 5.8

Sample of the advertising campaign promoting the use of condoms and aimed at women. Many TV networks and large magazines were reluctant to run such advertisements, which they viewed as too controversial at the time. Courtesy of the New York Department of Health and Mental Hygiene.

developed vaginal condoms. The basic theme of these campaigns is that AIDS is not solely a disease of homosexual men. The CDC has developed **Take Charge, Take Test (TCTT)**, a phase of the *Act Against AIDS* campaign designed to increase HIV testing among black/African American women ages 18 to 34.

In women, vaginal infections represent a unique problem that contributes to the risk of acquiring HIV infection. In addition to the well-known **sexually transmitted diseases (STDs)**, a microbial infection called bacterial vaginosis increases a woman's susceptibility to HIV infection because the disease interferes with the normal metabolism of cells lining the vaginal cavity. The bacteria produce offensive odors and cause internal pruritus (itching) and excessive vaginal discharge. Although there is little of the ulceration and inflammation associated with STDs such as gonorrhea or syphilis, bacterial vaginosis has a higher incidence rate and thus may be equally important to the spread of HIV. For example, researchers studied 1,196 pregnant women in Malawi and found that those with bacterial vaginosis were 3.7 times more likely to acquire HIV than those without vaginosis. Furthermore, the study showed that a man's risk of acquiring HIV rises when he has intercourse with an HIV-infected woman who has vaginosis. In the United States, rates of gonorrhea and syphilis are higher among women of color than among white women.

Researchers have also found that opportunistic diseases are equally prevalent in HIV-infected women and men, with the exception of cervical cancer and Kaposi's sarcoma. Erosion of the esophagus with *Candida albicans* and infections caused by herpes simplex virus and cytomegalovirus appear to occur in higher incidence in HIV-positive women who inject drugs (although the reason is not clear). Gynecologic complications associated with STDs, including abscesses of the ovaries and oviducts, also require special attention. Abnormalities of cells of the cervix can occur at any stage of HIV infection, but it is more frequent as the number of T-lymphocytes declines. Human papilloma viruses may cause the abnormalities and induce a progression to cervical cancer; however, Pap smears detect early signs of infection, and antiviral therapy for papilloma viruses is helpful. Antiretroviral therapy against HIV has also been shown effective for reducing papilloma infection.

In women who are pregnant, determining HIV status (**Box 5.1**) is particularly important because intervention with AZT can interrupt viral passage to the fetus (as we discuss in the next section). Moreover, during delivery, an obstetrician will postpone rupture of the amniotic membranes or other potentially invasive procedures that might encourage HIV transmission.

HIV and AIDS in Injection Drug Users

HIV is proportionally more prevalent among injection drug users because these individuals often share contaminated syringes and needles and consequently directly inject HIV and HIV-infected T-cell into their bloodstream. How is this done? Sharing needles and syringes (one's "works") is a common practice among groups of heroin addicts and other drug users. A syringe and needle become bloodstained when they are used to inject the drug into the veins (i.e., intravenously). Then the syringe and needle are passed to the next person, who uses them without the benefit of disinfection and is exposed to the first person's blood. Thus, the virus passes easily between the two individuals. Rinsing the syringe and needle in a bleach solution can eliminate the virus and interrupt transmission. Another practice often followed by injection drug users in a group is basically the same as a blood transfusion: One addict fills the syringe with the drug, inserts the needle, pulls back on the plunger,

BOX
5.1

Head, Tails, and the Truth

Studies linking disease to sexual behavior are difficult to perform because people are reluctant to reveal details of their private lives. For example, asking a man whether he has had a sexual encounter with anyone other than his wife often elicits a look of skepticism and an unreliable answer. Try as they might, researchers can rarely be certain they are getting the truth when they ask such questions.

In 1987, Joel E. Cohen, a mathematical biologist from Rockefeller University, suggested a statistical tool that could add reliability to the answer. The technique was invented in 1965 by Stanley Warner of Ontario's York University. It has been used periodically since then. Its basic premise is that people will reveal the truth if they believe their answers are secret.

The method involves a question and a coin flip. For example, a questioner asks a man, "Have you ever had a sexual encounter with anyone other than your wife?" The man then flips the coin in privacy. He answers no if he has not had an encounter and if the coin comes up tails. Otherwise, he must answer yes.

Using this method, a man can answer no because the flip resulted in tails, and he can answer yes, also because of the coin. A yes does not necessarily incriminate him; a no does not necessarily mean that he is lying.

Totaling the results, the researcher simply doubles the number of no answers to get an almost correct number of nos. The theory is simple: Given enough flips, the coin should come up tails half the time; this means that only half of the total number of men who actually had encounters said no; therefore, doubling the number of no answers gives a fairly accurate picture of the true number of nos. Subtracting the number of no answers from the total number of men questioned reveals the number of yes answers.

and draws blood into the syringe. Then he or she pushes some but not the entire drug–blood mixture back into the vein. The needle is then passed to the second individual who also draws blood back into the syringe to mix with the first person's blood. Now some of the drug–blood mixture is injected into the vein, and the syringe is passed to the third individual, who continues the pattern. The possibilities for HIV transmission during this practice are multifold.

In 2010, 9% of HIV infections were attributable to the use of injection drugs, and 45% of those testing positive were unaware of their infection. Although this does represent a dramatic drop from 2003, a year in which 27% of the HIV infections resulted from injection drugs, 9% remains a very high number and a significant problem due to the illegal aspects of this mode of transmission. Sex, the most common route of transmission, can be discussed and condoms can be legally obtained to prevent transmission. Although some states have legalized a needle exchange program, in most places, syringes and needles are hard to come by and even when it is available, taking advantage of an exchange program brings attention to an illegal behavior.

The disproportionate spread among blacks/African Americans that was noted above also carries over into the injection drug use population. **Table 5.3** shows that in 2010 there were 1,673 cases where HIV infections that were determined to have been transmitted by injection drug use alone. Of these cases in 2010, 40% were among the black/African-American population.

Because injection drug users and their sex partners are concentrated in the nation's inner cities, the problems created by the AIDS epidemic have been linked directly to

TABLE 5.3 Diagnosis of HIV Infection Attributed to Injection Drug Use Only*, by Race/Ethnicity, 2010–46 States

Race/Ethnicity	Cases	Percentage
American Indian/Alaska Native	11	0.7%
Asian	17	1.0%
Black/African American	665	40%
Hispanic/Latino	368	22%
Native Hawaiian/Other Pacific Islander	3	0.2%
White	586	35%
Multiple races	23	1.4%
Total	**1,673**	

*Data were compiled from males and females that participate in intravenous drug use alone excluding male-to-male sexual conduct and injection drug use. Data include persons with a diagnosis of HIV infection regardless of AIDS status at diagnosis. Data were from 46 states with confidential name-based HIV infection reporting.

Data from: CDC, *HIV/AIDS Surveillance Report*, Vol. 22, 2010.

problems of urban decay, drug traffic, and homelessness (**Figure 5.9**). Thus, the AIDS epidemic is gradually undergoing a face change and is taking on sociological, political, and economic overtones that differ from those of the past. In 2009, the CDC's National HIV Behavioral Surveillance System conducted interviews and HIV testing in selected metropolitan areas. This was the largest risk behavior assessment since 1993. **Table 5.4** lists the results of their survey of 10,073 participants used to compare selected behavior characteristics among injecting drug-users and HIV prevalence. The CDC's prevention strategy for intravenous drug users includes more HIV testing linked to treatment and access to sterile syringes.

HIV and AIDS in Gay and Bisexual Men and Transgender People

When the AIDS epidemic was revealed in 1981, the first cases were reported among sexually active men living a homosexual lifestyle. Epidemiologists were uncertain whether the frequent use of certain drugs was to blame or whether a series of sexually transmitted diseases might be depressing the immune system's function in these individuals. When it became apparent that a transmissible agent was involved, epidemiologists focused on the type and frequency of sexual activities among homosexual men. Those men having few sex partners were less likely to have AIDS; men having unusually high rates of sexual activity had a correspondingly high chance of acquiring AIDS.

For homosexual men, the primary mechanism of HIV transfer is anal intercourse. In this practice, the penis of one male (the **insertive partner**) is placed into the rectum of the second male (the **receptive partner**) and semen is ejaculated. During the process, the delicate lining of the rectum may tear and bleed because the rectal tissue is composed of columnar epithelial cells that are easily damaged. For this reason, HIV is more easily transmitted by anal intercourse than it is through the stratified vaginal epithelial cells, which are designed to sustain the friction of sex. Furthermore, the rectum is highly

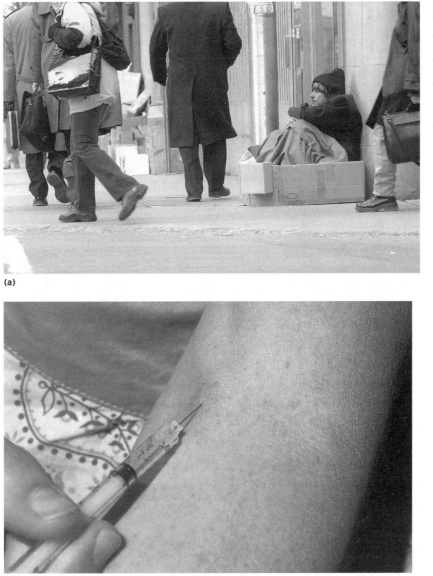

(a)

(b)

FIGURE 5.9

The percentage of total cases of AIDS reported since 1981 continues to rise among intravenous drug users and is expected to remain on that path well into the twenty-first century. (a) A homeless intravenous drug user. © AbleStock. (b) Heroin addict shooting up. © Laurin Rinder/ ShutterStock, Inc.

susceptible to abrasions and bleeding because, like the rest of the digestive tract, it is rich in blood capillaries for absorbing nutrients. Thus, if the insertive partner is infected with HIV and if the virus is present in the semen, the viruses can easily penetrate the rectal lining and enter the blood of the receptive partner. T-lymphocytes infected with HIV may also be present in the semen and may carry HIV in its provirus form into the receptive partner.

Anal intercourse also poses a risk for the insertive partner. It has been found, for example, that HIV-infected T-lymphocytes accumulate within the rectum (T-lymphocytes are normally found in this organ). It is possible that during anal intercourse, infected T-lymphocytes from the receptive partner can enter the urethra (the tube within the penis) of the insertive partner. If the cells make their way through

TABLE 5.4	HIV Infection among Injecting Drug Users by Selected Characteristics, 2009	
Characteristic	Percentage	% Positive HIV Status
Gender: Male	71%	9%
Gender: Female	29%	10%
18-29 years of age	11%	3%
30-39 years of age	19%	10%
40-49 years of age	32%	11%
≥ 50 years of age	38%	10%
Heroin injected most frequently	64%	7%
Other/multiple drugs injected most frequently	36%	14%
Education: Less than a high school diploma	36%	13%
Education: High school diploma	39%	8%
Education: More than a high school diploma	25%	7%
Financial status: At or below poverty level	81%	10%
Financial status: Above federal poverty level	19%	7%
Race/Ethnicity: Hispanic/Latino	22%	12%
Race/Ethnicity: Black	42%	11%
Race/Ethnicity: White	31%	6%
Race/Ethnicity: All other	4%	-
U.S. region: Northeast	34%	12%
U.S. region: South	27%	11%
U.S. region: Midwest	8%	5%
U.S. region: West	28%	6%

Modified from CDC, *Morbidity and Mortality Weekly Report*, vol. 61/no. 8, March 2, 2010.

an eroded urethral lining into the circulation, the insertive partner is infected. Again, this localization of T-cells in the anus makes anal sex a greater risk for HIV transmission compared with vaginal sex, as T-cells are generally not found in the healthy vaginal tract. Moreover, the insertive partner may have external lesions of the penis because of such diseases as syphilis and genital herpes. When the penis enters the rectum, free viruses or infected T-lymphocytes can pass from the rectal fluid through the open lesion into the insertive partner's blood.

Public health epidemiologists point out that those who engage in anal intercourse are at extraordinarily high risk for AIDS. Practicing a homosexual lifestyle is immaterial, they note, but engaging in anal intercourse is of great consequence ("It's not who you are. It's what you do"). For that reason, this book specifies "sexually active" homosexual men (**Figure 5.10**). To minimize the risk of transfer by semen, epidemiologists recommend using condoms. To further reduce the risk, they suggest reducing the number of sexual partners, thereby reducing the possibility of coming in contact with an HIV carrier. These precautions also hold true for heterosexual sex.

The early high prevalence of HIV in the active homosexual population suggests that HIV entered the United States homosexual population very early in the epidemic here. The early infection rates were nearly 100% in homosexuals. Gay and bisexual men are still the most severely affected by HIV than any other group in the United States. From 2006 to 2009, HIV infections among young black/African American gay and bisexual men increased by 48%. In 2009, men who have sex with men accounted for 61% of all new HIV infections. (**Figure 5.11**). In 2010, men who had sex with men accounted for 78% of estimated HIV diagnoses among males aged 13 and older. At the end of 2009, an estimated 784,701 individuals were living with an HIV diagnosis. Of these, 396,810 (51%) were men who have sex with men. About 48% of men who have sex with men were white, 30% were black/African American and 19% were Hispanic/Latino.

The CDC recommends that all men who have sex with men get tested for HIV at least once a year. The CDC supports behavioral health interventions such as **Many Men, Many Voices (3MV)**, *d-up: Defend Yourself!*, and *Mpowerment* for men who have sex with men. *Act Against AIDS* is a five-year national campaign launched by the CDC and the White House to create an awareness among all Americans in an effort to reduce the risk of HIV infection among the hardest-hit populations, gay and bisexual men, blacks/African Americans, and Hispanics/Latinos.

Transgender communities in the United States are among the groups at highest risk for HIV infection. **Gender identity** refers to how a person identifies their gender as male, female, or some other gender such as transgender. **Transgender**

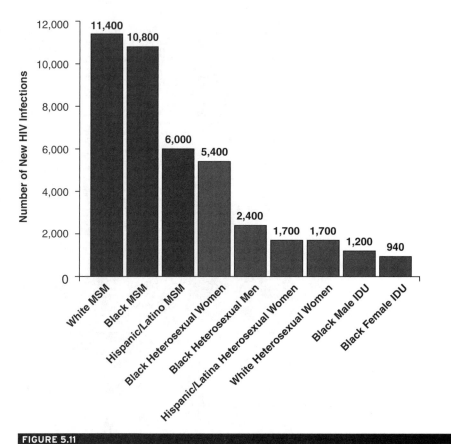

FIGURE 5.11

Graphical analysis of the estimated new HIV infections in the United States, 2009, for the most-affected groups. MSM = men who have sex with men; IDU = intravenous drug users. Reproduced from CDC.

refers to individuals who do not conform to the traditional definition of gender based on external genitalia or their sex assigned at birth. Data are lacking on how many transgender people in the United States are infected with HIV. However, there are some data collected by CDC-funded HIV testing programs, local health departments, and scientists that show high HIV infection rates among transgender people.

In a 2009 survey, about 4,100 of 2.6 million HIV testing events identified at least one participant who identified themselves as transgender. Based on these surveys, there were 2.6% newly diagnosed HIV infections among transgender people compared to 0.9% for males and 0.3% for females. More than half (52%) of the testing events with transgender persons occurred in nonclinical settings. A study in New York City from 2005 to 2009 had 206 new HIV diagnoses among transgender people, 95% of which were among transgender women. Among the newly diagnosed people with HIV infection, 50% of the transgender women had medical records that documented substance abuse, prostitution work, homelessness, incarceration, and/or sexual abuse as compared with 31% of other people who were not transgender.

Identifying transgender people can be challenging. High levels of HIV risk behaviors have been reported among transgender people. HIV infection among transgender women is associated with having multiple sex partners and unprotected receptive or insertive anal intercourse. No studies have reported links between behaviors and HIV infections among transgender men. The CDC is developing HIV-related behavioral surveys to monitor current HIV-related risk behaviors and prevention experiences among transgender women.

HIV and AIDS in Heterosexuals

For individuals in a heterosexual relationship, the risk of contracting AIDS can be substantial, depending on circumstances. Since the epidemic began, more than 80,000 people with AIDS were infected through heterosexual sex have died, including 4,424 in 2009. Heterosexuals generally practice vaginal intercourse. Normally, the lining of the vaginal tract will not erode or tear during intercourse. If, however, there are wounds, lesions, or abrasions along the lining, then viruses can enter a woman's bloodstream through the opening. An infection such as gonorrhea, syphilis, or genital herpes can be the source of such lesions or abrasions. Moreover, should vaginal intercourse take place at the beginning of or during a woman's menstrual period, the viruses could pass through the lining of the uterus, which are more fragile because the blood vessels are breaking down and rebuilding at this time. Researchers have also suggested that macrophages infected with HIV can pass through the tissue of the cervix at the opening to the uterus. Additional information on AIDS in women is presented in the next section.

In a heterosexual relationship, the male may also be at risk. It may happen, for instance, that the male has lesions on the outside surface of the penis, perhaps because of a herpes simplex infection. If the female is infected, her vaginal fluid may contain free viruses and/or HIV-infected T-lymphocytes, and these could pass through the lesions on the penis into the circulation. It is also possible that viruses or infected T-lymphocytes could enter the urethra of the penis at the conclusion of ejaculation and penetrate an internal lining eroded by a sexually transmitted disease such as gonorrhea.

From the foregoing, it follows that abstinence from sexual intercourse with individuals of unknown health status is the most sensible and efficient way of minimizing the risk of transmitting AIDS. For those who are sexually active, public health officials advise using condoms to reduce the possible passage of semen and viruses. They also recommend limiting the number of sexual partners. As we have noted, more than a million Americans are currently carrying the virus. Having multiple sex partners dramatically increases the odds of coming in contact with HIV.

The truth of this last statement was demonstrated among young people living in a small town in rural Mississippi. During 2000, public health investigators studied a group of individuals who had shared sex partners and identified seven who had become HIV-infected. Forty-four other adolescents and young adults were pinpointed in the network, and their HIV status was either negative or unknown at the time of the study (Figure 5.12). Antiretroviral therapy was recommended for all those found to be infected. The study underscores the importance of HIV counseling and testing services in rural areas.

In 1999, the CDC reported that 15% of all cases reported that year could be traced to vaginal intercourse among heterosexuals who had no other risk factors (such as injection drug use). Two years previously, the percentage was 14%. This statistic has changed appreciably since the 1980s, when the percentage was closer to 5%. Indeed,

FIGURE 5.12

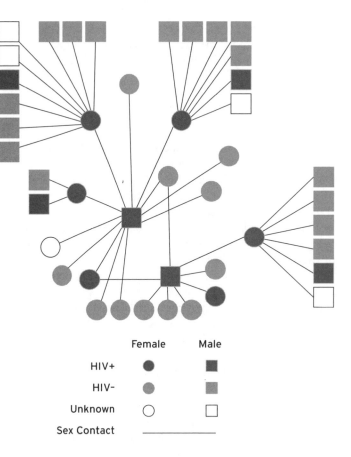

A network of young people in a rural Mississippi town who shared sex partners and spread HIV among others in the network. Data from: CDC, 2000. *HIV Surveillance Report: Diagnoses of the HIV Infection and AIDS in the United States and Dependent Areas,* 1999. Vol. 12.

	Female	Male
HIV+	●	■
HIV-	●	■
Unknown	○	□
Sex Contact	———	

of the 1,163,575 cumulative HIV/AIDS cases reported to the CDC from the epidemic's beginning up to 2010, almost 206,612 cases were transmitted by heterosexual contact, with no other risk factor involved. Heterosexual transmission accounted for 27% of estimated new HIV infections in 2009. Some cases related to injection drug use may, in fact, have occurred by heterosexual transmission, and some of the apparently homosexual men may have been bisexual and may have acquired HIV during vaginal intercourse with women.

Epidemiologists also point out that the characteristics of the heterosexual transmission group are changing. Before 1985, most cases linked to heterosexual contact occurred in people born in other countries (hence, HIV was "imported"). Since 1985, most such cases have occurred in persons in a sexual relationship with an infected person or with a person at high risk for AIDS (e.g., an injection drug user). The HIV was, therefore, "domestic."

HIV and AIDS in Seniors

Older individuals or seniors (55 and older) represent a special subgroup among the heterosexual population. Researchers point out that the older persons are far less likely to practice "safer sex" methods (such as using condoms); moreover, they have weaker immune systems with which to battle HIV than younger people do, and their HIV infection is less likely to be diagnosed by doctors, especially as fatigue, pneumonia, dementia, weight loss, and other AIDS symptoms are often related to advanced age or illnesses associated with aging. Older women may be especially at risk because

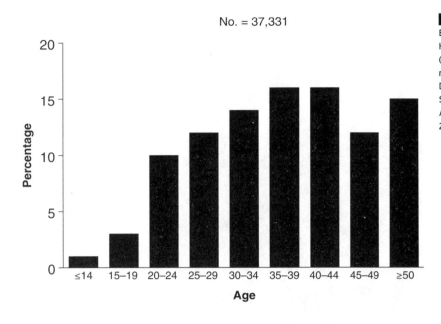

No. = 37,331

FIGURE 5.13

Estimated numbers of cases of HIV/AIDS by age, 2005. Seniors (age 50 or older) represent a significant number of HIV infections. Data from: CDC HIV/AIDS Facts Sheet: HIV/AIDS among Persons Aged 50 and Older, February, 2008.

age-related vaginal thinning and dryness can cause tears in the vaginal area during sexual activity. Nevertheless, older Americans, especially males, represent a significant number of the total AIDS cases in the United States. In 2010, 24% of new HIV/AIDS diagnoses were persons aged 50 and over (**Figure 5.13**).

During the 1980s, most older Americans with AIDS contracted HIV as a result of receiving a transfusion of contaminated blood, but currently, the vast majority has become infected as a result of heterosexual contact. Public health epidemiologists point out that the number of cases and deaths in this group is expected to increase further because of the new treatments that prolong life. Good nutrition, exercise, and an upbeat attitude also contribute to extending one's life. For many older patients, overcoming the denial that AIDS can strike them is a major obstacle in treatment. The stigma of HIV/AIDS may be more severe among older persons, leading them to hide their diagnosis from family and friends. Failure to disclose their HIV status may limit potential emotional and practical support.

HIV and AIDS in Correctional Facilities

The correctional setting is often the first place incarcerated men and women are diagnosed with HIV and provided treatment. More than 2 million people are incarcerated in the United States. At the end of 2010, 1,612,395 people were in federal and state prisons (**Figure 5.14**). At midyear of 2010, 748,728 people were in local jails. In 2008, 20,449 state prisoners and 1,538 federal prisoners (1.4% of the total prison population) were reported to be living with HIV or AIDS. In 2007, the rate of confirmed AIDS cases among state and federal prisoners was 2.4 times the rate in the general U.S. population. Of the 120 AIDS-related deaths in state prisons in 2007, nearly two thirds (65%) were among black/African American inmates, compared with 23% whites and 12% Hispanic/Latino inmates.

Healthcare providers and staff at correctional settings are up against challenges related to implementing HIV testing, treatment, prevention programs, and support services for prisoners after their release. The CDC recommends HIV screening upon entry into prison and before release in addition to voluntary HIV testing periodically during

FIGURE 5.14

Each year one in seven people living with HIV pass through a correctional facility. © Comstock/Thinkstock.

incarceration. Inmates are hesitant to be tested because of the stigma associated with a positive diagnosis and concern that medical confidentially will not be maintained.

AIDS in Schools and Colleges

Although the incidence of AIDS in school-age children and teenagers is relatively low, the potential exists for young people to be the next victims of the AIDS epidemic because young people experiment with drugs and sexuality, two modes for the transmission of HIV. As the number of sexual experiences and sex partners increases, the odds increase that an individual will come in contact with an HIV-infected person. Public health officials suggest that having sexual contact with a certain individual essentially brings a person in contact with all other individuals with whom that individual has had sexual contact. It is quite conceivable that HIV-infected T-lymphocytes or free viruses could be moving through any person-to-person chain, despite the fact that all appear to be in good health.

In 2009, 8,294 young persons aged 13 to 24 were diagnosed with HIV infection in the 40 states. These cases represent about 20% of the persons diagnosed during 2009 (**Figure 5.15**). Seventy-five percent (6,236) of these diagnoses occurred in young people aged 20 to 24 years (Figure 5.15), the highest number and rate of HIV diagnoses of any age group.

Ignorance and feelings of invincibility are but two of the obstacles that must be overcome in AIDS education programs for schools and colleges. Research has shown that a large proportion of young people are not concerned about becoming infected with HIV. This lack of awareness can translate into taking measures that could protect their health. Young people in the United States use tobacco, alcohol, and drugs at high rates. The CDC's National Youth Risk Behavior Survey found that 24.2% of high school students had had five or more drinks of alcohol in a row on at least one day during the 30 days before the survey, and 20.8% had used marijuana at least once

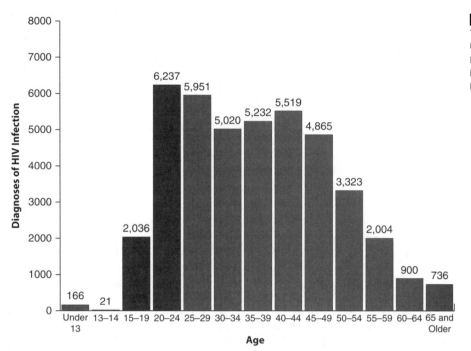

FIGURE 5.15
Young people aged 13 to 19 in the United States are at high risk for HIV infection. Data from: CDC HIV among Youth Fact Sheet, December, 2011.

during the 30 days before the survey. Casual and chronic users of drugs or alcohol are more likely to engage in high-risk behaviors, such as unprotected sex, when they are under the influence of drugs or alcohol. HIV prevention outreach and education efforts including programs on abstinence and safe sex are required to reduce the incidence of HIV infection and AIDS in young people. The CDC is working on a multifaceted approach to interventions, including the use of prevention messages on cell phones, gaming systems, websites, and social media.

HIV and AIDS in Newborns

Women can pass HIV to their babies during pregnancy, while the baby is being delivered, or through breast-feeding. Among the more tragic cases of AIDS were those in infants born to women infected with HIV. By December 2006, the CDC had reports of more than 9,000 cumulative confirmed cases of so-called pediatric (age < 13) AIDS in the United States and nearly 5,000 deaths. In many cases of transmission of HIV from mother to child, the mother was exposed to HIV through the use of intravenous drug injection, either by the mother or through infection of the mother by a partner who contracted HIV through injection drug use (**Table 5.5**). Indeed, until 1999, this link to injection drug use was the most common mode of transmission to newborns. However, between 2000 and 2006, mothers who transmitted HIV to their babies more often than not contracted HIV from sexual activity with partners whose exposure route to HIV was uncertain or the mothers themselves contracted HIV through uncertain means (Table 5.5). As was observed previously here with HIV/AIDS rates among the various ethnicities, pediatric HIV/AIDS has always been and continues to be most prevalent among blacks (Table 5.5).

Research studies indicate that HIV is able to pass the placental barrier and infect the fetus while it is still developing in the uterus. (A 1999 study concluded that the amount of HIV in a woman's bloodstream is a critical factor in whether transmission can occur.) Infection can therefore take place within the woman's body. In addition,

The Epidemiology of AIDS

	Year of Diagnosis					Cumulative (1981–2006)
	2002	2003	2004	2005	2006	
Race/Ethnicity						
White, not Hispanic	14	12	7	4	4	1,599
Black, not Hispanic	70	46	33	38	30	5,654
Hispanic	18	10	9	8	3	1,748
Asian/Pacific Islander	1	0	1	1	1	54
American Indian/Alaska Native	1	0	1	0	0	31
Transmission Category						
Hemophilia/coagulation disorder	0	0	0	0	0	226
Mother with documented HIV infection or one of the following risk factors:	104	70	53	52	37	8,508
Injection drug use	12	8	5	3	4	3,220
Sex with injection drug user	4	6	3	2	1	1,397
Sex with bisexual male	2	0	3	1	1	209
Sex with person with hemophilia	0	0	0	0	0	35
Sex with HIV-infected transfusion recipient	0	0	0	0	0	22
Sex with HIV-infected person, risk factor not specified	36	20	20	25	13	1,530
Receipt of blood transfusion, blood components, or tissue	2	1	0	0	0	144
Has HIV infection, risk factor not specified	47	34	21	21	17	1,951
Receipt of blood transfusion, blood components, or tissue	2	0	0	0	0	374
Other/risk factor not reported or identified	0	0	0	0	0	36
Total*	106	70	53	53	38	9,144

*Includes children of unknown race or multiple races. Cumulative total includes 58 children of unknown race or multiple races. Because column totals were calculated independently of the values for the subpopulations, the values in each column may not sum to the column total.

Reproduced from CDC, *HIV/AIDS Surveillance Report*, vol. 18, 2006.

public health officials have received reports of women who contracted HIV infection after giving birth and then infected their infants. Because virtually all such infected infants were breast fed, the findings suggested the passage of HIV in mother's milk.

Indeed, in 1999, researchers confirmed HIV infection via the breast milk. They studied 672 infants born without HIV infection to women in the developing African nation of Malawi. Forty-seven cases of HIV infection were apparently due to breast milk, nearly half occurring within the first five months after birth. Where inflammation of the mammary tissues occurred, the passage of HIV particles was higher. Researchers noted that discontinuing breast feeding would lessen the possibility of HIV passage, but it would also reduce the ability of mothers to pass helpful antibodies and nutrients to their offspring.

For many years, the study of pediatric AIDS was hampered by the lack of a diagnostic test to detect HIV in the newborn's tissues. Antibody tests were the staple means of diagnosis, but positive antibody tests could have reflected either the newborn's antibodies or those passing across the placenta from the mother. Then, during the mid-1990s, the viral load test was introduced, and physicians could detect the RNA or DNA associated with HIV. Availability of this test encouraged studies to determine whether AZT or other anti-AIDS drugs could be given to the mother to prevent prenatal transmission of HIV.

The studies indicated that rapid implementation of AZT treatment can effectively block HIV's transmission to the fetus, and a steep decline of prenatally acquired AIDS occurred during the last half of the 1990s and leveled off at very low levels by 2006, as **Figure 5.16** illustrates. The decline was further encouraged by improved treatments for HIV-infected newborns, treatments that delay the onset of AIDS; however, it is the prophylactic treatment of the mother with AZT and other anti-HIV drugs early in pregnancy that best reduces the risk of spread to the fetus. Improved surveillance of pregnant women early in pregnancy and subsequent treatment accounts for the rapid and dramatic drop in maternal–fetal spread of HIV. Indeed, greatest efficacy of anti-HIV drugs and therapy in stopping AIDS cases has been observed in the reduction of the maternal–fetal spread of HIV.

It is unclear why these drugs are more effective in blocking the spread to the infant compared with the effect on the adult, and this is an area of active research. Without treatment or breastfeeding about 25% (1 in 4) of pregnant women infected with HIV will transmit the virus to their babies. *If pregnant women take antiretroviral drugs, HIV transmission is reduced to less than 2% (fewer than 2 in 100).* Regularly testing

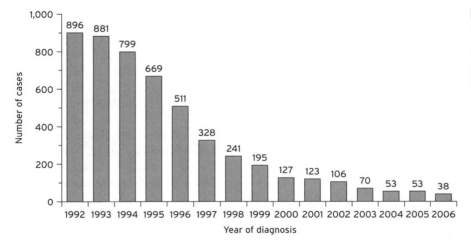

FIGURE 5.16
Estimated numbers of AIDS cases in children < 13 years of age, by year of diagnosis, 1992-2006. Reproduced from CDC, *HIV/AIDS Surveillance Report*, vol. 18, 2006.

Mapping HIV

Africa

The average life expectancy in sub-Saharan Africa is now 54.4 years and is below 49 years in countries heavily affected by HIV/AIDS.

- 22.9 million people were living with HIV in sub-Saharan Africa in 2010.

- 90% of the 16.6 million children orphaned by AIDS live in sub-Saharan Africa.

- South Africa had the highest number people living with HIV (5.6 million or 17.8%), followed by Nigeria (3.3 million) and Kenya (1.5 million) in 2009.

- 1.2 million people died from AIDS in sub-Saharan African and 1.9 million people became infected with HIV in 2010.

- West Africa has been less affected by HIV/AIDS but some countries are experiencing rising HIV prevalence rates (e.g., Gabon and Nigeria).

- HIV prevalence in East Africa exceeds 5% in Uganda, Kenya, and Tanzania.

- Antiretroviral drugs reached 64% of HIV-infected pregnant women in Eastern and Southern Africa and 18% in West and Central Africa in 2010.

Data from: *UNAIDS Report on the Global AIDS Epidemic, 2010.*

pregnant women for HIV and providing antiretroviral drugs if they are infected dramatically reduce the number of children born with HIV. The success of great reductions in the spread of the virus to the infant can be enjoyed by families as well as lessen the emotional and financial healthcare burdens from the local to national levels.

Cases of pediatric AIDS in the United States were concentrated in the New York City area, probably because of the high number of female residents who are injection drug users or sex partners of injection drug users. No other state reported a number close to New York's. Through June 2000, for example, 1,997 cases of pediatric AIDS were reported in the New York area since the beginning of the epidemic. The next closest metropolitan area was Miami, Florida (473 cases), and then Newark, New Jersey (320 cases). At the end of 2008, 3,022 children under the age of 13 were living with HIV in the U.S. Epidemiologists are quick to point out that the cases noted are only those that fit the definition for AIDS.

An interesting aspect of AIDS in newborns emerged in 1996 when the highly respected journal *Science* discussed new data under the provocative headline, "Can Some Infants Beat HIV?" The thrust of the article was that a collaborative study in Europe indicated that some HIV-infected newborns apparently cleared the HIV from their tissues. Investigators studied 2,319 children who had HIV antibodies at birth but subsequently tested negative. Tests performed for the presence of the virus showed that nine of the children cleared the viruses as well as the antibodies from their bloodstreams. The results confirmed an earlier French study in which 12 of 188 children cleared HIV from their bodies. These studies strengthened the possibility that in some newborns antibodies might be able to neutralize HIV before it becomes entrenched in the tissues; however, the concept that infant can clear HIV has been set aside. It now appears that these infants never had genuine HIV infections but simply possessed maternal antibodies that eventually cleared the neonates' system.

HIV and AIDS in Africa

In 2010, there were 34 million people globally living with HIV infection, a reduction of nearly 14% compared with the estimated 39.4 million in 2006 (**Figure 5.17**). Annual new HIV infections fell 21% between 1997 and 2010. The single biggest contribution to this reduction was an intensive effort to improve access to treatment. The most dramatic increases in antiretroviral therapy coverage have occurred in sub-Saharan Africa, with a 20% increase between 2009 and 2010 alone (**Figure 5.18**). It is estimated that at least 6.6 million people in low- and middle-income countries are receiving HIV treatment. In low- and middle-income countries, 47% of the 14.2 million eligible people living with HIV were on antiretroviral therapy at the end of 2010, compared to 39% at the end of 2009.

Africa remains the hardest-hit continent of the world. In 2010, for instance, as much as 68% of the approximately 36.1 million people with HIV infection or AIDS were living in sub-Saharan Africa. Sub-Saharan Africa accounted for 70% of new HIV infections in 2010. Nearly 50% of the deaths from AIDS-related illnesses in 2010 occurred in southern Africa. AIDS has claimed at least 1 million lives annually in sub-Saharan Africa since 1998. The total number of new HIV infections in sub-Saharan Africa has dropped by more than 26%, down to 1.9 million from an estimated 2.6 million at the height of the epidemic in 1997.

Since the first reports of AIDS in Africa were published in 1983, it has become apparent that the disease exhibits a different pattern of spread than that seen in Western Europe and the United States. In Africa, most transmissions of HIV take place during heterosexual vaginal intercourse, and currently, the percentage of women living with HIV in sub-Saharan Africa is 59%. In contrast, there is an approximate three-to-one ratio of infected men to infected women in the United States.

Epidemiologists cannot fully account for the majority of transmissions by heterosexual intercourse, but they believe that lesions and sores from STDs provide a gateway to the bloodstream. African men and women suffer from a high number of STDs,

> **Focus on HIV**
>
> Botswana was the first African country to implement a free national antiretroviral therapy program. Treatment for HIV is now available in 30 hospitals and 130 satellite clinics countrywide.

FIGURE 5.17

Regional estimated numbers of adults and children living with HIV in 2010.
Data from: *UNAIDS World AIDS Day Report, 2011.*

20%–39%

Algeria
Angola
Armenia
Azerbaijan
Bangladesh
Bhutan
Bolivia
Bulgaria
Burundi
Cameroon
CAR
Chad
China
Colombia
Côte d'Ivoire
Equatorial Guinea
Fiji
Gambia
Ghana
Hungary
India
Indonesia
Kazakhstan
Lebanon
Liberia
Lithuania
Malaysia
Mauritania
Mongolia
Morocco
Myanmar
Niger
Nigeria
Panama
Poland
Rep. of Moldova
Russian Fed
Sao Tome and Principe
Serbia
Sierra Leone
Sri Lanka
Uzbekistan

40%–59%

Belarus
Belize
Benin
Burkina Faso
Cape Verde
Congo
El Salvador
Eritrea
Gabon
Guatemala
Guinea
Guinea-Bissau
Haiti
Honduras
Jamaica
Lao PDR
Lesotho
Malawi
Mali
Mozambique
Oman
Papua New Guinea
Peru
Philippines
Senegal
South Africa
Suriname
Togo
Turkey
Uganda
UR Tanzania
Venezuela
Viet Nam
Zimbabwe

0%–19%

Afghanistan
DR Congo
Djibouti
Egypt
Iran
Kyrgyzstan
Latvia
Madagascar
Maldives
Mauritius
Nepal
Pakistan
Somalia
Sudan
Tajikistan
Tunisia
Ukraine

60%–79%

Argentina
Brazil
Costa Rica
Dominican Rep
Ecuador
Ethiopia
Georgia
Kenya
Mexico
Paraguay
Romania
Swaziland
Thailand
Uruguay
Zambia

>80%

Botswana
Cambodia
Chile
Comoros
Croatia
Cuba
Namibia
Nicaragua
Guyana
Rwanda
Slovakia

FIGURE 5.18

Low- and middle-income population countries receiving antiretroviral therapy at the end of 2010.
Reproduced from *UNAIDS World AIDS Day Report, 2011.*

including syphilis, chancroid, gonorrhea, and genital herpes. Each of these diseases are accompanied by lesions of the genital organs, and contact with these lesions and sores may permit viral transmission by infected persons to sex partners. In one study, 54% of African sex workers (prostitutes) with genital lesions also were infected with HIV, whereas only 17% of sex workers without an apparent STD were HIV positive.

Another possible avenue for heterosexual transmission in Africa may arise from the refusal to circumcise males. It has been suggested that leaving the penile foreskin in place encourages inflammation and formation of sores under the foreskin, thereby providing a passageway for HIV to enter or leave the body. Kenya could prevent an estimated 73,000 new HIV infections from 2011 to 2025 if 80% of uncircumcised males (860,000) were circumcised by 2015. South Africa has introducing voluntary medical male circumcision programs. By the end of 2010, there were 232,237 circumcisions performed, which is 27% of the national target goal to be reached by 2015. Young men have been the most willing to take the option of circumcision, but the older men also need to be reached to achieve the population-wide prevention benefit. If 20 million more men in Eastern and Southern Africa are circumcised, it could avert about 3.4 million new HIV infections by 2015.

There is also the cultural practice of "dry sex" performed in certain African communities. In this process, a woman dries her vagina with powders or cloths before having sex, a practice that can increase friction during intercourse and heightens sexual pleasure but also leads to abrasion of the vaginal wall and entry of free HIV and HIV-infected T-lymphocytes. The practice of dry sex probably also destroys many of the helpful bacteria in the vaginal tract and increases the likelihood that a condom will tear.

Although it is clear that condom use can interrupt the spread of HIV, traditional attitudes often mitigate against their use. For example, if a woman insists that a man wear a condom, she may imply that she is a sex worker; thus, she will press a man to have sex without using a condom as proof of her virtue. Yet in some communities, if she becomes infected with HIV, a woman will be divorced by her husband and ostracized from the community. Alternatively, if a woman suggests condom use, her husband may become angry because she is suggesting something he should have thought of first. For such instances, counselors have devised scripts that women can follow to allow their partners to save face. Using microbicides may be a useful alternative.

In addition, AIDS also can be related to practices such as ritual scarring with contaminated needles, transfusions of contaminated blood to treat diseases such as malaria, reuse of contaminated needles during immunization programs, and ritual removal of the clitoris from women. Anal intercourse among homosexual men probably accounts for some cases of AIDS; however, admitting to this practice often incurs social disgrace in Africa, and homosexual men are reluctant to reveal their lifestyle.

The typical African patient with AIDS exhibits a wasting syndrome known as "slim disease." Its symptoms are similar to those seen in the United States and other parts of the world: Weight is lost rapidly. Intractable diarrhea develops. Fever runs high, and the patient abandons interest in eating. Within a period of two months, 30% of the body weight may be lost (**Figure 5.19**). Many patients have infection with the protozoal parasites *Cryptosporidium* and *Isospora*. These parasites attack the intestinal cells and encourage the body to pour out huge volumes of fluid. Researchers have also located HIV in intestinal macrophages and have suggested that viral infection of these macrophages may reduce the defense normally available to the body and exacerbate the vigorous parasitic infection.

Among the opportunistic diseases, tuberculosis is the most prevalent in Africa (most AIDS wards in African hospitals are tuberculosis sanitaria). Kaposi's sarcoma is found in both HIV-infected and noninfected individuals. *Pneumocystis* pneumonia is relatively rare in Africa. Subtype C of HIV accounts for most infections in South Africa.

FIGURE 5.19

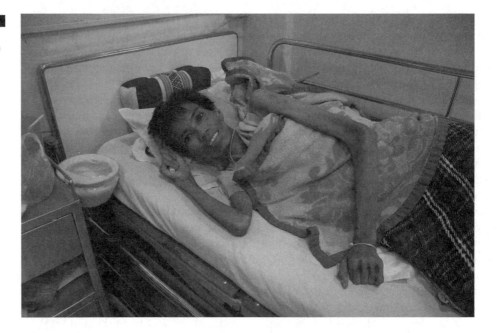

A woman suffering from AIDS displaying the symptoms of "slim disease," the wasting phase of HIV/AIDS. This photograph was taken at a hospital in Thailand. © FOTOSEARCH RM/age fotostock.

The AIDS epidemic in Africa is largely confined to urban centers, where infectious disease is common. Malaria is widespread in the cities, and few Africans reach the age of 30 years without suffering an attack of malaria or some other disease. Constant barrages of these diseases place the immune system under stress, a factor that may reduce its ability to resist HIV. This phenomenon, in which two diseases affect the body simultaneously, is called co-infection. The co-infection theory has gathered support in recent years and is offered as still another explanation for the pattern of distribution displayed by AIDS in Africa.

HIV and AIDS in the World

To international epidemiologists, the World Health Organization (WHO) and the United Nations Programme on AIDS (UNAIDS) are preeminent. These specialized agencies of the United Nations work to promote physical, mental, and social health in peoples of the world. Among their duties are collecting and distributing data on epidemics and establishing international programs for dealing with epidemics.

Worldwide, 34 million adults and children are living with HIV; of these, 2.7 million became infected in the year 2010, and almost 1.8 million individuals died of AIDS in 2010. Moreover, 269 countries are now involved in the pandemic. The year 2007 heralded changing regional rates of new HIV infections. The annual new HIV infections fell 21% between 1997 and 2010. The HIV epidemic has turned the corner. The proportion of women living with HIV has remained stable at 50% globally (with the exception of sub-Saharan Africa at 59% and the Caribbean at 53%). Rapid scale-up of access to antiretroviral therapy averted 700,000 AIDS-related deaths in 2010 alone.

However, several countries do not fit the declining trend of HIV infections. Sub-Saharan Africa remains the region of highest rates and greatest concerns by far. The number of people living with HIV rose 250% from 2001 to 2010 in Eastern Europe and Central Asia. While the Caribbean has the second highest HIV prevalence after sub-Saharan Africa, the epidemic has slowed significantly since the mid-1990s. In

Asia, prevalence of HIV is among populations at higher risk of infection: sex works, intravenous drug users, and men who have sex with men.

The transmission of HIV appears to fall into three major patterns, corresponding to three different portions of the world. In Central Europe, the pattern of transmission is similar to that in the United States, with the great majority of cases occurring in homosexual men and injection drug users. In Eastern Europe, the Russian Federation and Ukraine account for almost 90% of the region's epidemic. Injecting drug use remains the leading cause of HIV infection in this region. In contrast, in Northern Europe, the Middle East, North Africa, and Asia, the incidence of AIDS is relatively low. In these regions, cases are generally related to heterosexual contact with travelers who acquired HIV elsewhere, or they are associated with imported contaminated blood or implements. The third pattern of transmission is seen in Latin America and in areas of Sub-Saharan Africa. HIV probably spread through these areas in the 1970s. (Tests on blood stored in Zaire from 1959 showed that the virus was present at that time.) Most transmission in these regions is by heterosexual contact, as we noted previously. AIDS has particular significance to the world's developing countries because persons with AIDS are often young and middle-aged business workers who represent the country's future. A substantial economic impact on certain countries is predicted for the years ahead.

As of 2003, India was estimated to be second to South Africa in the total number of cases of HIV infection, and over 75% of those cases were believed to be acquired by heterosexual contact. Unfortunately, the disease remains highly stigmatized in India, especially for women—even if monogamous women are infected by their husbands, they often bear the blame for their own illness.

The rate of HIV infection in China was low as the new century began, and most of the confirmed cases were related to injection drug use. Concern has been expressed for the growing sex industry in China, especially because condom use is unpopular. Infected blood is another possible mode of HIV transfer because before it was banned, selling one's blood at a commercial collection center was a common way of earning money.

Nevertheless, the WHO and UNAIDS are committed to preventing new HIV infections through education, blood screening, and treatment for injection drug users. They also seek to bring support and care to the afflicted while linking together national and international efforts to break the chains of transmission. According to many public health epidemiologists, AIDS has no precedent in history. As long ago as 1988, Jonathan Mann, the former AIDS program director of the WHO, suggested that AIDS ". . . is changing the way the world thinks. It is becoming a key piece in the history of our time."

LOOKING BACK

Discovering that an epidemic is in progress and defining the circumstances under which it spreads are tasks for epidemiologists. For the AIDS epidemic, epidemiologists have concluded that AIDS is a bloodborne disease due to a fragile virus that is passed among individuals by blood-to-blood or semen-to-blood contact. Pregnant women can also pass the virus to the developing fetus or after birth and through breast milk.

In homosexual men, HIV is passed primarily by anal intercourse because bleeding of the rectal tissue permits entry of the virus. In this case, a risk exists for both the insertive and receptive partner. Injection drug users pass HIV by allowing their blood to mix with that of another person in a contaminated needle and syringe when "works" are shared. Over 12% of HIV infections in the United States occur in this way. Heterosexual couples may be at risk because during vaginal intercourse viruses may enter through lesions or abrasions on the genitals. Both males and females are at risk, and special risks exist among the older population. Newborns acquire HIV from their mothers via placental transfer, but drug therapy can reduce the possibility of such transfer.

In Africa, the major method for spreading the virus among heterosexuals is by vaginal intercourse. Epidemiologists believe that ulcers from sexually transmitted diseases allow entry into the bloodstream. The number of reported cases in sub-Saharan Africa far exceeds that in North America (1.3 million new infections in 2010), and the major manifestation of AIDS in African patients is the wasting syndrome. In the world at large, 269 countries had reported AIDS cases by 2002, and over 40 million people had AIDS. Travelers and imported blood may be involved in the spread.

Healthline Q & A

Q1 Why are sexually active homosexual men at risk for AIDS?

A Cases of AIDS among homosexual men are associated with the practice of anal intercourse. When the penis enters the rectum, the surface tissue often tears and bleeds, thereby allowing HIV-infected semen to enter the blood of the sexual partner.

Q2 Can infected heterosexuals spread AIDS if they practice anal intercourse?

A Yes. If the man's penis causes abrasions or bleeding in the lining of the woman's rectum, then semen-to-blood contact can take place. Should the semen contain free HIV or HIV-infected T-lymphocytes, the possibility of viral transmission is substantial.

Q3 Can lesbians (female homosexuals) spread AIDS from one to another?

A A lesbian relationship does not involve anal intercourse or blood-to-blood contact. Because these behaviors are excluded, transmission of HIV is not likely.

Q4 Why is AIDS more likely to be transmitted from men to women than from women to men during vaginal intercourse?

A Most studies indicate that a woman is more likely to contract AIDS from a male than the reverse. The reason is probably linked to the fact that the concentration of HIV is higher in the semen than in the vaginal secretions. Thus, an uninfected woman is exposed to more viruses than an uninfected man. In addition, the semen can remain for a long time in the vagina, thereby increasing the length of exposure to HIV.

REVIEW

This chapter has dealt with the spreading of the AIDS epidemic among different groups of individuals and in different parts of the world. To test your knowledge of these concepts, select the letter of the phrase that best completes each of the following statements. The correct answers are listed elsewhere in the text.

_____ 1. All of the following are part of the job of an epidemiologist except
    a. discovering that an epidemic is in progress.
    b. developing a drug for use in the epidemic.
    c. defining circumstances under which a disease spreads.
    d. studying the pattern of distribution of a disease.

_____ 2. In a male, the presence of a sexually transmitted disease (STD) such as gonorrhea or chlamydia can contribute to HIV transmission because
    a. the bacterial agents enhance the aggressiveness of HIV.
    b. the body's antibacterial antibodies react with HIV.

    c. STDs can cause erosion of the urethra and provide an entry path for HIV.

    d. HIV attaches to the bacteria and is transmitted when they are transmitted.

_____ **3.** Whereas the percentage of AIDS cases involving homosexual men is dropping, the percentage continues to rise in

    a. teenagers.

    b. heterosexuals.

    c. health workers.

    d. doctors and surgeons.

_____ **4.** Direct blood-to-blood or semen-to-blood transfer of HIV is necessary for the transmission of AIDS because

    a. the AIDS virus is a very fragile virus.

    b. blood and semen contain cofactors that stimulate the development of AIDS.

    c. the AIDS virus is a very resistant virus.

    d. contact with oxygen in the atmosphere kills the AIDS virus.

_____ **5.** The transmission of HIV can be limited by all of the following methods except

    a. using a condom during sexual encounters.

    b. disinfecting needles used to inject drugs.

    c. limiting the number of sexual partners.

    d. using mosquito repellents.

_____ **6.** In the Western Hemisphere, confirmed cases of AIDS have been reported

    a. in nearly every country.

    b. only in the United States and Canada.

    c. primarily in Mexico and the Caribbean islands.

    d. in all countries except those in South America.

_____ **7.** The typical AIDS patient in Africa exhibits symptoms that typify

    a. the AIDS-dementia complex.

    b. the HIV wasting syndrome.

    c. an extreme case of cytomegalovirus disease.

    d. an asymptomatic HIV infection.

_____ **8.** During vaginal intercourse, the risk of HIV infection increases if

    a. the woman has vaginal lesions.

    b. HIV antibodies are present in the semen.

    c. the man uses a condom.

    d. the woman uses birth control pills.

_____ **9.** In Africa, the most predominant method for HIV transmission appears to be

    a. ritual scarring and reuse of needles.

    b. heterosexual vaginal intercourse.

    c. imported contaminated blood.

    d. social kissing.

Brennan, M. et al., 2009. *Older Adults with HIV: An In-Depth Examination of an Emerging Population*. New York: Nova Science Publishers, Inc.

Centers for Disease Control and Prevention (CDC), 2010. "Preventing exposures to bloodborne pathogens among paramedics." DHHS (NIOSH) Publication No. 2010-139.

Centers for Disease Control and Prevention, 2010. "Workplace solutions. Preventing exposures to bloodborne pathogens in paramedics." Washington, DC: National Institute for Occupational Safety and Health. Retrieved November 14, 2012 from www.cdc.gov/niosh/docs/wp-solutions/2010-139/pdfs/2010-139.pdf.

Centers for Disease Control and Prevention (CDC), 2012. *HIV Surveillance Report, 2010*. Vol. 22. Atlanta: U.S. Department of Health and Human Services, Centers for Disease Control and Prevention.

Cholewinska, G. 2010. "Healthcare and antiretroviral treatment in HIV-infected detained persons at the penitentiary units." *HIV & AIDS Review* 7(4):10–12.

Denning, P.H., et al., 2011. "Characteristics associated with HIV infection among heterosexuals in urban areas with high AIDS prevalence—24 cities, United States, 2006–2007." *MMWR* 60:1045–1049.

Dowling, T., et al., 2007. "Rapid HIV testing among racial/ethnic minority men at gay pride events—nine U.S. cities, 2004–2006." *MMWR* 56(24):602–604.

Jasny, B., et al., 2010. "HIV/AIDS: Eastern Europe." *Science* 329:159.

Joint United Nations Programme on HIV/AIDS (UNAIDS), 2011. *UNAIDS World AIDS Day Report*. Geneva, Switzerland: United Nations.

Valdiserri, R. O., 2011. "Thirty years of AIDS in America: A story of infinite hope." *AIDS Education and Prevention* 23:479–494.

Thorne, E., et al., 2010. "Central Asia: Hotspot in the worldwide HIV epidemic." *Lancet Infectious Diseases* 10:479–488.

Wejnert, C., et al., 2012. "HIV infection and HIV-associated behaviors among injecting-drug users—20 cities, United States, 2009." *MMWR* 61:133–138.

Youngleson, M.S., et al., 2011. "Improving a mother to child HIV transmission programme through health system redesign: Quality improvement, protocol adjustment and resource addition." *PLoS ONE* 5:e13891.

As the AIDS epidemic continues to spread, individuals can take certain measures in their personal lives and at work to protect themselves and others from contracting the human immunodeficiency virus (HIV). This chapter describes some of the available measures. On completing the chapter, you should be able to . . .

- Recognize STD trends and why undetected and untreated STDs can increase a person's risk for HIV and other serious health conditions such as infertility.
- Recognize how behaviors during sexual activity can be altered to prevent the passage of HIV.
- Understand the methods for preventing HIV transmission among those who use injection drugs.
- Describe the universal precautions recommended for all healthcare workers and be familiar with particular precautions suggested for surgical procedures, dentistry, embalming, and medical laboratories.
- Specify the precautions recommended by public health officials for those who work in public safety professions.
- Discuss the reasons why AIDS education is necessary in schools and survey the methods for implementing such education.

INTRODUCTION

In the 1840s, a **cholera** epidemic broke out in Europe. It spread quickly and reached England in June 1849. The district near Golden Square in London was among the worst affected areas—more than 500 people died in one 10-day period that August. A physician, John Snow, lived near Golden Square. Snow had studied cholera for years, and he saw in the epidemic an opportunity to test some of his ideas. Most scientists of his time believed that infected air was the probable cause of cholera, but Snow believed water was the culprit. With meager resources but much determination, he began a systematic study of the source of the cholera.

Snow discovered that most people living near Golden Square drew their water from a well on Broad Street. A hand pump operated the well, and anyone could work it. Snow examined the well closely and found that it was being contaminated by sewage overflowing from a nearby tenement, and a person living in the tenement had died recently of cholera. The relationship between the well and the disease was unmistakable.

On September 7, 1849, Snow presented his findings to the local community council. The members listened attentively but were skeptical. How, they asked, did Snow intend to interrupt the epidemic? Snow thought for a moment and then replied with the now-classic solution: "It's simple—take the handle off the Broad Street pump." The next day the handle was gone, and soon the epidemic subsided in that area.

Of course, not all epidemics are quite so easy to interrupt. For AIDS, scientists know what causes the disease (in fact, Snow did not know the cause of cholera), and they know how HIV spreads through a population; however, halting the spread of AIDS requires more than knowledge alone. It demands changes in social customs and cultural practices, and it requires the willingness of people to adapt their behaviors. As we see in this chapter, there are many ways of interrupting the spread of AIDS, but they require more than simply removing the handle of a water pump.

At the 2008 XVII International AIDS Conference in Mexico City, the international community used the conference to emphasize the goal of HIV prevention. The conference focused on the complications of globalization and discrimination in the prevention of the spread of HIV and providing universal access to effective care. Indeed, in his opening remarks, Dr. Pedro Cahn, the president of the IAS (International AIDS Society) stated the following to the 2008 International AIDS Conference:

> . . . *"I challenge you and member states to work with all the populations at risk for HIV some of which, 27 years into this epidemic, they are still to name. It is time to challenge the tyranny of ignorance and denial. It is time for leaders of all kinds, political, religious and community to help move this group in from the margins of society to the center of the global response to HIV. It is time for political courage, not time for political experience. And that means fighting the gender inequality, homophobia and poverty, which continue to drive this epidemic. We can and we must do better."*

June 5, 2011 marked the 30th anniversary since the first cases of AIDS were reported in 1981. Worldwide, more than 33.3 million people were living with HIV in 2010, including 1.2 million in the United States. HIV/AIDS has been responsible for the deaths of an estimated 30 million individuals. Although the rate of HIV infections is declining or stabilized in most parts of the world, the disease continues to take an enormous toll: 1.8 million deaths in 2010 alone, grief and hardship on countless families and communities, and harmful effects on the economy as those in the prime of their lives are lost in the workforce and as parents.

Anthony Fauci (Director) and Gregory K. Folkers (researcher) at the National Institute of Allergies and Infectious Diseases at the National Institutes of Health in Bethesda, Maryland, stated their viewpoints in the *Journal of the American Medical Association (JAMA),* July 25, 2012. They commented that,

> . . . *"despite these daunting statistics, the fight against HIV/AIDS is currently viewed with considerably more optimism than in years past because powerful interventions have been developed, scientifically proven effective, and refined. If these tools are made widely available to those who need them, an AIDS-free generation may be possible—that is, today's children could one day live in a world in which HIV infections and deaths are rare."*

Personal Protection

Before World War II, doctors had to contend with a variety of **sexually transmitted diseases (STDs)**, including **syphilis, gonorrhea, chancroid,** and **lymphogranuloma;** however, with the widespread use of antibiotics after the war, the incidence of syphilis and gonorrhea declined, and that of most other STDs dropped as well.

Then came the 1960s and the sexual revolution. It was a time of affluence and defiance of traditional values. Birth control pills, contraceptive devices, and vasectomies offered sexual liberation to go along with social and economic freedoms, and antibiotic resistance surfaced. Not surprisingly, the incidence of STDs soared. Today, the United States is in the grip of an epidemic of STDs. The CDC estimates there are 19 million new infections every year, costing $19 billion in health care and those affected face life-long health consequences. In 2010, more than 300,000 cases of gonorrhea were reported to the CDC. However, the CDC estimates that more than 700,000 people in the United States have been infected by the bacterium that causes gonorrhea each year. In 2012, the CDC changed treatment guidelines because of significant increases in the percentage of *Neisseria gonorrhoeae* bacterial isolates (the cause of gonorrhea) that were resistant to the effects of antibiotics used to treat gonorrhea.

Moreover, the list of STDs has continued to expand. In the United States, the STDs with the most troubling trends in 2010 are gonorrhea, syphilis, and chlamydia (**Figure 6.1**). Other STDs reported to the CDC today include genital herpes, trichomoniasis viral hepatitis, bacterial vaginosis, pelvic inflammatory disease, papillomavirus infections, lymphogranuloma and chancroid, and the latest addition to the list is HIV/AIDS. Untreated STDs have long-lasting effects that are serious for young people. For example, untreated gonorrhea and chlamydia can cause infertility in women. Untreated syphilis can lead to long-term complications, including brain, heart, and organ damage. Syphilis in pregnant women can result in congenital syphilis, which can cause stillbirths, physical and neurological deformities in children, and infant death. Studies have shown that people with gonorrhea, chlamydia, or syphilis are at increased risk for HIV.

The common denominator among STDs is the fragility of the infectious agent. The responsible infectious agent—whether it causes gonorrhea, syphilis, chlamydia, or AIDS—prefers the warm, moist environment of the human body and rapidly dies or is inactivated when exposed to an environment outside the body. The agent is able to infect only when passed directly from person to person in a moist environment, and because it infects organs of the reproductive system, the usual method of transfer involves physical contact between these organs during vaginal intercourse or anal intercourse.

Public health officials advise that couples should not have to worry about AIDS so long as they have been mutually faithful since the 1970s (when HIV is believed to have entered the United States) and avoid other risk factors, such as contaminated needles associated with injection drug use. In addition, individuals at risk through body fluid contact can avoid HIV infection as well as other STDs by practicing sexual abstinence.

In comparison, those who continue to have multiple sexual contacts place themselves and their partners at risk for HIV infection and AIDS. Sexual contact with sex workers (prostitutes) increases the risk further because prostitutes have innumerable sexual contacts. The presence of another STD pushes the risk even higher, as lesions and sores from the STD permit entry of HIV. Epidemiologic studies indicate that vaginal intercourse permits transmission of HIV and that oral–genital contact may be a risky practice as well. Anal intercourse involves the highest risk because tearing and

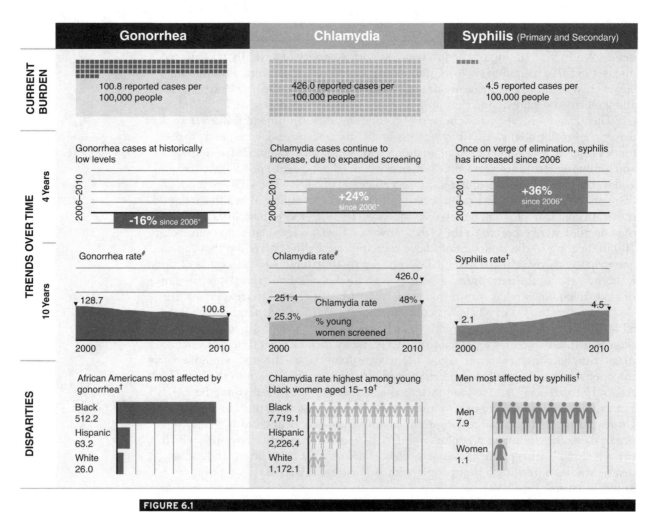

	Gonorrhea	Chlamydia	Syphilis (Primary and Secondary)
CURRENT BURDEN	100.8 reported cases per 100,000 people	426.0 reported cases per 100,000 people	4.5 reported cases per 100,000 people

TRENDS OVER TIME

4 Years (2006–2010)

Gonorrhea cases at historically low levels — **-16%** since 2006*

Chlamydia cases continue to increase, due to expanded screening — **+24%** since 2006*

Once on verge of elimination, syphilis has increased since 2006 — **+36%** since 2006*

10 Years (2000–2010)

Gonorrhea rate# — 128.7 (2000) → 100.8 (2010)

Chlamydia rate# — 426.0; Chlamydia rate 251.4 → 48%; % young women screened 25.3% (2000–2010)

Syphilis rate† — 2.1 (2000) → 4.5 (2010)

DISPARITIES

African Americans most affected by gonorrhea†
Black 512.2
Hispanic 63.2
White 26.0

Chlamydia rate highest among young black women aged 15–19†
Black 7,719.1
Hispanic 2,226.4
White 1,172.1

Men most affected by syphilis†
Men 7.9
Women 1.1

FIGURE 6.1

Snapshot of the 2010 Trends in the top three STDs in the U.S.: gonorrhea, chlamydia, and syphilis. Reproduced from CDC, *Sexually Transmitted Diseases Surveillance*, 2010.

bleeding of the rectal tissue permits HIV access to the bloodstream. The next paragraphs explore how the possibility of HIV transmission can be reduced.

Safer Sex Practices

The foolproof way to avoid contact with HIV through sexual activity is to avoid sexual activity with infected persons or to have a lifelong monogamous relationship. Because this suggestion is not realistic for many people, public health officials trying to prevent the spread of AIDS have encouraged modifications of sexual behaviors. These modifications are collectively known as safer sex practices. The implication is that they can reduce the risk of HIV transmission, but they cannot eliminate the risk completely (they are *safer* sex practices, *not* safe sex practices, the latter term implying a 100% guarantee).

Among the most important safer sex practices is the use of condoms. Anything that creates a barrier between the semen and the internal tissues of sexual partners can lessen the risk of HIV transfer. The purpose of a condom is to form such a barrier. A condom (or "rubber") often made of latex that is a soft, stretchable sheath is placed over the penis during sexual intercourse, whether vaginal, oral, or anal. The condom acts as a barrier to collect semen and prevent it from entering the sexual partner's

BOX
6.1

No Excuses

Excuses for not using condoms are usually so persuasive that couples find it hard to answer them effectively, and they often give in to unprotected sex. Here are a few excuses for not using condoms and some reasonable replies:

She: "I use the pill; you don't have to use a condom."

He: "Let's use one anyway; either of us could be infected with something and it will help prevent passage."

She: "You carry condoms around? Was this all a trick to get me into bed?"

He: "I keep one with me because I care about your health and mine, and you never know where a night is going to end up."

He: "I don't have any infection, and I haven't had sex for months."

She: "Thanks for letting me know. I don't have anything either, but one of us could be harboring something without having any symptoms. Let's use a condom to be sure."

She: "It's hard to feel anything when you wear a condom."

He: "It's hard for me too, but there are lots of other feelings—and the protection's worth it."

He: "I might lose my erection if we stop right now."

She: "Not necessarily; I'll help you put it on. We'll make it a part of the whole thing."

He: "It's messy, and it smells bad."

She: "A little mess isn't so bad when you consider the alternative—AIDS!"

She: "They're unnatural and a fake."

He: "Maybe so, but getting sick is even worse."

She: "I'm insulted! Do you think I'm some sort of slut?"

He: "I didn't say that. I'm just trying to be careful for both of us. I want you to be sure I don't give you anything."

She: "Real men aren't afraid of getting sick; they don't use condoms."

He: "Maybe I'm a little more caring than those other guys."

He: "I love you. I wouldn't make you sick. I don't need a condom."

She: "I believe you love me and that you wouldn't intentionally make me sick, but AIDS can be in your body without your knowing it. So if you really love me you'd be aware of my needs and care for me."

He: "Just this once, no condom."

She: "All it takes is once; AIDS is forever!"

body (**Figure 6.2**). Extensive studies have demonstrated conclusively that condoms can decrease the risk of STDs, including HIV infection, by decreasing the transmission of infectious agents. In 1993, for example, the Centers for Disease Control and Prevention (CDC) reported that HIV failed to be transmitted among 123 couples who used condoms when one person was infected. In laboratory tests, infectious agents as small as viruses (including HIV) failed to pass the barrier (**Box 6.1**).

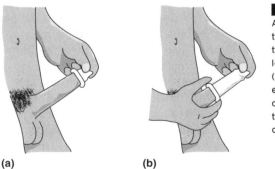

(a) (b)

FIGURE 6.2
A condom creates a barrier between the semen and the internal tissues of the sexual partner and lessens the risk of HIV transfer. (a) The condom is put on when erection is achieved. (b) The condom is rolled directly onto the penis, leaving a reservoir for catching the semen.

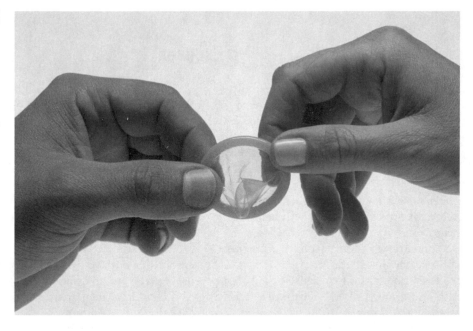

Several types of condoms are available commercially (**Figure 6.3**). In the laboratory, studies under mechanical conditions simulating sexual intercourse indicate that
latex condoms can prevent the passage of all infectious agents, including HIV. Natural skin (membrane or "lambskin") condoms are believed to be less effective. This is
probably because latex condoms are consistently thick and of quality (although the
latex can cause allergic reactions), whereas natural skin condoms made from lamb's
intestinal tissue can vary in quality and are generally more porous. In addition, fresh
condoms are preferable to older ones because heat and aging can weaken the material. To alert consumers, boxes of condoms are labeled with an expiration date.

For greatest effectiveness, the condom must be used correctly and in every sexual
encounter when there is doubt about the health status of the partner. Some helpful
guidelines for condom use to decrease the risk of HIV infection are as follows:

1. A condom must be worn each time one has genital, anal, or oral sexual contact,
   and the condom must be put on as soon as erection is achieved and before the
   penis is inserted into the partner. Care should be taken to avoid damage with the
   fingernails or other sharp objects.
2. The condom should be rolled directly onto the penis. It should not be unrolled
   from the ring then stretched over the penis. If a reservoir tip for catching semen is
   not present, a small amount of space should be left at the tip.
3. A spermicidal cream or foam containing nonoxynol-9 or a water-based jelly (such
   as KY) can be used to lubricate the tip of the condom before it is put on (however,
   nonoxynol-9 kills helpful bacteria in the woman's vaginal tract and may change the
   chemical environment there, and thus, pathogenic bacteria could thrive and cause
   urinary tract infections). The use of petroleum lubricants such as petroleum jelly,
   vegetable oil, or massage or body lotions is not advised because they can promote
   condom failure by reducing its strength.

4. The condom should be removed soon after ejaculation to avoid leaking of semen from the condom as the penis becomes limp. It is also a good idea to hold the rim of the condom during withdrawal to prevent its slipping off.

5. Condoms are for single-time use. They are not intended for reuse and should be discarded properly. Washing the hands and the penis after sexual intercourse is also recommended.

Condoms reduce the risk of infection for both the wearer and his sex partner. For the wearer, condoms lessen the exposure of the penis to infectious secretions and infected cells of the partner's reproductive organs or rectum. For the receptive sex partner, condoms prevent deposit of infectious semen, contact with infected discharge from the male reproductive tract, and exposure to lesions on the penis. Although the actual effectiveness of condom use in prevention of STDs is difficult to assess, several case-control studies indicate a lower rate of STDs in condom users and thus fewer health-related problems. Indeed, many colleges have decided to remove all of their cigarette machines and replace them with condom machines. "Why kill people," said one administrator, "when you can help save their lives?"

Other safer sex practices involve activities where no body fluids are exchanged, that is, activities taking place outside the body. Among these practices are massaging various sensitive parts of the body (the back, scalp, neck, and face are examples); rubbing the bodies together and stroking the partner's genital organs, as long as the penis does not come in contact with the vagina or rectum; masturbation, either of self or of the sexual partner; and kissing various body parts. That deep, or "French," kissing, however, could involve some slight risk if mouth or gum sores are present and if large amounts of saliva are exchanged. The effectiveness of safer sex practices is underscored by the results of the CDC study cited earlier and by results of a study published in 1988 by Margaret A. Fischl of the University of Miami. In the study, Fischl reported on 58 heterosexual married couples. In each couple, one spouse was infected with HIV. Over the duration of the study, 13 couples abstained from sexual intercourse, and none of the 13 spouses transmitted HIV; 23 couples engaged in sexual intercourse using condoms and HIV was transmitted in only three cases, and 22 couples engaged in sexual intercourse without condoms, with the result that 14 spouses were infected with HIV (**Table 6.1**). Methods proven to prevent HIV infection in men include condoms, male circumcision, and behavioral

TABLE 6.1	The Relationship of Sexual Activity to Development of HIV Antibody Among 58 Spouses of HIV-Positive Individuals*			
Sexual Activity	**Total Number**	**HIV+**	**HIV-**	**Percentage Converted**
Abstinence	13	0	13	0
Sexual contact with condoms	23	3	20	13
Sexual contact without condoms	22	14	8	64

* Spouses did not display HIV antibodies at the beginning of the study.

Source: Fischl, N. A., et al. "Heterosexual Transmission of Human Immunodeficiency Virus (HIV): Relationship of Sexual Practices to Seroconversion." Third International Conference on AIDS. Washington, D.C., 1988, p. 178.

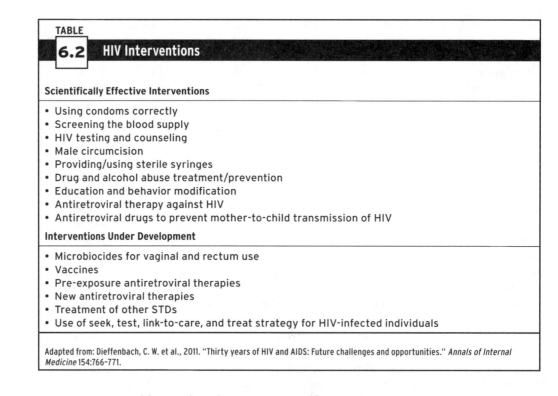

TABLE 6.2	HIV Interventions

Scientifically Effective Interventions

- Using condoms correctly
- Screening the blood supply
- HIV testing and counseling
- Male circumcision
- Providing/using sterile syringes
- Drug and alcohol abuse treatment/prevention
- Education and behavior modification
- Antiretroviral therapy against HIV
- Antiretroviral drugs to prevent mother-to-child transmission of HIV

Interventions Under Development

- Microbiocides for vaginal and rectum use
- Vaccines
- Pre-exposure antiretroviral therapies
- New antiretroviral therapies
- Treatment of other STDs
- Use of seek, test, link-to-care, and treat strategy for HIV-infected individuals

Adapted from: Dieffenbach, C. W. et al., 2011. "Thirty years of HIV and AIDS: Future challenges and opportunities." *Annals of Internal Medicine* 154:766–771.

interventions (**Table 6.2**) but these are not sufficient enough to protect women. Women in Sub-Saharan Africa have one of the highest risk factors for HIV infection where condom use is lower.

Condoms are not practical for women who want to conceive children or who cannot persuade their partners to use them. Topical **microbiocides** under development offer an approach to safer sex practices for women. Vaginal microbicides can be foam or gels that women can use to protect themselves against HIV and other sexually transmitted microorganisms. They are designed to be applied either before sexual activity or on a daily basis. Such a product would have to be nontoxic and nonirritating; it would have to withstand changing acidity levels in the vagina; it would need to fulfill aesthetic requirements for female consumers; and it could not kill the lactobacilli and other beneficial bacteria normally found in the vaginal tract. A microbiocide must be affordable to at-risk populations.

Early microbiocides under development were simple, non-drug based gels that could be provided over-the-counter without a prescription. Some contained surfactant-containing spermicides because early observations suggested spermicides could inactivate HIV *in vitro*. Other approaches were developing microbicides that would lower the pH of the vagina, which could inactivate HIV or polyanionic gels that inhibit HIV from binding to host cells. Corporate excitement for such products diminished because these microbiocides were not sufficiently potent to prevent HIV infection.

Current interventions under development focus on microbiocides that contain one or more antiretroviral drugs that target key events of HIV replication. A number of Phase III clinical trials are underway or about to be initiated. The major ongoing issue will be that of compliance. Researchers are hopeful that microbiocides will serve as another avenue through which sex can be made safer.

For Injection Drug Users

Injection drug users who share drug paraphernalia are at high risk for HIV infection because they are exposing their blood to the blood of an infected individual when they share needles and syringes. HIV need not pass through any cell layers or penetrate a lesion to reach the bloodstream (as it must during passage by sexual activity).

Ironically, the social structure of injection drug users may promote the sharing of syringes and needles. For example, the ethics of cooperation within small friendship groups is applied to the sharing of drug paraphernalia, which drug users call their "works." To refuse to share one's works can damage the reputation and reliability of the owner. Limited supplies of drug paraphernalia can also encourage sharing. Sharing is promoted by the rental of "houseworks" at places where injection drug users congregate (the so-called shooting galleries). After being used, the houseworks are returned to the owner for the next rental. Reusing contaminated needles and syringes under such conditions easily spreads HIV from a small group of friends into a larger population of injection drug users. An infected individual might then have sexual intercourse with a nonuser and transfer the virus to that person.

Clearly, the best way to avoid HIV from contaminated needles is to avoid using injection drugs. For those who choose to inject drugs, the safer practice is to avoid sharing drug paraphernalia. Although this practice may lead to social isolation, the benefits may outweigh the risk of contracting HIV.

It is also possible to decontaminate the syringe and needle with household bleach and hot water. Bleach contains chlorine, and the chlorine reacts with proteins in the HIV capsid and envelope to alter the chemical structure of these structures and thereby inactivate the virus. The syringe should be filled with bleach and then flushed out a few times. After this filling and flushing, the syringe should be filled with hot (preferably boiled) water a few more times to rinse it thoroughly (**Figure 6.4**). A 10% bleach solution (one part bleach added to nine parts water) inactivates HIV. More concentrated solutions may also be used, but they may leave a chlorine residue in the syringe that could be toxic to the body. Also, using a lighted match to heat the needle is not advised because HIV can be trapped within clots of blood left in the syringe. Furthermore, placing the needle and syringe in boiling water is inadequate because the hot water may not completely fill the implements.

Among the more controversial issues of contemporary society is whether municipalities should participate in needle exchange programs. In such a program, used needles (and syringes) are traded at no cost for sterile ones at designated locations and times. Opponents of these exchange programs argue that they condone and encourage drug use; proponents counter that they help interrupt the AIDS epidemic among injection drug users. Since the beginning of the HIV epidemic, injection drug use has accounted for 36% of AIDS cases in the United States and 57% of the cases of HIV infections in women.

One of the first needle exchange programs in the United States was put into operation in New Haven, Connecticut in 1992. A van, painted with vivid stripes and a rising sun, plied the streets of the city four times a week. Identifying themselves by code names, drug users came to the van and exchanged their needles for "survival kits" containing clean needles, bottles of bleach and water, and condoms. Rather than accelerating inner-city drug abuse, the program has apparently slowed the rate of infection, increased addict referrals to treatment programs, increased police empathy for addicts (police cars are steered clear of the van to avoid giving the impression of entrapment), and gained the trust of addicts. The van has become a visible expression of the city's desire to help its underprivileged population (**Figure 6.5**). To avoid subjectivity,

FIGURE 6.4

Eight steps for disinfecting a needle and syringe following injection drug use. All eight steps should be followed to ensure HIV destruction. The bleach solution can have a 10% concentration; the water should be hot (preferably boiled).

Bleach   Bleach

1. Fill syringe   **2.** Empty syringe   **3.** Fill syringe again   **4.** Empty syringe

Water   Water

5. Fill syringe   **6.** Empty syringe   **7.** Fill syringe again   **8.** Empty syringe

FIGURE 6.5

Needle exchange emergency distribution (NEED) program vehicle used in Berkley, California in 2007. Courtesy of Scott B. Fisher.

the city used scientists at nearby Yale University to evaluate the program. Experimental programs in New York City and Washington, D.C. soon followed.

The number of needle exchange programs continued to grow through the 1990s and into the new century as moral and philosophical support came from the CDC, the National Institutes of Health (NIH), and the American Medical Association. By 2009, more than 184 programs were operating (including sidewalk tables, health clinics, cars, and storefronts) in 36 states, and organizations were actively campaigning for more programs (**Figure 6.6**). Over 20 million syringes were being exchanged each year, and onsite counseling and testing for HIV were being offered at most of the

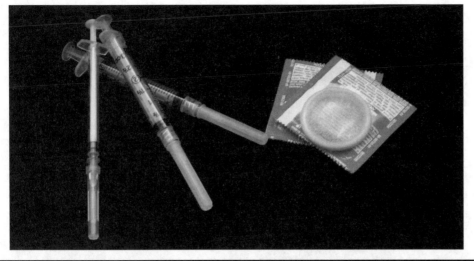

AIDS activists continue to campaign for condom supply and needle exchange programs as a way of reducing HIV transmission among injection drug users. By 2000, about 150 needle exchange programs were operating in the United States. © Jones & Bartlett Learning. Photographed by Christine McKeen.

sites. Although many of the programs receive state, county, and/or city funding, the major stumbling block to their growth is that federal funding is not permitted by law. In response, program supporters continue to point out that preventing HIV infection among injection drug users costs about $20 per user per year, a notable difference from the estimated $120,000 it costs to treat an HIV-infected person until death.

Needle exchange programs continue to face lack of funding and perennial resistance from communities that do not want such programs to operate in their backyards. Supporters point out that the latter objection could be addressed by using mobile vans, as in New Haven, Connecticut. Also, while the federal funding debate continues, programs could be helped by state approval of syringe sales at local pharmacies. In New York, for example, the sale of syringes won approval during 2001; a 10-pack costs $2 to $6 at local pharmacies. That year, New Hampshire and Rhode Island also removed the requirement for a prescription to purchase syringes and needles.

Interrupting the spread of AIDS among injection drug users has a corollary benefit to society because injection drug users are the major link by which HIV enters the heterosexual population. It was reported in the 2012 UNAIDS Report on the Global AIDS Epidemic that in the 49 countries with available data, the HIV infection rate is 22 times higher for people who inject drugs than for the population as a whole. Clearly, the education campaigns directed at injection drug users merit public support. Such campaigns include counseling about "safer shooting" practices, distribution of bleach vials for cleaning syringes, and expedited admission of infected individuals to drug treatment.

Globally, only 82 countries have needle-exchange programs. For example, in Germany there are 250 needle/syringe exchanges, yet drug users receive only an average of 2 needles/syringes per year. Experts recommend that in order to make an impact on HIV transmission, a distribution rate of 200 needles/syringes per drug injection user is needed. China has increased the number of needle exchange programs from 92 in 2006 to 901 in 2010, but syringe distribution remains low, an average of 32 needles/user per year. Australia needle exchange programs have by far the greatest rate of distribution at 213 needles/syringes/user per year (**Table 6.3**). There remains an urgent need to improve coverage of these services in this at-risk population.

Mapping HIV

Eastern Europe and Central Asia

- The number of people living with HIV rose 250% from 2001 to 2010 in Eastern Europe and Central Asia.

- The Russian Federation and Ukraine account for almost 90% of the HIV epidemic in this region of the world.

- Injecting drug use remains the leading cause of HIV infection in this region, but significant transmission also occurs to the sexual partners of those who inject drugs.

- According to interview-based research, many dealers who prepare a homemade heroin called **chernaya** sell prefilled used syringes picked off of the street that are contaminated with blood (the blood may be from an HIV-infected injector).

- Contrary to most other regions, AIDS-related deaths continue to rise in Eastern Europe and Central Asia.

- The incidence of HIV infections was slowing during the 2000s but has been accelerating since 2008.

Data from: *UNAIDS Report on the Global AIDS Epidemic*, 2010.

TABLE 6.3	Needle and Syringe Programs (NSPs), Selected Countries		
Country/Region	Number of Intravenous Drug Users Accessing NSPs in a Year	Average Number of Needles-Syringes Distributed per Intravenous Drug User per Year	Number of Intravenous Drug Users Receiving Antiretroviral Therapy
Ukraine/Eastern Europe	94,583–132,361	32	1,860
Czech Republic/Eastern Europe	27,200–34,000	151	12
United Kingdom	NK	188	623
Germany/Western Europe	NK	2	3,000
China	>38,000	32	9,300
Indonesia	49,000	3	5,406
Iran	55,000	41	580
Mexico	12,819	NK	NK
Canada	NK	46	NK
United States	NK	22	NK
Australia	NK	213	518

NK: not known.

Adapted from Mathers, B. et al., 2010. "HIV prevention, treatment, and care services for people who inject drugs: a systematic review of global, regional, and national coverage." *The Lancet* 375:1014–1028.

Protecting Healthcare Workers

Epidemiologists reviewing the scientific literature of recent years have concluded that HIV transmission within healthcare settings is extremely rare but nevertheless possible. As the AIDS epidemic continues to spread, the possibility increases that healthcare workers will be exposed to blood and body fluids from patients infected with HIV, and the number of healthcare workers is considerable. Health care is the fastest growing sector of the U.S. economy, employing over 18 million workers including students and trainees whose activities involve contacts with patients or with blood or other body fluids from patients. These workers may be involved in surgery, emergency room care, dentistry, laboratory, or morgue service. Even laundry service workers are considered healthcare workers.

The transmission of HIV among healthcare workers can occur by any of several means. These include cuts with **sharps** (scalpels, razors, needles, broken ampules, or any object that can cause a wound or punctures), contamination of open wounds, exposure of mucous membranes, and of particular concern, self-injection with blood-contaminated needles ("needlestick").

Epidemiologists point out that **infection control procedures** significantly reduces the number of accidental occupational exposures to HIV. Fundamental to infection control is the assumption that all patients are potentially infected with HIV in their blood and other body fluids. An assumption such as this must be made because a medical history and routine examination cannot reliably identify all patients infected with HIV and because testing for HIV may require the consent of the individual. This consent is not necessarily forthcoming from the patient.

Universal Precautions

Recommendations for protecting healthcare workers were first made by the CDC in 1983. As the years passed, CDC officials realized that the wisest course for these workers was to consider all patients as potentially infectious. Public health officials therefore promulgated a series of "universal blood and body fluid precautions," known simply as **"universal precautions."** Universal precautions are employed to protect the healthcare worker from infection by the patient as well as the reverse, and to prevent the spread of HIV among patients through contaminated devices or surfaces. The precautions apply to blood, semen, and other body fluids, but they do not apply to feces, saliva, tears, and urine because the amount of HIV is extremely low in most cases. Among the universal precautions are the following:

1. Healthcare workers should use barrier precautions such as gloves to prevent skin and mucous membrane exposure to blood or other body fluids from a patient. Gloves should be worn for touching blood and body fluids, mucous membranes, or nonintact skin of all patients; for handling items or surfaces soiled with blood or body fluids; and for performing venipuncture and other procedures where access to the patient's blood is required. Gloves should be changed after contact with each patient. Masks and protective eyewear or face shields should be worn during procedures that are likely to generate airborne droplets of blood or other body fluids, to prevent exposure of mucous membranes of the mouth, nose, and eyes. Gowns or aprons should be worn during procedures that are likely to generate splashes of blood or other body fluids.
2. Hands and other skin surfaces should be washed immediately and thoroughly if contaminated with blood or other body fluids. Hands should be washed immediately after gloves are removed.

Focus on HIV

Based on an Exposure Prevention Information Network and National Surveillance System for Hospital Healthcare Workers, an estimated 385,000 sharps-related injuries occur annually in hospitals in the United States, with nurses representing the largest percentage of healthcare workers sustaining injuries.

Focus on HIV

The average risk of HIV transmission after a needlestick exposure to HIV-infected blood has been estimated to be approximately 0.3%.

3. Healthcare workers should take precautions to prevent injuries caused by needles, scalpels, and other sharp instruments or devices when performing procedures, cleaning used instruments, disposing of used needles, and handling sharp instruments after procedures. To prevent needlestick injuries, needles should not be recapped, purposely bent or broken by hand, removed from disposable syringes, or otherwise manipulated by hand. After they are used, disposable syringes and needles, scalpel blades, and other sharp items should be placed in puncture-resistant containers for disposal; the puncture-resistant containers should be located as close as practical to the use area. Large-bore reusable needles should be placed in a puncture-resistant container for transport to the reprocessing area. The container should be placed out of reach of small children.

4. To minimize the occurrence of mouth-to-mouth resuscitation, mouthpieces, resuscitation bags, or other ventilation devices should be available for use in areas in which the need for emergency resuscitation is predictable.

5. Healthcare workers who have open lesions or fluid-yielding skin inflammations should restrain from all direct patient care or handling of patient-care equipment until the condition resolves.

6. Pregnant healthcare workers should be especially familiar with the precautions and adhere to them strictly to minimize the chances of HIV transmission.

The universal precautions promulgated by the CDC are intended to supplement (but not replace) recommendations for routine infection control, such as hand washing and glove use to prevent microbial contamination of the hands. Epidemiologists at the CDC point out that specifying the type of barrier protection for every clinical situation is impractical, and thus, healthcare workers must exercise their own good judgment.

Invasive Procedures

Invasive procedures are those that involve surgical entry into the tissues, cavities, or organs of a patient. They also involve repair of major traumatic injuries in such areas as the delivery room, the emergency room, or the physician's or dentist's office (**Figure 6.7**). In such a case, universal precautions are advised, with the addition of the following precautions:

1. All healthcare workers who participate in invasive procedures must routinely use appropriate barrier precautions to prevent skin and mucous membrane contact with blood and other body fluids of all patients. Gloves and surgical masks must be worn for all invasive procedures. Protective eyewear or face shields should be worn for procedures that commonly result in the generation of droplets, splashing of blood or other body fluids, or the generation of bone chips. Gowns or aprons made of materials that provide an effective barrier should be worn during invasive procedures that are likely to result in the splashing of blood or other body fluids. All healthcare workers who perform or assist in vaginal or cesarean deliveries should wear gloves and gowns when handling the placenta or the infant until blood and amniotic fluid have been removed from the infant's skin and should wear gloves during postdelivery care of the umbilical cord.

2. If a glove is torn or a needlestick or other injury occurs, the glove should be removed and a new glove used as promptly as patient safety permits; the needle or instrument involved in the incident should also be removed from the sterile field.

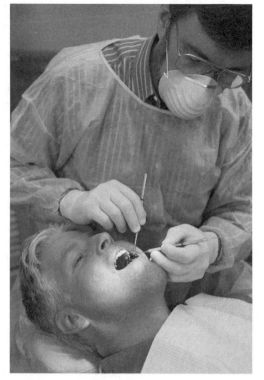

FIGURE 6.7
Among the chief recommendations in the universal precautions is the use of barrier precautions when participating in invasive procedures such as surgery and dentistry. Gloves, gowns, and masks are essential barrier precautions. © Photodisc.

Dentistry

Epidemiologists at the CDC have recommended that special precautions be implemented for dentistry because contamination of the saliva with blood from the gums is predictable. Moreover, trauma to the hands of dental professionals is common, and blood spattering may occur. Infection control procedures also protect the patient's oral membranes from possible exposure to blood from breaks in the skin of the dental worker's hands. Among the special precautions for dentistry are the following:

1. All dental workers should wear gloves for contact with oral mucous membranes of all patients and should wear surgical masks and protective eyewear or chin-length plastic face shields during dental procedures in which splashing or spattering of blood, saliva, or gingival fluid is likely. Rubber dams, high-speed evacuation, and proper patient positioning, when appropriate, should be used to minimize generation of droplets and splatter.

2. Handpieces should be sterilized after use with each patient since blood, saliva, or gingival fluid may be aspirated into the handpiece or waterline. Handpieces that cannot be sterilized should be flushed; the outside surface should be cleaned and wiped with a suitable chemical germicide and then rinsed. Handpieces should be flushed at the beginning of the day and after use with each patient. Manufacturers' recommendations should be followed for use and maintenance of waterlines and check valves and for flushing of handpieces. The same precautions should be used for ultrasonic scalers and air/water syringes.

3. Blood and saliva should be thoroughly and carefully cleaned from material that has been used in the mouth (e.g., impression materials, bite registration), especially before polishing and grinding intraoral devices. Contaminated materials,

impressions, and intraoral devices should also be cleaned and disinfected before being handled in the dental laboratory and before they are placed in the patient's mouth.

4. Dental equipment and surfaces that are difficult to disinfect (e.g., light handles or x-ray unit heads) and that may become contaminated should be wrapped with impervious-back paper, aluminum foil, or clear plastic wrap. The coverings should be removed and discarded, and clean coverings should be put in place after use with each patient.

As a general principle, the CDC recommends that dental workers consider the blood, saliva, and gingival fluid from all patients to be potentially infectious. The special precautions are formulated on this basis.

That dental practices can transmit HIV was vividly demonstrated in the early 1990s when six patients of a Florida dentist were identified with HIV infection. Investigators found that all six patients had the same strain of HIV as the dentist, and epidemiologists could not identify any risk factors (such as injection drug use or heterosexual contact with an infected individual) that could account for HIV transmission. Five of the patients had root canal therapy or dental extraction. One patient, Kimberly Bergalis, was the first to link her infection to treatment via a series of letters to the Florida Health Department. Bergalis died in 1992; two years before, the dentist had died. Three other cases of transmission of HIV from a healthcare worker to a patient were reported; a nurse in France, an orthopedic surgeon in France, and an obstetrician in Spain.

Until December 2002, there have 344 published cases worldwide in which healthcare workers were infected with HIV as a result of their profession. Of these, 106 were documented to result from occupational exposure. For dental professionals, eight possible occupationally acquired HIV infections were published, although no cases were confirmed.

Autopsy and Embalming

Autopsy and embalming procedures carry special risk because of the workers' exposure to HIV-infected tissue of internal organs. For all persons performing or assisting in postmortem procedures, the CDC recommends adherence to the universal precautions, with special reference to the wearing of gloves, masks, protective eyewear, gowns, and waterproof aprons. Particular attention should be paid to decontamination of all instruments and surfaces with appropriate chemical germicides both during and after postmortem procedures.

Epidemiologists have suggested additional precautions during procedures that involve transporting and embalming the body of an HIV-positive person or AIDS patient. Funeral directors, for example, are advised to spray body wrappings liberally with disinfectant and then add a layer of plastic wrapping or place the body in a pouch. At the funeral home, disinfectant should be sprayed and swabbed on the entire body, especially the orifices and genital organs. A stronger-than-normal solution of embalming fluid is recommended, and the drainage vessel should remain closed for an extended time to ensure that the fluid has contacted any viral particles in the blood. During drainage, care should be taken to avoid blood splash on the embalming table and, if possible, in the waste sink. Before treating the body cavity, epidemiologists suggest that several hours pass to maintain embalming fluid pressure within the blood vessels and keep the fluid within the blood. When the cavity is treated, the use of an extra measure of fluid is recommended in both thoracic and abdominal cavities. Extreme caution must be used when handling any visceral organs or tissues, and the embalmer should exercise care to avoid abrasions and scratches from such things as cut bones of the rib cage. After suturing, all incisions should be thoroughly sealed.

Medical Laboratories

A medical laboratory is a specialized facility where urine, feces, blood, and other patient specimens are analyzed biologically and chemically by skilled technologists and technicians. Laboratory workers receive blood and other specimens on a regular basis and are expected to perform the protocols desired by attending physicians and to report the results. CDC epidemiologists recommend that laboratory workers follow the universal precautions, supplemented with additional specified precautions, as follows:

1. All specimens of blood and body fluids should be put in well-constructed containers with secure lids to prevent leaking during transport. Care should be taken when collecting each specimen to avoid contaminating the outside of the container and the laboratory from accompanying the specimen.
2. All persons processing blood and body fluid specimens (such as removing tops and vacuum tubes) should wear gloves. Masks and protective eyewear should be worn if mucous membrane contact with blood or body fluids is anticipated. Gloves should be changed and hands washed after completion of specimen processing.
3. For routine procedures, such as histologic and pathologic studies or microbiologic culturing, a biological safety cabinet is not necessary; however, biological safety cabinets (Class I or Class II) should be used whenever conducting procedures that have a high potential for generating airborne droplets. These include activities such as blending, treating with ultrasonic vibrations, and vigorous mixing.
4. Mechanical pipetting devices should be used for manipulating all liquids in the laboratory. Mouth pipetting must not be done.
5. The use of needles and syringes should be limited to situations in which there is no alternative, and the recommendations for preventing injuries with needles outlined under universal precautions should be followed.
6. Laboratory work surfaces should be decontaminated with an appropriate chemical germicide after a spill of blood or other body fluids and when work activities are completed.
7. Contaminated materials used in laboratory tests should be decontaminated before preprocessing or placed in bags and disposed of in accordance with institutional policies for disposal of infective waste.
8. Scientific equipment that has been contaminated with blood or other body fluids should be decontaminated and cleaned before being repaired in the laboratory or transported to the manufacturer.
9. All persons should wash their hands after completing laboratory activities and should remove protective clothing before leaving the laboratory (**Figure 6.8**).

Decontamination Procedures

In the vast majority of healthcare settings, the standard sterilization and disinfection procedures are adequate for decontaminating patient care equipment. Because HIV is fragile, it is inactivated rapidly after exposure to commonly used chemical germicides, even when the germicides are used in lower concentrations than normal.

Hand washing is one of the simplest yet most effective methods of preventing the spread of HIV. Any healthcare activity should always be followed by hand washing, regardless of how major or minor the activity has been. If there were a single rallying cry among public health officials attempting to stem the spread of AIDS and other diseases, it would be this: "Wash your hands!" Courtesy of Kimberly Smith/CDC.

In addition to commercially available germicides, a solution of household bleach (sodium hypochlorite) is an inexpensive and effective germicide. The solution should be prepared daily in a concentration from 10% (one part bleach to nine parts water) to 1% (one part bleach to 99 parts water). The concentration depends on how much blood, mucus, or other organic material is present—the higher concentration for the higher measure of contamination. Unfortunately, many surfaces and instruments are corroded by bleach solutions, and thus, commercially available germicides may be preferable.

Extensive studies of the survival of HIV in the healthcare environment indicate that the virus rapidly becomes inactive on exposure to the air. Of course, conditions vary (e.g., temperature, pH, amount of blood present, number of viruses present), but the general principle is that no changes in sterilization, disinfection, or housekeeping procedures are warranted when HIV is present. Standard decontamination procedures assume "worst case" conditions of extreme viral, bacterial, or other microbial contamination; therefore, extraordinary attempts to eliminate HIV are not required. For example, surfaces in patient care settings are usually cleaned on a regular basis as well as when spills occur. These decontamination procedures help prevent HIV transmission. **Table 6.4** summarizes the methods for preparing equipment used in healthcare settings to limit HIV contamination.

For decontaminating spills of blood and other body fluids, a chemical germicide approved for use as a "hospital disinfectant" is recommended. The germicide should be applied to the spill by first surrounding the spill and then working in toward the center. Paper toweling should be placed on top of the liquid, and a period of several minutes should elapse before the area is wiped clean to give the chemical time to react with the microorganisms. Fresh germicide should be used to treat the area a second time. Gloves should be worn at all times. For laundry, normal hygienic and commonsense procedures should be followed when dealing with soiled linens. Linen, for example, should be bagged where it was used, and bags that prevent leakage should be used. Similarly, infected gauze pads and other forms of waste that may contain HIV require no special procedure but should be handled as they would be to deal with any infectious microorganisms.

TABLE 6.4	Suggested Methods for Processing Patient Care Equipment in Healthcare Settings		
Sterilization:	Destroys:	All forms of microbial life, including high numbers of bacterial spores.	
	Methods:	Steam under pressure (autoclave), gas (ethylene oxide), dry heat, or immersion in EPA-approved chemical "sterilant" for prolonged period of time, e.g., 6–10 hours or according to manufacturers' instructions. Note: liquid chemical "sterilants" should be used only on those instruments that are impossible to sterilize or disinfect with heat.	
	Use:	For those instruments or devices that penetrate skin or contact normally sterile areas of the body, e.g., scalpels and needles. Disposable invasive equipment eliminates the need to reprocess these types of items. When indicated, however, arrangements should be made with a healthcare facility for reprocessing of reusable invasive instruments.	
High-Level Disinfection	Destroys:	All forms of microbial life **except** high numbers of bacterial spores.	
	Methods:	Hot water pasteurization (80–100°C, 30 minutes) or exposure to an EPA-registered "sterilant" chemical as above, except for a short exposure time (10–45 minutes or as directed by the manufacturer).	
	Use:	For reusable instruments or devices that come into contact with mucous membranes (e.g., laryngoscope blades, endotracheal tubes).	
Intermediate-Level Disinfection:	Destroys:	*Mycobacterium tuberculosis,* vegetative bacteria, most viruses, and most fungi, but does **not** kill bacterial spores.	
	Methods:	EPA-registered "hospital disinfectant" chemical germicides that have a label claim for tuberculocidal activity; commercially available hard-surface germicides or solutions containing at least 500 ppm free available chlorine (a 1:100 dilution of common household bleach—approximately $1/4$ cup bleach per gallon of tap water).	
	Use:	For those surfaces that come into contact only with intact skin, e.g., stethoscopes, blood pressure cuffs, and splints, and have been visibly contaminated with blood or bloody body fluids. Surfaces must be precleaned of visible material before the germicidal chemical is applied for disinfection.	
Low-Level Disinfection	Destroys:	Most bacteria, some viruses, some fungi, but not *Mycobacterium tuberculosis* or bacterial spores.	
	Methods:	EPA-registered "hospital disinfectants" (**no** label claim for tuberculocidal activity).	
	Use:	These agents are excellent cleaners and can be used for routine housekeeping or removal of soiling in the **absence** of visible blood contamination.	
Environmental Disinfection:		Environmental surfaces that have become soiled should be cleaned and disinfected using any cleaner or disinfectant agent that is intended for environmental use. Such surfaces include floors, woodwork, ambulance seats, countertops, etc.	

IMPORTANT: To assure the effectiveness of any sterilization or disinfection process, equipment and instruments must first be thoroughly cleaned of all visible soil.

Source: Guidelines for Environmental Infection Control in Health-Care Facilities: Recommendations of CDC and the Healthcare Infection Practices Advisory Committee (HICPAC). 2003.

Exposures to HIV

Several medical centers have conducted ongoing studies of healthcare workers exposed to possible infection through contact with blood or body fluids from AIDS patients. In virtually all of these studies, the risk of transmission was found to be extremely low. For instance, researchers at the NIH tested 983 healthcare workers who were exposed to possible HIV infection through contact with patients or infected materials: 137 had suffered needlestick injuries; 345 had experienced mucous membrane exposure; the remainder had miscellaneous exposures. None of the workers displayed evidence of HIV infection six months after exposure. Another study was conducted at the University of California, where 212 workers were tested after contact with infected blood; after six months, only one person had evidence of HIV infection.

One of the most comprehensive and long-range of such studies was performed by CDC researchers beginning on August 15, 1983, and terminating on April 20, 1989. In the study, 1,449 healthcare workers were followed for a minimum of one year and were tested for evidence of HIV exposure at intervals of six weeks and three, six, and 12 months after exposure. The workers came from throughout the United States and had suffered needlestick injuries (80%), cuts with sharp objects (8%), contamination of open wounds (7%), and mucous membrane exposure (5%). They suffered exposures in different locations in the healthcare facilities and under different working conditions (**Figure 6.9**). Of 1,172 healthcare workers with complete data as of April 1989, only four developed HIV infection, a total of less than 1% (Figure 6.9).

Continuing monitoring of these workers and additional workers with new exposures has been conducted by the National Institute for Occupational Safety (NIOS). Today the estimated risk of HIV infection from sharps is 1 in 300 (0.3%). The CDC reported 57 documented cases and 140 possible cases of HIV transmission to U.S. healthcare workers between 1981 and 2006. Of these, 48 out of 57 cases were associated with a puncture/cut injury (often referred to as a **percutaneous injury**). According to the NIOS 2004 report, of those exposures that convert to HIV infection, nurses continue to be the group of healthcare workers at greatest risk (24% of infection) for exposure to HIV infected body fluids, with laboratory workers coming in second with 19% of work-related infections (Figure 6.9).

In its surveillance report on AIDS for 1995, the CDC summarized the number of cases of HIV infection and AIDS occurring in healthcare workers as a result of occupational exposure. As of December 31, 1994, over 440,000 cases of AIDS had been reported in the United States, and 42 healthcare workers (less than one hundredth of 1% of the total) were infected in documented occupational transmission. Another 91 workers (two hundredths of 1% of the total) were possibly infected during their work. Fifteen of the 42 cases occurred in laboratory technicians and 13 in nurses, although Figure 6.8 shows that by 2004, the percentage of work-related infections was highest in nurses.

A CDC report issued in June 2001 noted that the risk of HIV transmission through contact among healthcare workers was about 0.3%; through mucous membrane exposure, it was about 0.09%. The CDC had also received reports of 56 healthcare workers who contracted HIV while at work and 138 who may have done so through occupational exposure. Data such as these indicate that the risk of HIV transmission to healthcare workers during the normal course of their duties is extremely low; however, the findings do not justify a cavalier disregard of prudent precautions. This point was driven home when CDC officials reported three cases of HIV transmission via skin exposure to blood. In one case, a hospital worker was exposed to blood when she pressed gauze against the arm of an AIDS patient who was bleeding. In the second case, blood spilled onto the arms and hands of a worker

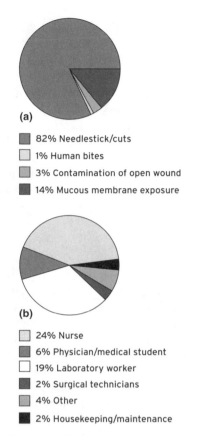

(a)

- ■ 82% Needlestick/cuts
- □ 1% Human bites
- ■ 3% Contamination of open wound
- ■ 14% Mucous membrane exposure

(b)

- □ 24% Nurse
- ■ 6% Physician/medical student
- □ 19% Laboratory worker
- ■ 2% Surgical technicians
- □ 4% Other
- ■ 2% Housekeeping/maintenance

(c)

Case of HIV Infection Resulting from Exposure

Type of Exposure	Number Tested	Number Positive	Percentage of Positives
Needlestick or cut with object	1,031	4	0.39
Mucous membrane or wound contamination	141	0	0.00
Total	1,172	4	0.34

FIGURE 6.9

Risk of healthcare workers (HCW) for work-related infection with HIV. (a) Percentages of HCW exposures to patient body fluid by type of exposure. (b) Percentages of HCW exposures by occupation of the worker that have led to HIV infection, 1981–2002. (c) Results of a CDC study of 1,449 HCWs exposed to patients infected with HIV. The study was conducted between August 15, 1983 and April 20, 1989. Among 1,449 exposures, only four workers acquired HIV infection. Studies like this demonstrate that the risk of HIV transfer during the normal course of duties is extremely low. *Source:* CDC, *NIOSH and Health, Worker Health Chartbook*, 2004.

manipulating a machine used to separate blood components. In both of these cases, workers had skin lesions or chapped skin, and neither was wearing gloves. In the third case, a rubber stopper popped off a glass tube, splattering blood onto the skin and into the mouth of a worker. CDC recommendations for reporting an occupational exposure are listed in **Table 6.5.**

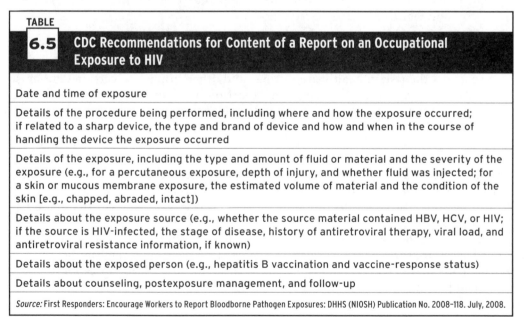

TABLE

6.5 **CDC Recommendations for Content of a Report on an Occupational Exposure to HIV**

Date and time of exposure
Details of the procedure being performed, including where and how the exposure occurred; if related to a sharp device, the type and brand of device and how and when in the course of handling the device the exposure occurred
Details of the exposure, including the type and amount of fluid or material and the severity of the exposure (e.g., for a percutaneous exposure, depth of injury, and whether fluid was injected; for a skin or mucous membrane exposure, the estimated volume of material and the condition of the skin [e.g., chapped, abraded, intact])
Details about the exposure source (e.g., whether the source material contained HBV, HCV, or HIV; if the source is HIV-infected, the stage of disease, history of antiretroviral therapy, viral load, and antiretroviral resistance information, if known)
Details about the exposed person (e.g., hepatitis B vaccination and vaccine-response status)
Details about counseling, postexposure management, and follow-up
Source: First Responders: Encourage Workers to Report Bloodborne Pathogen Exposures: DHHS (NIOSH) Publication No. 2008–118. July, 2008.

BOX
6.2

"An Ounce of Prevention . . . "

The report of HIV infection or AIDS in a fellow worker at a healthcare facility can be an unsettling experience. Other employees will need reassurance that the workplace is safe and that the infection was probably acquired in the worker's private life. Still, some fear may remain that the workplace is dangerous, especially if it is a healthcare facility in which AIDS patients receive treatment. To minimize this concern and to reduce the risk of infection, a plan can be implemented to deal with potential or actual HIV infection among employees. The plan could include the following:

Education
Guest speakers, including specialists at the facility, can be invited to address healthcare workers on the various ramifications of AIDS. Reading materials can be supplied on the cause, transmission, diagnosis, and treatment of AIDS with emphasis on precautions taken while handling blood-contaminated materials. Besides reassuring employees, the sessions demonstrate management's concern about their well-being and its willingness to confront AIDS-related issues.

Protective Materials
Management can provide protective supplies to further demonstrate its commitment to employee safety via materials expenditures. Needle drop boxes and gloves (in small, medium, and large) can be placed in all work areas, and masks, goggles, and gowns can be made available. Containers of 10% bleach solution can be placed throughout the facility in case of spills. And employees can be taught precautionary safeguards such as glove replacement and how to clean up a spill safely.

Precautions Policy
A policy can be developed that stresses prevention first, but also guidance and support when workers contract or believe they may have contracted an infectious disease. Incorporation of employee concerns, legal considerations, and suggestions from human resource groups is essential to such policy development. The policy should include such issues as AIDS education, how employees who test positive for HIV will be evaluated, precaution against infection, insurance coverage, and rights to privacy.

Working with AIDS patients is often a stressful situation, but healthcare workers can act more intelligently if they know enough to protect themselves. Both responsible management and employee compliance can contribute to stress reduction.

To assess the risk in healthcare settings and other situations, certain data must be available. For instance, the geographic location in which the incident took place should be known. (The chances of encountering contaminated blood are higher in urban San Francisco than in a rural Iowa community.) One should ask how exposure took place. (Airborne blood splashes are much less likely to transmit HIV than needlestick injuries.) What volume of blood was transmitted? (Was the skin scratched with blood present, or were there a deep penetration wound and a considerable amount of blood?) How much virus did the blood contain? (Blood from a person with HIV infection probably contains less of the virus than blood from a person with AIDS.) Data such as these lend individuality to each case and help determine the likelihood that the worker has been infected. The aforementioned cases are rare, but they illustrate the need for caution in the healthcare setting (**Box 6.2**). The take-home message is brief and straightforward: Observe universal precautions and be careful!

Protecting Public Safety Workers

The general infection-control procedures outlined by the universal precautions apply not only to healthcare settings but also to any environment where workers contact other individuals and where microbial transmission can occur. Public safety workers perform their assignments in such environments. They hold positions as emergency medical technicians, firefighters, and law enforcement and correctional facility officers.

Working in the public safety sector carries some risk of exposure to HIV. Exposure can be unpredictable, and thus, protective measures should be used, even when the risk is not obvious. Because public safety workers perform their duties under extremely variable conditions, control measures must necessarily be simple and uniform to ensure compliance.

Fire and Emergency Medical Services

Among the workers considered in the general category of fire and emergency medical services are firefighters, paramedics, emergency medical technicians, and advanced-life support personnel. Individuals in these professions perform their duties in pre-hospital, uncontrolled environments where time for decision making is extremely limited. Recommendations are made with these conditions in mind and under the basic premise that workers must be protected from exposure to blood and other body fluids that are potentially infectious. Accordingly, CDC epidemiologists recommend the following:

1. Disposable gloves should be a standard component of emergency response equipment and should be put on by all personnel prior to initiating any emergency care tasks involving exposure to blood or other body fluids to which universal precautions apply. Extra pairs should always be available. Considerations in the choice of disposable gloves should include dexterity, durability, fit, and the task being performed. For situations where large amounts of blood are likely to be encountered, gloves should fit tightly at the wrist to prevent contamination of hands by blood seepage around the cuff. For multiple trauma victims, gloves should be changed between patient contacts, if the emergency situation allows. More extensive personal protective equipment measures are indicated for situations in which broken glass and sharp edges are likely to be encountered, such as when extricating a person from an automobile wreck.

2. Masks, eyewear, and gowns should be carried in all emergency vehicles that respond to medical emergencies or victim rescues. These protective barriers should be used in accordance with the level of exposure encountered. Minor lacerations or small amounts of blood do not require the same barrier use as is required with victims who are bleeding heavily, either externally or internally. Masks and eyewear (including safety glasses) should be worn together, or face shields should be used. These should be put on by all personnel before beginning work in any situation where splashes of blood or other body fluids are likely to occur. If large splashes or quantities of blood are present or anticipated, impervious gowns or aprons should be worn. An extra change of work clothing should be available at all times.

3. For artificial ventilation of trauma victims, disposable airway equipment or resuscitation bags should be used. Disposable resuscitation equipment and devices should be used only once. If reusable, they should be thoroughly cleaned and disinfected

6.6 Examples of Recommended Personal Protective Equipment for Worker Protection Against HIV Transmission in Prehospital Settings*

Task or Activity	Disposable Gloves	Gown	Mask	Protective Eyewear
Bleeding control with spurting blood	Yes	Yes	Yes	Yes
Bleeding control with minimal bleeding	Yes	No	No	No
Emergency childbirth	Yes	Yes	Yes, if splashing is likely	Yes, if splashing is likely
Blood drawing	At certain times	No	No	No
Starting an intravenous (IV) line	Yes	No	No	No
Endotracheal intubation, esophageal obturator use	Yes	No	No, unless splashing is likely	No, unless splashing is likely
Oral/nasal suctioning, manually cleaning airway	Yes	No	No, unless splashing is likely	No, unless splashing is likely
Handling and cleaning instruments with microbial contamination	Yes	No, unless soiling is likely	No	No
Measuring blood pressure	No	No	No	No
Measuring temperature	No	No	No	No
Giving an injection	No	No	No	No

*Defined as settings where delivery of emergency health care takes place before arrival at hospital or other healthcare facility.

Source: CDC, 1989. Guidelines for the Prevention of Transmission of Human Immunodeficiency Virus and Hepatitis B Virus to Health-Care and Public-Safety Workers. *MMWR* 38: No. S-6.

after each use, according to the manufacturer's recommendations. In addition, mechanical respiratory assist devices (e.g., bag-valve masks, oxygen-demand valve resuscitators) should be available on all emergency vehicles and to all emergency response personnel during any medical emergency or victim rescue. Pocket mouth-to-mouth resuscitation masks designed to isolate emergency response personnel (the so-called double lumen systems) from contact with victims' blood and blood-contaminated saliva, respiratory secretions, and vomitus should be available to all personnel who provide emergency treatment. **Table 6.6** summarizes recommended personal equipment that should be used for protection in the prehospital setting.

Law Enforcement Officers

Law enforcement officers risk exposure to HIV if they encounter blood while performing their duties. At a crime scene, for instance, police officers may need to handle blood-contaminated materials or help remove a body. In correctional facilities, officers may encounter contaminated syringes and needles during searches, or they may have to subdue disruptive inmates. Epidemiologists make certain recommendations for law enforcement officers to reduce the risk of acquiring HIV in the normal course of a day's events. They point out, however, that an extremely diverse range of potential situations exists, and the informed judgment of the officer is paramount when unusual events present themselves. Among the guidelines are the following:

1. Law enforcement officers are exposed to a range of fights and assaults, during which they may be exposed to blood or blood-contaminated materials. In cases such as these, appropriate protection should be worn if feasible and as conditions permit. For instance, gloves should be put on and a change of clothing should be available if blood is splattered.
2. Exposure to HIV may occur during cardiopulmonary resuscitation (CPR), and thus, protective masks and airways should be available to officers and training should be given in their use.
3. During searches and evidence handling, an officer may come in contact with HIV, especially through a puncture wound with a blood-contaminated needle; therefore, individual discretion should be used to determine if a prisoner should empty his or her own pockets, and protective gloves should be worn for all body cavity searches. Long-handled mirrors and flashlights should be used to search hidden areas such as above ceilings or under car seats. Purses should be searched by turning them over and emptying the contents onto a table. Puncture-proof containers should be available for sharp objects.

Prison populations in all states include individuals at high risk for AIDS. For example, injection drug users often continue their drug use while in prison, and illegal tattoo machines are available among prisoners. Homosexual practices also occur in prisons. Corrections officers thus can come in contact with HIV-infected persons on a regular basis and should take appropriate precautions.

An interesting dilemma has arisen regarding the education of inmates. In the outside community, risk-reduction measures for HIV exposure are emphasized, but such measures are not feasible in prisons. Condom use, for example, is strongly advised in the community, but within prison confines, homosexual practices are prohibited and punished, and thus, the risk reduction afforded by the use of condoms cannot be encouraged. Similarly, needle disinfection is advocated in the community, but drug use in prison is not tolerated; thus, risk-reduction measures for needle users are not promulgated. The discrepancies between recommended public health measures and prison practices have evoked legal issues. For example, inmates argue that they should be taught how to avoid the risky consequences of their behavior, but prison officials often disagree. These questions are currently being tested in the court system.

Other Transmission Methods

Because HIV is present in blood and semen and, in some cases, in very low levels in saliva or other body fluids, it is remotely possible that one could acquire HIV by methods other than anal or vaginal intercourse or sharing of contaminated needles. For example, any instrument coming in contact with blood or body fluid can be a potential source of HIV if very large numbers of viruses are present and if large amounts of fluid are transmitted, among other factors. Acupuncture needles are an example of such an instrument, as are tattoo needles, razors, and implements used to puncture the ears during ear piercing; however, the risk of contracting HIV from these instruments is extraordinarily low because many concurrent factors, such as viral numbers, volume of fluid, health condition of the recipient, and inoculation site, must be taken into consideration.

For HIV transmission, ordinary social kissing is not considered a high-risk behavior, but the risk increases with open-mouthed, or "French," kissing. This is because of the considerable amounts of saliva exchanged. The risk increases still further if the uninfected individual has lesions or sores of the mouth (e.g., canker sores); it is important to note, however, the fragility of the virus argues against transmission by deep kissing. (Laboratory experiments indicate that HIV survives very poorly in saliva.) In addition, the level of HIV in saliva is extremely low: In one research study, only 1 of 83 HIV-infected individuals had HIV in the saliva.

Oral–genital contact may be a risky practice, depending on the circumstances. If there are wounds, abrasions, or lesions of the mouth, for example, free viruses or HIV-infected T-lymphocytes from the semen may enter the circulatory system. Few researchers would argue that HIV cannot be spread through oral–genital contact, but the actual risk has not yet been defined. Part of the reason is that most studies contain too few patients to yield truly convincing data, and these studies were rarely designed to assess the actual time of HIV infection. There is also the daunting task of evaluating the risk of a single sexual act in the context of an individual's entire sex history.

Improbable Transmissions

Epidemiologists have established that a multitude of common activities do not allow HIV to pass among individuals because such activities do not involve contact with blood or semen or involve maternal–fetal transfer (**Figure 6.10**). Most of these activities are listed under the broad category of casual contact. Social kissing is included here, as are shaking hands, hugging, and other forms of physical contact. HIV is not transmitted by contacting bodily secretions such as sweat, mucus (as in sneezing or coughing), or saliva (as in sharing eating utensils or a can of soda). One does not acquire HIV by using a telephone, toilet seat, or soap bar used by an AIDS patient. The virus, if once present, has probably disintegrated; even if it were still there, no opportunity exists for penetrating the recipient's bloodstream.

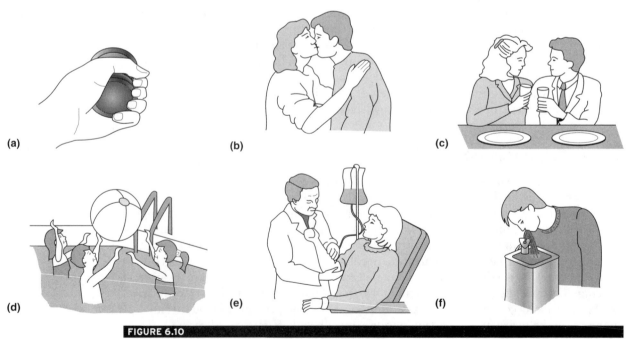

(a)
(b)
(c)
(d)
(e)
(f)

FIGURE 6.10

Methods that are not known to transmit human immunodeficiency virus. (a) Touching a doorknob; (b) a social kiss; (c) eating at a restaurant; (d) swimming at a pool; (e) donating blood; (f) drinking at a water fountain.

The list goes on. HIV is not contracted in restaurants or eating facilities or from an infected chef or waiter (unless, of course, one shared contaminated needles or had sexual contact with that individual). HIV does not remain active in the water of a hot tub, on the surface of a deodorant stick, or on the metal of a doorknob. Swimming pools do not transmit AIDS, nor do laundry facilities, drinking fountains, or public restrooms. It is not likely that one could become infected by donating blood, by caring for an AIDS patient, or by coming in contact with a child who has AIDS. Even people who touch the blood of an AIDS patient are at minimal risk. As one prominent researcher has suggested, "You need quite a slug of blood in your veins to become infected."

It should be understood that each case is unique and that making blanket statements is not advised. Based on all the best evidence to date, however, there is nothing to indicate that HIV can be transmitted by any of the aforementioned ways.

Rare Routes of Transmission

Organ donors are routinely screened for HIV infection in the United States. From 1994 to 2007, nearly 300,000 transplants were performed without any cases of HIV transmission. In 2009, the first documented case of HIV transmission through organ transplantation occurred in a New York City hospital. A living donor tested HIV negative 10 weeks before donating a kidney. Unfortunately the donor contracted HIV via sexual transmission some time during the 10-week period prior to surgery and the recipient contracted HIV. This rare occurrence prompted the CDC to recommend that living donors be rescreened by two different methods no longer than 7 days before surgery.

Three unusual cases of HIV transmission in African-American children aged 9 months to 39 months were reported in the journal *Pediatrics* in 2009. Researchers found compelling evidence that the three children were infected with HIV from mothers or caregivers primarily through pre-chewed food. Pre-chewing food is also known as **pre-mastication.** In developing countries, dental care is lacking and commercially available baby foods and blenders are not available. Mothers and caregivers pre-masticate or pre-chew foods for infants. The mothers or caregivers were not adhering to their antiviral therapy regime and their gums were bleeding in the mothers or caregivers as well as the infants because they were teething. All of these factors increased the risk for HIV transmission. This was the first report of HIV transmission through pre-mastication. Albeit it rare, the CDC advises against this practice for HIV-infected caregivers or parents.

The Mosquito Myth

Throughout recorded history, mosquitoes and other arthropods have been responsible for spreading many epidemics. Plague was spread by fleas, typhus by lice, and malaria by mosquitoes. In addition, ticks transmit the agents of Lyme disease and Rocky Mountain spotted fever.

In all of these instances, the disease organisms greatly increase their numbers within the arthropod and then exit from it efficiently. For malaria and yellow fever, the respective parasites (a protozoan and a virus) multiply in the mosquito and then concentrate in its salivary gland. When the mosquito takes its next blood meal, the parasites pass with its saliva into the victim's blood.

No such model exists for AIDS. Since AIDS has been known in the United States, researchers have been unable to locate multiplying HIV in mosquitoes or other arthropods, nor have they observed the viruses in a place where they could exit the arthropod host. Mosquitoes suck blood—they do not inject blood from one person into another. To be sure, residual blood might remain after a bite on a mosquito's needle-like injecting tube (the proboscis), but entomologists point out that a mosquito continually washes its proboscis with saliva, thereby cleansing it. Moreover, the amount of blood on the proboscis is roughly a thousandth of what might be encountered in a needlestick.

Additional evidence that no animal intermediate is involved in HIV transmission comes from Africa. It is well known that African children play outside in mosquito-infested areas. Children would constitute a large percentage of African AIDS cases if the disease were mosquito-borne. Statistics show, however, that children represent a relatively small percentage of Africans with AIDS. Many African children suffer from malaria, but comparatively few suffer from AIDS.

Scientists have shown that HIV can remain active within mosquitoes for a period of several hours after they have been fed blood with a high concentration of the virus. This observation, however, does not support the notion that mosquitoes transmit HIV in nature because the virus neither multiplies within a mosquito nor concentrates at a point where it can leave the mosquito. Although these activities occur with many microbial agents, they do not appear to do so for HIV. Barring some fundamental biochemical change in the virus or the mosquito, an HIV–mosquito connection is unlikely to emerge in future years.

AIDS Education in the Schools

Public health experts generally agree that education is the most valuable tool against the spread of HIV infection and AIDS, but they advise that an AIDS education program must be carefully constructed. It must, for example, deal with the public health threat of AIDS while taking positive health behavior into account. In addition, it must consider an array of moral, religious, and legal values.

Education programs can be effective in preventing the spread of HIV because the virus is transmitted almost exclusively by behaviors that individuals can modify. With this in mind, the CDC has promulgated a number of guidelines for school health education to prevent the spread of AIDS. It stipulated, however, that the specific scope and content of AIDS education in schools should be determined locally and should be consistent with parental and community values. Essentially, the proposed curriculum was designed to ensure that students understand the nature of the AIDS epidemic and the specific actions they can take to prevent HIV infection.

The CDC guidelines advocate that school systems advise students to abstain from sexual intercourse and intravenous drug use as ways of avoiding HIV. Those students who engage in sexual intercourse or inject drugs should be encouraged not to do so. For students unwilling to adopt less risky behaviors, the school program should advise them to (1) avoid sexual intercourse with infected persons, (2) use condoms, (3) seek treatment for drug addiction, (4) avoid sharing needles, and (5) seek HIV counseling if they suspect they are infected.

In the AIDS education curriculum, students should learn the biology of AIDS as well as the emotional and social factors that influence the spread of HIV (**Figure 6.11**). Accordingly, in the early elementary grades, teachers should allay the fears of young children that they will become infected. Emphasis should be placed on the facts that

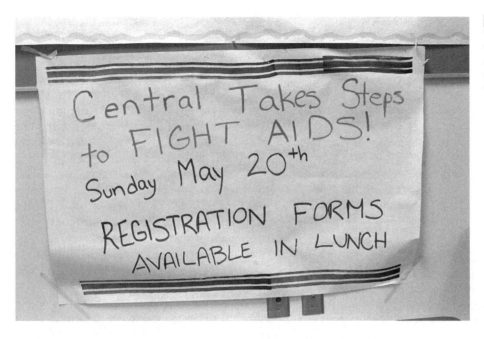

FIGURE 6.11
The breadth of AIDS education in the schools should be consider-able, especially in view of the susceptibility of students to HIV. © Steve Skjold/Alamy Images.

AIDS is hard to get, that AIDS primarily affects adults, and that researchers are actively seeking a cure for the disease.

Once students reach the late elementary and middle school grades, the curriculum could encompass discussions about the nature of viruses, including their transmission and disease potential. Brief information can be communicated about the symptoms of AIDS, the estimated infected population, the ubiquitous nature of the disease, methods for transmitting the virus, and how the virus is not transmitted.

For students at the junior and senior high school levels, the breadth of AIDS education can be considerable. The three major methods of transmission can be studied in depth, and risk factors can be explored, including how risks are increased and reduced. Protective devices for sexual intercourse and precautions for injection drug use can be discussed, and consideration for others can be emphasized. Counseling should be encouraged to increase student understanding of AIDS and to explain how testing can be performed to give students a clearer understanding of their HIV status.

Another AIDS-related educational issue in secondary school concerns the use of blood in school laboratories (**Box 6.3**). However, among the most controversial issues that secondary schools must address is whether to distribute condoms to students. Proponents point to their desire to save lives, but opponents are equally vociferous about sending the message that promiscuous sex is condoned. Proponents counter that students must have parental permission to obtain condoms and must take the initiative to get them in an appropriate place such as a nurse's office, whereas opponents counter that providing condoms will give the incorrect impression that wearing them makes sex completely safe. Proponents counter that parents can opt their children out of the program, whereas opponents ponder the psychological effects of such a decision on the child. As the debates continue, school officials continue to advocate AIDS education programs.

BOX
6.3

End of the Blood Typing Lab?

"I couldn't believe all the students in this A/P lab who took alcohol swabs and set them down on the lab tables instead of discarding them right into the hazardous waste bag... And then this one girl pricked herself with another girl's used lancet!! I asked her if she realized what she did and she said, 'Yes, but (Laura) is a good Christian girl.'"

Letter from a student

One of the outfalls of the AIDS epidemic has been a serious reconsideration of the use of human blood in the high school and college teaching laboratory, particularly in the exercise involving blood typing. In this exercise, blood is obtained from each student by pricking a fingertip with a lancet. Drops of blood are then placed on a slide and combined with anti-A, anti-B, and anti-Rh sera to determine the blood type and which antigens are present on the red blood cells.

Although it is statistically unlikely that a student is unknowingly carrying HIV, nevertheless, it is possible. If so, then HIV transmission during blood typing could conceivably occur in the following ways: Blood lancets may inadvertently be placed in wastebaskets where they pose a hazard to custodians and other students. Students may bleed excessively and deposit fresh blood on the desk or other surface where other individuals can contact it. Cotton or gauze is often used to wipe excess blood from the fingertip, and cotton or gauze dropped in a wastebasket can be dangerous to others. Slides containing blood may be dropped and broken, and the broken pieces may pierce the skin and inoculate HIV if present. The procedure of slide cleaning may bring the technical staff in contact with contaminated blood, and if skin blisters or cuts are present, HIV transmission could occur.

For teachers and professors who consider it essential to demonstrate the principles of blood typing, alternative methods may suffice. For example, screened blood from a local blood bank may be employed. Computer simulations are also available, and video software can be purchased for showing in class. Disposable blood typing cards can be obtained from biological supply companies to eliminate slide washing at the end of the exercise. Regardless of whether instructors choose to discontinue blood typing or be more cautious than in the past, they should at the very least consider the value of reducing the risk of HIV transmission.

It bears mention that any AIDS education campaign must be carefully thought out, as illustrated by a serious mistake made in the mid-1980s in Mexico. A government campaign was warning teens to avoid casual sex, but it also reassured them that casual contact would not spread AIDS. The campaign did not seem to be working until a psychologist found out why: In the local vernacular, "casual sex" and "casual contact" were both interpreted to mean the same thing—sex without commitment. What had begun as a campaign to stop AIDS instead was helping to spread it. The lesson was clear: Halting the AIDS epidemic requires attention to local customs, traditions, and language.

LOOKING BACK

Preventing the spread of AIDS is a task that can be accomplished by individuals at risk for acquiring HIV. Individuals, for example, can interrupt the passage of HIV by avoiding sexual intercourse with multiple partners and by not using injection drugs. Those who continue to have multiple sexual partners should use devices such as condoms to form a barrier between semen and the tissues of the sexual partner. Condoms protect both the wearer and his sexual partner.

Safer sex practices can also be used to prevent semen passage and, by inference, HIV passage. Activities taking place outside the body are considered safer sex practices. Microbiocides are under development for women as vaginal foams and gels that will target/inhibit HIV replication. Those who use injection drugs can interrupt the spread of HIV by disinfecting drug paraphernalia with a chemical such as bleach and by refusing to share drug paraphernalia. Needle exchange programs are also an efficient way of reducing HIV transfer.

To interrupt the spread of HIV among healthcare workers, the CDC has promulgated universal precautions that all such workers should observe. Barrier measures, hand washing, injury precautions, and care for abraded skin are among the key recommendations. Special precautions are also promulgated for invasive procedures, dentistry, autopsy and embalming, and medical laboratories.

Some risk of HIV transmission exists for public safety workers, especially when contact is made with blood. Risk reduction recommendations are therefore made by the CDC for firefighters, emergency medical personnel, and law enforcement and correctional officers. Gloves and other barrier protections are suggested. Informed judgment and common sense are paramount when unusual events present themselves. Although rare, HIV has been transmitted through organ transplantation and pre-mastication.

One method for preventing the spread of AIDS is through education in the schools, beginning in the elementary grades and proceeding through the high school years. Student surveys indicate a desire to know about AIDS and reveal certain misperceptions that require correction. Among the more controversial issues facing secondary schools is whether to distribute condoms to students.

Healthline Q & A

Q1 Should women carry condoms?

A AIDS poses a risk to women who are exposed to the infected semen of men. It is therefore in a woman's own best interest to insist that a man use a condom. Women who anticipate having sexual intercourse should carry condoms. They should also investigate female condoms as a method of protection.

Q2 If a couple has been engaging in unprotected sexual intercourse, is it too late to start using a condom?

A It is never too late to begin using condoms. If, for example, one partner becomes infected by sharing a contaminated needle today, HIV could be passed to the other partner the next time they have sexual intercourse. Using a condom would limit that possibility.

Q3 Can I kill HIV by heating my needle with a match?

A Using a match is better than not doing anything, but the heat will probably cause the blood to clot and trap the virus. Rinsing out the needle and the syringe in a 10% bleach solution is the more effective way of disinfecting it.

Q4 I've heard that when you're on drugs, you're more likely to spread AIDS by sexual contact than when you're not on drugs. Why?

A When injection drug users get high and have sex, they are less likely to think about using condoms. Also, prostitutes often pay for an expensive drug habit by trading sex for money. AIDS is therefore more likely to be spread by injection drug users than by nonusers.

Q5 Have any AIDS researchers contracted HIV infection during the course of their work?

A Yes. In the fall of 1987, federal officials reported that two laboratory researchers had contracted HIV, apparently in the course of their work. In both cases, the workers came in contact with liquid containing HIV. The first worker sustained a cut through a glove and was apparently contaminated with HIV; the second had numerous cuts and abrasions of the hand through which the virus may have passed.

Having completed this chapter, you should be familiar with methods by which the spread of HIV can be interrupted. To test your knowledge, place a T to the left of a statement if it is true or an F if it is false. Correct answers are elsewhere in the text.

_____ 1. A lighted match is an excellent way for injection drug users to sterilize their needles.

_____ 2. One method by which healthcare workers can be exposed to HIV-contaminated blood is through needlestick injuries.

_____ 3. The universal precautions promulgated by the CDC are recommendations for all healthcare workers.

_____ 4. A condom prevents the passage of bacteria from the semen to the man's sex partner but does not prevent the passage of viruses.

_____ 5. The agents of STDs, including AIDS, are generally unable to withstand exposures to the outside environment that other microorganisms can tolerate.

_____ 6. In controlled laboratory studies, natural lambskin condoms were found to be more effective for preventing HIV passage than latex condoms.

_____ 7. Law enforcement officers may be exposed to HIV-contaminated blood when they suffer needlestick injury while conducting a search.

_____ 8. AIDS education in the lower school grades should begin with descriptions of HIV and include a thorough grounding in sex education, including condom use.

_____ 9. One of the most effective germicides for destroying the AIDS virus is a solution of bleach in water at a concentration of 1% to 10%.

_____ 10. Healthcare workers should assume that all patients are infected with HIV in their blood and other body fluids.

_____ 11. Disposable gloves need not be carried in emergency response equipment because technicians rarely encounter accident victims who are infected with HIV.

_____ 12. One widely adopted way of interrupting the spread of AIDS among prison inmates is to distribute clean, sterile needles to drug addicts.

_____ 13. Medical laboratory technicians should pipet liquids by mouth because the probability of dealing with HIV-contaminated fluids is very low.

_____ 14. Funeral directors and embalmers should be concerned about exposure to HIV because the virus may be present in the blood, other body fluids, and organs of the deceased.

_____ 15. Abstinence from sexual intercourse and avoiding injection drug use are effective methods for avoiding exposure to HIV.

FOR ADDITIONAL READING

Carvalho, F. T. et al., 2011. "Behavioral interventions to promote condom use among women living with HIV." *Cochraine Database Syst Rev* 7:CD007844.

Centers for Disease Control and Prevention (CDC), 2011. "Sexually transmitted diseases surveillance, 2010." Atlanta: U.S. Department of Health and Human Services.

Ciesielski, C., et al., 1992. "Transmission of human immunodeficiency virus in a dental practice." *Annals of Internal Medicine* 116:798–805.

del Rio, C. et al., 2012. "Update to CDC's Sexually Transmitted Diseases Guidelines, 2010: Cephalosporins no longer a recommended treatment for gonococcal infections." *MMWR* 61:590–594.

Dieffenbach, C. W., et al., 2011. "Thirty years of HIV and AIDS: Future challenges and opportunities." *Annals of Internal Medicine* 154:766–771.

Johnson, S., 2007. The Ghost Map: The Story of London's Most Terrifying Epidemic–and How it Changed Science, in *Cities and the Modern World*. Riverhead Trade, New York.

Laheij, A. M. G. A, et al., 2011. "Healthcare-associated viral and bacterial infections in dentistry." *Journal of Oral Microbiology* 4:17659.

Lindholm, P. F. et al., 2011. "Approaches to minimize infection risk in blood banking and transfusion practice." *Infectious Disorders-Drug Targets* 11:45–56.

Mathers, B., et al., 2010. "HIV prevention, treatment, and care services for people who inject drugs: A systematic review of global, regional, and national coverage." *The Lancet* 375:1014–1028.

Matseke, G. et al., 2012. "Inconsistent condom use among public primary care patients with tuberculosis in Africa." *Scientific World Journal* 2012:501807.

Obiero, J., et al., 2012. "Topical microbiocides for prevention of sexually transmitted infections." *Chochrane Database System Review* June 13:6: CD007961.

Renaud, T. C., et al., 2009. "The free condom initiative: Promoting condom availability and used in New York City." *Public Health Reports* 124:481–489.

Shattock, R. J., Rosenberg, Z., 2012. "Microbiocides: Topical prevention against HIV." *Cold Spring Harbor Perspectives in Medicine* 4:a007385.

van Wijk, P. T., et al., 2012. "The risk of blood exposure incidents in dental practices in the Netherlands." *Community Dentistry and Oral Epidemiology* doi:10.1111/j.1600-0528.2012.00702.x.

LOOKING AHEAD

In contemporary medicine, laboratory tests provide essential information relative to a physician's diagnosis and help to ensure that the diagnosis is correct. This chapter surveys some of the procedures used in laboratory testing for HIV infection and AIDS. On completing the chapter, you should be able to . . .

- Summarize the general approach used by physicians to diagnose HIV infection and AIDS.
- Describe the various laboratory tests available to assist the diagnosis of HIV infection and understand the scientific basis for each test.
- Conceptualize how different laboratory tests detect patient antibodies produced against HIV.
- Summarize how the different nucleic acid tests work and their applications in detecting HIV.
- Discuss the meaning of viral load, and describe how the viral load is determined and how it correlates with the disease progression to AIDS.
- Understand the advantages and disadvantages of laboratory tests for HIV infection.
- Summarize the controversies surrounding interpretation and use of the results from laboratory tests with special reference to voluntary testing and mandatory testing.
- List high-risk behaviors for HIV infection and why it is recommended for high-risk groups to be routinely tested for HIV infection.
- List what U.S. groups undergo mandatory HIV testing.
- Discuss the importance of HIV education and the need for HIV testing programs.

INTRODUCTION

For many centuries, physicians practiced heroic medicine to save patients from the ravages of disease. They prescribed frightening courses of blood-lettings and purges, and they subjected patients to enormous doses of strange medicines and concoctions, ice-water baths, starvation, and other drastic remedies. These treatments probably made an already bad situation worse by reducing the body's natural defenses to the point of exhaustion.

Then, during the 1820s, a group of physicians in Boston and London experimented to see what would happen if they withheld treatment from patients and let nature take its course. Surprisingly, they found survival rates of untreated patients similar to and sometimes better than those of treated patients. As word of their experiments gradually spread, many of the worst features of heroic medicine began to disappear. A more conservative, nonmeddling approach to disease developed, and physicians sharpened their skills as diagnosticians. Their job was to recognize a certain illness, distinguish it from other illnesses, explain it to the family, and attempt to predict what would happen in the future. Then they would care for the patient within the limits of what was known.

Today's physicians understand the role of microorganisms in disease, and they have available numerous therapies and preventatives to help control disease. Nevertheless, their skills as diagnosticians must remain acute because they are usually the first healthcare providers we seek out when we are ill. In the effort to interrupt the AIDS epidemic, HIV testing is a key diagnostic tool because physicians can identify those who harbor the human immunodeficiency virus and counsel them on measures to prevent its spread. Also, partner notification may follow diagnosis when circumstances warrant, and infected individuals can be evaluated for drug treatment for HIV and preventative therapy for opportunistic diseases. Finally, the statistics generated from HIV testing and diagnosis help agencies plan medical and social support programs while providing a framework for interrupting the spread of the AIDS epidemic.

In 2006, the CDC released revised recommendations for HIV testing of adults, adolescents, and pregnant women in healthcare settings. These recommendations are listed in **Table** 7.1. The changes were prompted by the fact that 1 in 5 (20%) or about 240,000 people in the United States do not know that they are infected, yet often individuals with HIV infection visit healthcare settings (e.g. hospitals, acute-care clinics, and STD clinics) before they are diagnosed or tested for HIV. Routine screening of pregnant women identifies the infection so that counseling and treatment can begin as soon as possible.

TABLE

7.1   **2006 CDC HIV Screening Recommendations in the United States**

For patients in all healthcare settings
- HIV screening is recommended for patients (aged 13–64) in all health care settings unless the patient declines (opt-out screening).
- Persons at high risk for HIV infection should be screened for HIV at least annually.
- NOTE: Healthcare providers should initiate screening unless prevalence of undiagnosed HIV infection in their patients has been documented to be < 0.1%. In the absence of existing data for HIV prevalence, healthcare providers should initiate voluntary HIV screening until they establish that the diagnostic yield is < 1 per 1,000 patients screened, at which point such screening is no longer warranted.

For pregnant women
- HIV screening should be included in the routine panel of prenatal screening tests for all pregnant women.
- HIV screening is recommended after the patient is notified that testing will be performed unless the patient declines (opt-out screening).
- Repeat screening in the third trimester is recommended in certain jurisdictions with elevated rates of HIV infection among pregnant women.

Adapted from CDC, *Morbidity and Mortality Weekly Report*, vol. 55/RR-14, Sept. 22, 2006.

Adolescents aged 13 to 19 are at high risk. The 2005 Youth Risk Behavior Survey indicated that 47% of high school students reported they had sexual intercourse at least once, and 37% of sexually active students had not used a condom during their most recent act of sexual intercourse. In 2007, the CDC launched an Expanding HIV Testing Initiative in the United States. That same year, the World Health Organization/UNAIDS guidelines recommended routine HIV screening in healthcare settings.

In this chapter, we discuss the laboratory methods used for detecting HIV and helping physicians identify HIV infection and AIDS. Considering the advanced state of biotechnology, these methods tend to be complex. We shall also explore the implications of HIV testing and the conflicting views on how test results should be used. These are among the more controversial issues associated with the AIDS epidemic.

HIV Antibody Tests

To even the casual observer, it is obvious that the most direct diagnostic method for HIV infection is identifying the human immunodeficiency virus in the body tissues. As we discuss here, such a test is available, but the more cost-effective way of detecting HIV infection is to locate antibodies produced against the virus rather than the virus itself. These antibody methods are accurate and reliable, within certain limitations.

Any laboratory test for detecting antibodies is called a **serological test** because it involves a patient's **serum.** The patient who tests positive has the antibodies sought by the test and is said to be **seropositive.** In contrast, the patient who tests negative lacks the antibodies and is termed **seronegative.** One other bit of diagnostic jargon: A patient **seroconverts** when the serum tests positive.

In most cases, serologic tests are used to detect the presence of disease. They are based on the supposition that when the body is exposed to an infectious agent, such as HIV, the immune system reacts through the process of either antibody- or cell-mediated immunity. Certain tests, such as the tuberculosis skin test, are designed to detect the products of cellular immunity, but most bacterial and viral tests identify the products of antibody-mediated immunity—the **antibodies.** Because serologic tests do not detect the infectious agent itself, they are indirect tests.

For the diagnosis of HIV infection and AIDS, a number of antibody-based serologic tests are available to the physician. Two such tests, the enzyme-linked immunosorbent assay and the Western blot analysis, are routinely used to detect and confirm HIV antibodies in the blood. Although the theories behind the tests are somewhat involved, performing the tests is routine for trained, licensed technicians.

Enzyme-Linked Immunosorbent Assay

Currently, the most widely used diagnostic test for HIV infection is an antibody-based test called the **enzyme-linked immunosorbent assay (ELISA).** Through July 2010, the U.S. **Food and Drug Administration (FDA)** had licensed seven different manufacturers to produce ELISA HIV-1 kits, one ELISA HIV-2 kit, and twelve ELISA HIV-1 and 2 test kits for the U.S. market. Laboratories may use any of the approved kits for their diagnostic tests.

Each of the ELISA kits uses the same general procedure: A sample of a patient's serum is added to HIV antigens (protein fragments of HIV). These antigens are bound to beads or coated along the walls of tiny wells (**Figure 7.1**). The mixture is incubated for a period of time to permit antibodies in the serum to attach to (immunosorb) to the HIV antigens. Next, enzyme molecules linked to antibodies are added to the mixture. The antibodies, in this case, are produced by horses injected with human antibodies. The horse's immune system views the human antibodies as foreign substances and produces antibodies against them. These "antihuman" antibodies produced in horses

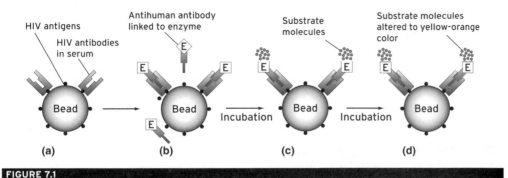

A positive ELISA test. (a) HIV antigens are bound to microscopic beads. A serum sample containing HIV antibodies is added to the beads, and the mixture is shaken. The HIV antibodies bind to the HIV antigens. (b) A sample of enzyme-linked antihuman antibodies is added. The enzyme-linked antibodies bind to the HIV antibodies on the surface of the beads. (c) Substrate molecules are added. (d) Enzyme molecules on the bead surface alter the substrate molecules and change their color to a yellow-orange that can be detected by a color-sensing instrument. The reaction shows that HIV antibodies are in the serum sample.

are subsequently isolated from the horse's blood and chemically linked to a highly specific enzyme for use in the test.

After the enzyme-linked antibodies have been mixed in, the reagents are permitted to react. Then the residual enzyme-linked antibodies are washed away. Next, the technician adds a **substrate**, that is, a chemical compound that will react with the enzyme. When the substrate reacts with the enzyme, a color change takes place; depending on whether the color change occurs, the technologist can infer whether the serum had HIV antibodies or lacked them. Figure 7.1 displays how a positive test works.

The ELISA test is complex, but you should make the effort to understand what is taking place. Given the medical and social significance of a positive ELISA test, the results must be accurate (i.e., it must be possible to unambiguously distinguish negative results from positive results), and interpretation of the results must be correct. In one survey, when 601 test laboratories ("participating laboratories") performed the ELISA test on samples identical to those tested by 15 established laboratories ("referee laboratories"), the results showed a very close correlation. This observation indicates that the ELISA test can be performed by different laboratories with accurate results.

Clinical data submitted by ELISA kit manufacturers to the FDA indicate that both the sensitivity and the specificity of their kits exceed 99%. Sensitivity refers to the probability of a positive test resulting when the serum sample has antibodies; for ELISA, the sensitivity is greater than 99%. Specificity refers to the probability of a negative test when the serum sample lacks antibodies; for ELISA, the specificity is also greater than 99%. Indeed, the American Red Cross Blood Services laboratories have reported that a 99.8% specificity was consistently achieved during testing of donated blood for transfusion purposes.

In current laboratory practice, a sample of serum that tests negative (or "nonreactive") is considered free of HIV antibodies. If a sample tests positive (or "reactive"), the results are reported as "initially reactive," and the sample is retested twice (**Figure 7.2**). If both retests yield negative results, the sample is reported as "nonreactive for HIV antibodies;" however, if either or both retests yield positive results, the sample is reported as "repeatedly reactive for HIV antibodies." Under these conditions, the ELISA test results should be validated by an independent supplemental test. In the United States, the validation test most often used is the Western blot analysis. We examine that test next. The section on **False Positives and False Negatives** will discuss the **"window period"** between infection with HIV and the ability to detect it with antibody-based tests such as the ELISA, which can vary from person to person.

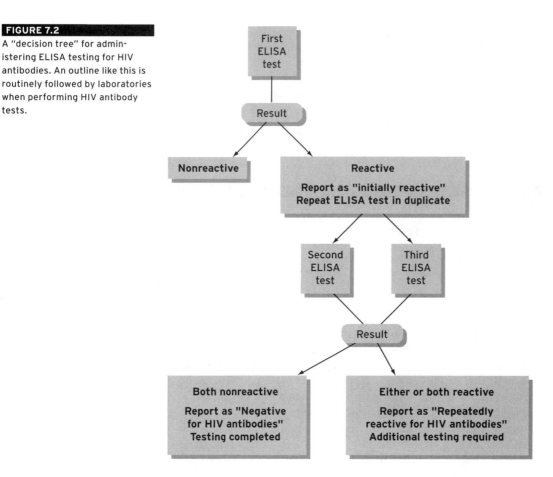

Western Blot Analysis

The Western blot analysis is a variation of a procedure devised in the 1970s by British investigator E. M. Southern. Southern used his procedure to separate fragments of DNA and identify them. He began with a standard laboratory technique called **gel electrophoresis**. The basic principle of gel electrophoresis is that molecules dissolved in a gel of agarose will respond to an electrical field and move through the gel according to fragment size (**Figure 7.3**). Usually, the smaller fragments move faster than the larger fragments. Once separated, the DNA fragments are transferred to filter paper by placing the gel in contact with the paper. The paper draws the fragments out of the gel, literally blotting them exactly as they were positioned in the gel. This procedure came to be known as the Southern blot technique.

When this technique was adapted to study fragments of RNA instead of DNA, it was whimsically named the "Northern" blot procedure. Further adaptation was needed when analyzing protein molecules such as those from viruses, and thus, researchers modified the technique and devised the **"Western" blot analysis**. (There is no "Eastern" blot analysis yet.)

The Western blot analysis is used as a confirmatory test when an ELISA test has given a positive result. In the laboratory, HIV proteins from laboratory cultures are separated by gel electrophoresis. Such proteins as gp41 and gp120 are used. Then the proteins are transferred to a special nitrocellulose paper by blotting the paper against the gel. Next, a person's serum is diluted and added to the paper

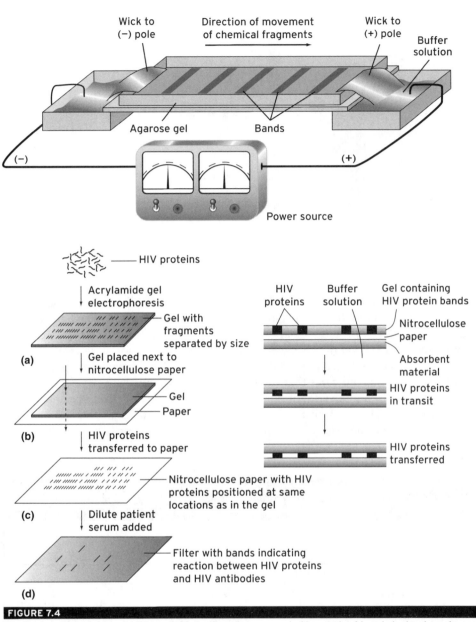

Wick to (−) pole Direction of movement of chemical fragments Wick to (+) pole Buffer solution

Agarose gel Bands

(−) (+)

Power source

FIGURE 7.3

An electrophoresis apparatus such as that used in the Western blot analysis. An electrical power source maintains positive (+) and negative (−) charges in two buffer solutions. Wicks connect the agarose gel to the buffer solutions. Chemical fragments move through the electrically charged gel according to their sizes. Bands form where the movements of different size fragments come to an end.

HIV proteins

Acrylamide gel electrophoresis

Gel with fragments separated by size

(a) Gel placed next to nitrocellulose paper

Gel
Paper

(b) HIV proteins transferred to paper

Nitrocellulose paper with HIV proteins positioned at same locations as in the gel

(c) Dilute patient serum added

Filter with bands indicating reaction between HIV proteins and HIV antibodies

(d)

HIV proteins Buffer solution Gel containing HIV protein bands

Nitrocellulose paper

Absorbent material

HIV proteins in transit

HIV proteins transferred

FIGURE 7.4

The Western blot analysis. (a) HIV proteins (antigens) are separated by acrylamide gel electrophoresis, as shown in Figure 7.3. The proteins form bands in the electrophoresis gel. (b) When the gel is placed next to a nitrocellulose paper in a buffer solution, the solution carries the proteins to the paper and deposits them there (right side of diagram). (c) The result is a paper containing the proteins positioned in the same place as they are on the gel. (d) Finally, a sample of diluted patient serum is added. If HIV exposure has taken place, different HIV antibodies will react with different HIV proteins, and when a staining reagent is added, the stain will gather where the antibody–protein reaction has taken place and form distinctive bands of color. The bands indicate that the person is seropositive. No bands of color will develop if the person is seronegative.

(**Figure 7.4**). If HIV antibodies are present, they bind to their complementary viral proteins (i.e., gp41 antibody binds to gp4l protein, whereas gp120 antibody binds to gp120 protein).

Because these reactions are invisible, the product of the reaction is made visible by adding a staining reagent. The reagent consists of stain combined with commercially prepared antihuman antibodies. These antibodies react with human HIV antibodies.

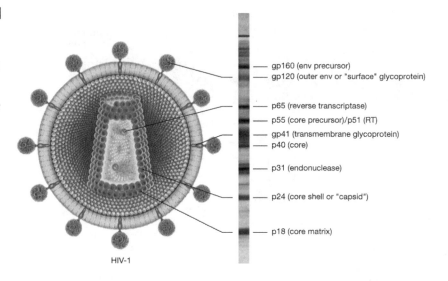

gp160 (env precursor)
gp120 (outer env or "surface" glycoprotein)

p65 (reverse transcriptase)
p55 (core precursor)/p51 (RT)
gp41 (transmembrane glycoprotein)
p40 (core)

p31 (endonuclease)

p24 (core shell or "capsid")

p18 (core matrix)

HIV-1

The stain gathers where HIV antibodies have accumulated, and distinctive bands of color appear on the paper when the person is seropositive (**Figure 7.5**). The bands correspond to different HIV proteins separated from one another in the electrophoresis step. Obviously, if no HIV antibodies were in the serum, no reaction with HIV antigens would take place, and no stain would be attracted to the nitrocellulose paper. The absence of stained bands indicates that the person is seronegative.

For the Western blot analysis, antibodies reacting with at least three specified HIV antigens must be detected for the test to be considered positive (**Table 7.2**). Thus, the Western blot analysis is more specific than the ELISA test (where a mixture

TABLE 7.2	A Comparison of Responses by Referee and Participant Laboratories on Samples Tested for HIV-1 Antibodies by Western Blot Testing

Organization	Criteria
Association of State and Territorial Public Health Laboratory Directors/CDC	Any two of: • p24 • gp41 • gp120/gp160*
FDA-licensed DuPont test	p24 and p31 and gp41 or gp120/gp160
American Red Cross	≥ 3 bands; 1 from each gene-product group: • gag and • pol and • env
Consortium for Retrovirus Serology Standardization	≥ 2 bands: p24 or p31, plus • gp41 or • gp120/gp160

*Distinguishing the gp120 band from the gp160 band is often very difficult. These two glycoproteins can be considered as one reactant for purposes of interpreting Western blot test results.

Data from: Recommendations for HIV testing of adults, adolescents, and pregnant women in health-care settings. *MMWR* 55 (RR-14):1–17.

of "generalized" HIV antibodies is detected). For Western blot analyses that are inconclusive (e.g., only one or two HIV antigens detected), the results are reported as "indeterminate."

As with the ELISA test, the sensitivity and specificity of the Western blot analysis exceed 99% as long as certain criteria are met; however, if a laboratory chooses to use different Western blot reagents or unlicensed tests or less stringent interpretive methods, then the 99% levels of sensitivity and specificity may not hold, as these are determined under carefully controlled conditions. Laboratories can also ensure the reliability of test results by training their personnel carefully, establishing quality controls, and participating in performance evaluation programs administered by public health agencies. Standardization of test kits and attention to technical proficiency also boost confidence in test results.

False Positives and Negatives

Since the ELISA test was introduced in 1985, its results, validated by the Western blot analysis, have been used to make many medical and personal decisions. For example, millions of donated pints of blood have been screened using the tests, and only those units testing negative for HIV antibodies against p24 are used for transfusion. Donated organs have also been tested and assumed safe for transplant if HIV antibody tests are negative, and the shaping of personal behaviors (e.g., whether to use a condom) has been influenced by the results of HIV tests.

Nevertheless, it is common knowledge that the antibody tests are not absolutely perfect. Since 1985, several cases of AIDS were linked to transfused blood previously tested negative for HIV antibodies. Moreover, several individuals who believed they had HIV infection on the basis of antibody tests were later found to have no trace of the virus when their tissues were examined. It is possible, therefore, that HIV antibody tests (ELISA and Western blot analysis) can give false-negative and false-positive results. A false-negative result (or false negative) is one that indicates HIV is absent when, in fact, it is present in the body. A false-positive result (false positive) occurs when evidence indicates the presence of HIV in the body when, in fact, none is present.

How is it possible for an individual to be seronegative yet harbor HIV? It can take some time for a person's immune system to produce enough antibodies for the antibody-based tests to detect. This "window period" between the time of infection with HIV and the ability to detect it with antibody tests can vary from person to person. Most people will develop detectable antibodies that can be detected by the most commonly used tests in the United States within 2 to 8 weeks (average is 25 days) of their infection. Should an infected person go for HIV testing two or three weeks after being infected, the test results will probably be negative, even though HIV is present.

Ninety-seven percent (97%) of people infected with HIV-1 will develop detectable antibodies in the first 3 months. There is a small chance that some individuals will take longer to develop detectable antibodies. A person should consider a follow-up test more than 3 months after their last potential exposure to HIV. In rare cases, it can take up to 6 months to develop antibodies against HIV-1.

Another possible reason for a negative test may be the heterogeneity of HIV. Studies have shown that HIV undergoes numerous mutations in the lymphoid tissues and that numerous strains of HIV may be present in a single individual. In response, the body will produce numerous types of antibodies with different molecular configurations, and these may not be detected by the ELISA test. The test may therefore give a negative result because it is not detecting the type(s) of antibodies that are present.

False positives develop when tests signal the presence of HIV antibodies even though none are present. ELISA tests may give false positives, but the Western blot analysis eliminates virtually all of them. A possible reason for a false positive may lie in the test itself. The viruses for the ELISA test are cultivated in human cells, and a certain amount of cellular debris can remain with the viral fragments when they are isolated. This debris can attract human antibodies; when the latter accumulate on the bead, they can attract the enzyme-linked antibodies. A positive result ("reactive") will erroneously be observed. Another reason for a false positive may be the presence of other antibodies in a person's blood. For example, certain non-HIV antibodies are known to react slightly with HIV antigens. A person's serum may also give a positive result because of a previous HIV infection that is now gone.

False positives and negatives may also be associated with technical errors in the tests or with the person performing the tests. In the ELISA test, for example, improper washings carried on during test performance may contribute to errors, as contaminations of samples or color reagents or both may occur. Receiving news of a positive test can be extremely traumatic. Thus, the need for more reliable test results has encouraged public health officials to urge the development of tests that detect the virus itself.

Home and Simplified Tests

At the other end of the spectrum of HIV diagnostic testing is a simplified home test. Sold as "The Home Access HIV-1 Test System" or "The Home Access Express HIV-1 Test System," this test is manufactured by Home Access Health Corporation and requires that a person obtain three drops of blood using an antiseptic wipe and lancet supplied in the kit. The individual blots the blood onto a test card, which is coded with a unique identification number. The card is then mailed to a laboratory for ELISA testing; the person calls a 1-800 telephone number a few days later and uses the identification number to learn the result. If the result is positive, the call is always rotated to a counselor.

The efficacy of the home test has been demonstrated in numerous research studies. In one study, subject-drawn blood spot samples were compared with professionally drawn samples, and nearly 100% of the subject samples were found adequate. The test has been approved by the FDA since 1996 and is available in most pharmacies. Its accuracy, the availability of HIV therapies for those testing positive, and the public health benefits of knowing who is HIV infected have contributed to its use.

It should be stressed that the home test made by Home Access Health Corporation was the only FDA-approved home HIV test kit that could be legally sold in the United States. In September 1997, the FDA warned about the illegal sale of the "Lei-Home Access HIV test" distributed by Lei-Home Access Care located in Synovial, California, which had been advertised in magazines and on the Internet. Because of the illegal nature of this marketing, the businessmen responsible for the sale of these test kits were sentenced to 5 years in prison. Fraudulent sales of kits not approved by the FDA have also come from outside the U.S. borders. In February 2005, the FDA released a warning regarding home HIV test kits being sold by Globus Media devices that are not FDA approved. These kits were found to be on sale through certain websites; the FDA alert regarding Globus Media has provided authority to detain and refusal to admit these products into the United States and also advises U.S. Customs officials about these products.

In the clinical setting, there are tests that do not require the use of blood. HIV tests are available that simply require a urine sample for analysis. The test has been available to physicians since its approval by the FDA in 1996. Patients provide a urine sample, and a physician sends it to a laboratory for antibody testing. Although the level of accuracy is lower than for blood tests, the urine test is believed to increase the number of people found to be HIV-infected because more individuals are willing to be tested (because no blood is drawn). Thus, it will help epidemiologists track the AIDS epidemic and get patients into therapy sooner. Also, the test does not require the assistance of a person trained in venipuncture, which reduces the risk of occupational exposure as well as cost. Research studies indicate that a combination of blood and urine tests detects more HIV-positive individuals than either test alone, possibly because antibodies may be present in one body fluid but not in the other.

A number of additional simplified tests that do not require blood samples are in use. One such test, the **OraQuick Advance Rapid HIV-1/2 Antibody Test**, requires an oral fluid sample and was approved by the FDA as a rapid home-use rapid HIV-1/2 test kit in 2012 (**Figure 7.6**). It is the first and only over-the-counter **rapid home test** licensed by the FDA on the market. The collection of oral fluid samples is much easier and the OraQuick Advance HIV-1/2 test sample is not sent to a laboratory for testing. The OraQuick test provides a result in 20 to 40 minutes in the subject's own home. It works similar to an over-the-counter pregnancy test. Anyone who has a positive test result should have a follow-up laboratory test performed. The OraQuick Advance HIV-1/2 test is also suitable for clinical settings including private offices, college campuses, hospitals, and public health settings.

The ability to detect HIV at the point of care could be useful in inner-city emergency rooms, where patients who do not have regular physicians could be tested and informed quickly. The tests would also find use in HIV-infected women who are pregnant and whose first contact with the healthcare system comes at the time of delivery. Indeed, one CDC model has shown that in a single year, health officials could learn the HIV status of nearly 100,000 more people if rapid-screening tests were widely available. Moreover, as **Table 7.3** shows, if clinic personnel were encouraged to recommend

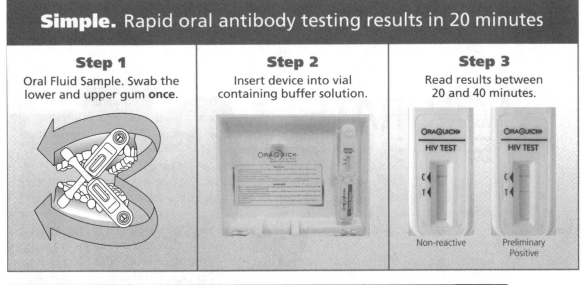

Simple. Rapid oral antibody testing results in 20 minutes

Step 1
Oral Fluid Sample. Swab the lower and upper gum **once**.

Step 2
Insert device into vial containing buffer solution.

Step 3
Read results between 20 and 40 minutes.

Non-reactive

Preliminary Positive

FIGURE 7.6
Product information about OraQuick, the first rapid home-use kit for self-testing approved by the FDA in 2012.
© Jones & Bartlett Learning.

TABLE 7.3	Number of Individuals Tested for HIV at a Georgia Clinic When Testing Was Not Routinely (1999) and Routinely (2000) Recommended		
Test Process	Testing Not Routinely Recommended (1999)	Testing Routinely Recommended (2000)	Increase from 1999 to 2000
Clinic visits	19,626	19,911	285
HIV tests conducted	1,100	2,787	1,687
Newly detected infections*	47	74	27
HIV-positive patients who learned they were infected	28	55	27
HIV-positive patients who entered into care	13	26	13

*Positive HIV test result (Western blot).

Modified from CDC, *Morbidity and Mortality Weekly Report*, vol. 50/no. 13, June 29, 2001.

HIV testing for all urgent-care patients, more HIV infections could be detected. Mobile testing units in urban settings and in developing countries with primitive clinical testing facilities also stand to benefit from these rapid tests.

HIV Viral Load Nucleic Acid Testing

Although serological tests are widely used to assist diagnoses and to screen blood, they are indirect tests of HIV infection and, as such, fail to give the clearest picture of viral presence. Furthermore, they are subject to false negatives because a person can remain seronegative for weeks or longer after exposure to HIV, and they do not establish how much virus is in the body (the viral load) or which tissues the virus has infected or which variant of the virus is present. For these reasons, diagnosis is better served by tests directed specifically at HIV or chemical fragments of HIV, commonly known as *HIV antigens*.

Detecting HIV and/or HIV antigens is not easy, however. The standard technique for pinpointing retroviruses, for example, is to identify the presence of the enzyme reverse transcriptase, but this technique has not been used as a diagnostic test because it is very complicated, expensive, and it identifies all retroviruses, not just HIV. Applying other established diagnostic techniques to detect HIV antigens has also proven difficult because HIV may exist as a provirus in T-lymphocytes, and the number of infected cells may be too low. In addition, tests available to detect infected cells are not sensitive enough to give confidence in their results when screening large numbers of blood samples.

The diagnostic tests that detect HIV antigens are directed at both proteins and nucleic acids. Proteins such as p24 are a logical target of diagnostic tests (as we discuss presently), but investigators have shown that HIV-infected cells may not be synthesizing the viruses; thus, viral proteins may not be present. They also caution that infected T-lymphocytes cultured in the laboratory and tested for viral proteins do not always reflect the level of viral infection in the individual. Thus, a test for detecting viral nucleic acids is considered more dependable. Such a test is called the **viral load test**. They are also referred to as **nucleic acid tests** (**NATs**). NATs are used for blood donor screening to detect the RNA of HIV. NATs use technologies of gene probes, gene amplification, and the polymerase chain reaction (PCR), the viral load test has gained wide acceptance among the FDA, physicians, and scientists. We discuss this test next.

Gene Probes

Gene probes and the PCR are central to the viral load test. They are outgrowths of research in molecular biology that began in the 1950s. With the discovery of new enzymes, new apparatus, and new technologies, scientists found they could reproduce DNA in a test tube, fragment it, determine its composition, change its structure, exchange pieces of it, and map its genes. As the age of molecular biology unfolded, scientists applied newfound principles to diagnosis, devising the antibody tests we have discussed previously, and developing the gene probes and viral load tests we explore here.

A **gene probe** is a single-stranded segment of DNA that can recognize and bind to a complementary segment of DNA on a large DNA molecule. The probe may be labeled with a radioactive isotope that will signal when binding has taken place (**Figure 7.7**). Underlying the technology is the fact that DNA (such as the DNA of an HIV provirus) exists as two strands opposing one another much like the sides of a ladder. To perform the viral load test and hunt for **proviral DNA**, infected T-lymphocytes are secured and broken open. All of the cellular DNA (including the proviral DNA) are then isolated and split apart, thereby separating the two strands. Now the radioactive gene probe is added. Like a left hand seeking its unique matching right hand, it mingles among all the DNA strands until it locates a complementary strand. Binding to the complementary DNA strand, the probe brings along its radioactive label, and the signal is given that proviral DNA has been found. In an uninfected individual, no such union takes place (there is no proviral DNA). Hence, no radioactivity accumulates, and no signal is sent.

The viral load test typifies the important relationship between basic research and practical applications. The test requires a gene probe, which is, as we have seen, a single-stranded DNA molecule to complement the HIV proviral DNA. Synthesizing such a probe mandated that scientists analyze and decipher the chemical structure

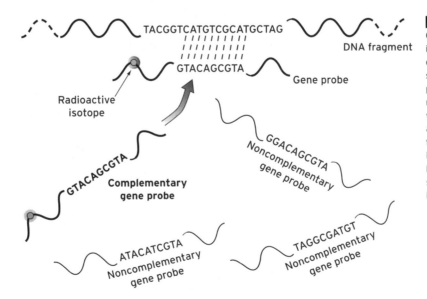

FIGURE 7.7

Gene probe activity. A gene probe is a single-stranded segment of DNA. When combined with separated DNA strands, the gene probe will seek out its complementary DNA segment and bind with it. If the probe is attached to a radioactive isotope, the radioactivity will accumulate at the binding site and signal that a reaction has taken place. The diagram shows only the nitrogenous bases involved in the union of the DNAs.

of HIV's DNA, a task that required a considerable expenditure of time and research funds. Gene probes now exist for the RNA in HIV as well as for the proviral DNA. Thus, gene probes can be used to determine the viral load in the bloodstream and other body fluids, as we see here.

Polymerase Chain Reaction

One major problem with the viral load test is securing enough DNA to perform the test (e.g., only 1 in 10,000 T-lymphocytes may be infected with HIV). This problem has been addressed by using a technology called gene amplification to increase the amount of DNA.

Gene amplification employs a procedure called the **polymerase chain reaction (PCR)**. The technology was developed by Nobel laureate Kary Mullis at the Cetus Corporation in California in 1984. Gene amplification is performed with double-stranded DNA obtained from T-lymphocytes (and presumably containing HIV proviral DNA). The DNA is separated to yield two strands by carefully heating the material to a specific temperature. After strand separation, a mixture of nucleic acid building blocks (nucleotides) is added, together with an enzyme called DNA polymerase and a short piece of DNA "primer." The primer specifies the segment to be copied. Then the temperature is reduced slightly, but not enough for the two DNA strands to recombine (**Figure 7.8**). Now the DNA polymerase synthesizes new complementary strands of DNA, using the building blocks, the primer, and the single strands of DNA as templates. A copy of each separated DNA strand is created. Where there were two DNA strands, there are now four.

The PCR is repeated dozens of times, each cycle taking 1 to 2 minutes. It is performed in a highly sophisticated automated processor and results in a million-fold amplification of the genes because each copied DNA segment serves as a source for millions of additional copies. By incorporating this technology with that of a gene probe, the medical equivalent of a needle in a haystack can be located.

The PCR systems available today are able to amplify RNA as well as DNA and to provide enough RNA to do a viral load test for free HIV particles. To perform the test, the RNA (if present) is used as a template to synthesize DNA, with the enzyme reverse transcriptase serving as a catalyst. Then the same DNA amplification and gene probe procedures are performed as mentioned previously here. The results are commonly expressed as copies of HIV RNA per milliliter (mL). In 1996, the FDA approved the first commercial test to measure viral loads in patients.

Advantages of the Viral Load Test

Because the viral load test identifies viral nucleic acid, it helps clinicians determine which babies are actually HIV-infected and assists epidemiologists in following the AIDS epidemic in newborns. Antibodies found in newborns are most likely derived from the mother via the placenta. The viral load test makes it possible to detect infected cells directly. In one study, for example, CDC investigators tested blood samples from 200 infants born in New York. Half of the infants had mothers who were positive for HIV antibodies. By using the viral load test, researchers succeeded in locating HIV infection at or near the time of birth in all but one of the children. In adults as well, tracking the viral load has become as valuable as following the clinical signs of AIDS.

A safer **blood supply** is another benefit from the viral load test. Because the test can detect as few as 100 HIV particles in a milliliter of blood, it is far superior to hunting for HIV antibodies. Moreover, the viruses can be detected as few as 10 days after infection. Unfortunately, the current high cost of the viral load tests precludes

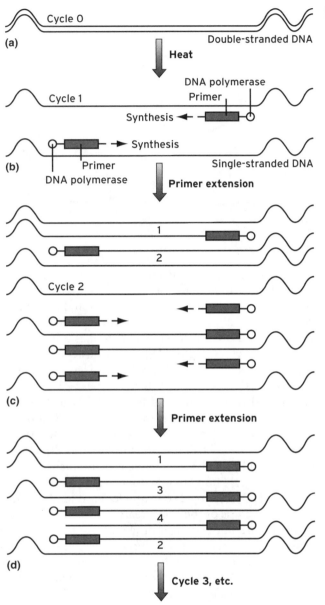

FIGURE 7.8

The PCR. (a) Heat is used to separate the double-stranded DNA molecule. In Cycle 1, nucleotides, an enzyme called DNA polymerase, and a piece of DNA primer are mixed together and added to the two molecules. (b) The DNA polymerase extends the primer using the available nucleotides to yield two double-stranded DNA molecules. (c) The process is repeated, and at the end of Cycle 2, four double-stranded DNA molecules are produced. Repeating the process in Cycle 3 yields a total of eight double-stranded DNA molecules.

their use on every unit of donated blood. To resolve this dilemma, blood banks employ a process called **mini-pooling.** In one example of mini-pooling, a blood bank combines samples from 16 blood donations in a primary pool and then combines samples from eight such primary pools in a master pool (thereby bringing together 128 samples). A sample from the master pool is subjected to the viral load test, and if it is positive for HIV RNA, each of the primary pools is tested until the infected blood donation is located. If the test on the master pool sample returns a negative result, all 128 units are declared safe for transfusion. In this way, the viral load test can be used, costs can be controlled, and the risk of HIV transfer through donated blood can be significantly lowered.

Measuring viral loads has also helped revolutionize our understanding of HIV's replication inside of cells. For example, by using the viral load test, scientists have

FIGURE 7.9

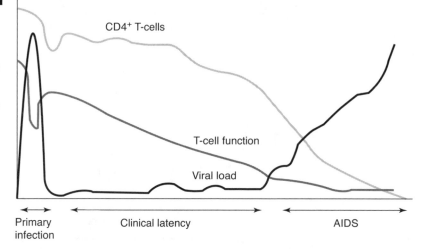

The relationship between viral
load and the T-lymphocyte count
in an AIDS patient. Early in the
disease, the viral load rises
dramatically as the T-lymphocyte
count drops. This same pattern
is seen in the later stages in the
disease as the patient progresses
to AIDS. Adapted from Wolthers,
K.C., Miedema, F, *Trends in
Microbiology*, Volume 6, 1998.

shown that soon after infection there is no true latent period, as once believed, but an explosion of viral activity in the bloodstream; then, with the onset of the immune response, viral multiplication stabilizes and remains constant for a long period of time (**Figure 7.9**). Furthermore, researchers have discovered that the viral load in a person with HIV infection can be used as a barometer of how quickly he or she will progress to AIDS (Figure 7.9). For instance, experimental results obtained in 1996 by John Mellors and colleagues at the University of Pittsburgh indicate that if the viral load is less than 4,500 HIV RNA copies per mL (cpm) the median progression time to AIDS is 10 years. If the viral load is 4,501 to 13,000 cpm, the median time to progress to AIDS is 7.7 years. If the viral load is 13,001 to 36,300 cpm, the median time is 5.3 years, and if the viral load is more than 36,300 cpm, the median time is 3.5 years. These results establish that the viral load is directly related to the rate of disease progression and provides a valuable measurement for discriminating disease stages.

Moreover, knowing how to measure the viral load has helped define new principles of HIV therapy. For example, viral load measurements can be used to determine the effectiveness of the HIV treatment (e.g., the viral load drops as treatment proceeds) or the failure of the treatment (e.g., the viral load returns to pretreatment levels). A viral load of 400 cpm was arbitrarily established as the level at which HIV is "undetectable" in the patient (although there are calls to reduce that number), and treatment guidelines aim to maintain the HIV level below 400 cpm. Finally, viral load measurements provide researchers in drug development with an easier way of evaluating clinical outcomes of their tests. For these reasons, viral load determinations have been paired with T-lymphocyte counts for guiding the decision-making process in diagnosis and therapy. Indeed, in 1996, the first evidence of the effectiveness of combination therapies came as a result of viral load studies. Indeed, NATs are exquisitely sensitive assays that can quantitate HIV present in patient serum/plasma. They are not used as diagnostic assays. Instead they are used to monitor patient viral load and screen donor plasma, blood and/or organs.

The FDA approved ten viral load test kits for use in the United States between 1999 and 2007. All of these test kits are performed on plasma or cadaveric serum (to screen nonliving organ donors). Half of the tests utilize PCR; the others involve gene probes or DNA sequencing to do genotyping. Genotyping is used to monitor the mutations in the genomes of viruses present in the patient to determine if the viruses have developed antiretroviral drug resistance.

Protein and T-Lymphocyte Tests

The HIV antibody and viral load NATs predominate among diagnostic or patient drug therapy applications, but a number of other tests are in use as well. In 1996, for example, the FDA approved the first blood test to detect HIV antigens rather than HIV antibodies. The test is known as the Coulter HIV-1 p24 Antigen Assay. It screens blood for the presence of the **p24 antigen,** a protein associated with the capsid of HIV. Research indicates that this protein is detectable about a week earlier than antibodies are typically detected, thus narrowing the so-called window period when a false negative may be obtained.

Among other available tests is an improved technique for measuring T-lymphocytes. A test for these cells has gained significance since January 1993 when the case definition for AIDS was changed to reflect the CD4 T-lymphocyte count in patients. The count of T-lymphocytes in a patient was traditionally determined by a process known as **flow cytometry.** In this test, live blood cells flow in single file through a measuring device where they scatter laser light, allowing a count to be taken. Only one blood sample can be measured at a time in a flow cytometer, and samples of blood must be no more than two days old. In 1995, the FDA approved a new test that employs blood samples up to five days old. The test passes blood cells over a plate coated with antibodies that react with the CD4 receptor sites. A visualization reaction similar to that in ELISA is then used to complete the test.

Implications of HIV Testing

One of the more volatile current issues concerns who should be tested for HIV infection. On the federal level, testing has been mandatory for members of the armed forces since 1985, and Congress passed a bill in 1987 requiring testing of immigrants. In state legislatures, more than 100 bills are introduced each year mandating various forms of HIV testing. Premarital testing appears to be the most popular approach to mass screenings, but there are many questions to be resolved: Should the infected person be notified of the presence of HIV or simply informed that the marriage application has been turned down? What should the intended spouse be told?

Health officials have used such terms as "universal," "mandatory," "compulsory," "routine," and "voluntary" to describe various plans. Of these adjectives, universal, mandatory, and compulsory evoke the strongest reaction. Proponents of universal testing maintain that it will help define the extent of the AIDS epidemic. They also point out that such testing will help identify those infected so that they and their partners can take precautions to limit the spread of HIV. In addition, universal testing can help discover who is in need of early treatment to slow the progression of AIDS and HIV-related diseases and indicate where available resources should be concentrated.

Opponents of universal testing are equally adamant about why it would be of little value. They admit that testing high-risk groups makes sense, but, they ask about how one defines a person at high risk. For example, is a male homosexual automatically at risk? Not necessarily, because a homosexual man with a very limited number of partners is at considerably less risk than a heterosexual man with multiple partners. They also see the value of testing people who use venereal disease clinics but point out that the threat of an AIDS test might scare away people who need the care offered by the clinic. Moreover, they note, testing must be done repeatedly if it is to be effective because a negative test is only valid until the next sexual encounter with a possibly infected individual.

There is also the important discrimination factor. For example, testing of hospital patients might yield valuable information on each person's health picture and relieve stress felt by healthcare workers, but a positive HIV test might also make it impossible for a person to secure health or life insurance. Proponents of testing counter that patients could be tested anonymously with the results available only to physicians. This might help epidemiologists while minimizing the possibility of discrimination, but opponents counter that failing to alert those who test positive would be unethical.

Most public health officials agree that identifying HIV-positive individuals can help curtail the spread of AIDS, but only if testing is combined with counseling. Those testing positive are often in a state of shock on learning the test results. Guilt, depression, and attempted suicide may follow. These individuals need to know how to live with the implications of the diagnosis. Counseling should emphasize what the test results mean and should suggest lifestyle changes necessary to prevent infecting others, including safer sex practices.

The next question that arises is this: Who else is entitled to know that an individual has tested positive for HIV? For the control of sexually transmitted diseases (STDs), including syphilis, gonorrhea, chlamydia, and others, a traditional public health approach has been to notify and offer treatment to sexual contacts of the infected person. The logical extension would be to apply the same approach in cases of HIV infection, but opponents point out that STDs are generally curable, whereas HIV infection is not; plus, they note, STDs do not carry the same stigma as HIV infection and AIDS. Thus, cooperation might not be forthcoming from the person who has tested positive. Health officials also point out that pressing an individual for names of sexual partners often elicits false names. Eventually, discrimination may surface, and the person testing positive may be denied work, housing, or insurance.

A similar dilemma exists for U.S. immigrants whose application for admission to the United States is denied because of HIV infection or AIDS. Who should be notified, and what should be done with the test results? Questions such as these continue to confront public health officials as they grapple with the question of who should be tested.

Mandatory Testing

As the controversial issues concerning HIV testing are debated in the United States, laws have been instituted to mandate testing for certain groups. Since 1990, for example, all of those desiring to donate blood, plasma, or organs must undergo an HIV test, as must all individuals wishing to donate sperm for artificial insemination or organs or tissues for transplantation purposes. The purpose is to prevent HIV transmission. Blood donors are asked to read an informational pamphlet describing the test, answer a series of medical questions, and sign a consent form to donate blood and be tested. Donors are notified only if their test results are positive, and they are invited for counseling and additional evaluation. For living donors of tissues and organs, a similar procedure is followed.

HIV testing is also mandatory for military recruits anticipating active duty and for those entering the foreign service or a federal service agency such as the Peace Corps or Job Corps. Health officials reason that military personnel must receive numerous immunizations, and these can be dangerous to the receiver's health if immune suppression caused by HIV is taking place. There also is a possibility that duty in a remote part of the world may expose the recruit to exotic organisms that could attack an HIV-infected body with unusual severity. Moreover, someone in the military might be called on in an emergency to donate blood to a wounded comrade, and the situation might not lend itself to pretransfusion HIV testing. For those entering the foreign service, similar reasons apply, plus there is a diplomatic responsibility to ensure that American representatives do not pose a danger to people in the countries in which they serve.

As of 2010, HIV testing is no longer mandatory for those wishing to emigrate to the United States or for refugees. Federal prison inmates are tested by law under certain circumstances. Persons testing positive receive counseling and medical care while in prison, and probation officers counsel inmates on appropriate behaviors when release is anticipated. Most prisons now provide special care facilities for those with AIDS, but not for those with HIV infection.

Another important issue concerns the mandatory testing of newborns. Beginning in 1988, federal legislation required that all newborns be tested, but the law stipulated that the results were not to be made available to mothers. Approximately 2 million infants were tested annually, and the results were used to give a map of where HIV was spreading among heterosexuals (without violating any confidentiality), as newborns have the same antibodies as their mothers. Then, in 1995, a congressional bill was introduced requiring that the results be given to the mother, in effect making the test mandatory for both mother and child. The response by the government was to cancel the national newborn testing program.

In the late 1980s, laws requiring mandatory testing were enacted in the states of Illinois and New York for those contemplating marriage. The spouses to be were each informed of the partner's results, but the couple could marry whether or not they tested positive for HIV infection. Vocal proponents of the test pointed out that in 1988, 23 cases of HIV infection were detected in Illinois, cases that otherwise might have been missed. Equally vocal opponents noted that a total of approximately 150,000 people took the test that year, and the low number of positives (23) showed the low risk of acquiring HIV in that population. They also pointed out the fact that couples were using their feet to vote their opposition to the law: The number of marriage ceremonies taking place in states neighboring Illinois and New York rose sharply. By 1991, the laws were repealed in both states.

Since 1990, all 50 states and the District of Columbia require healthcare providers to report new cases of AIDS to their state health departments. In addition, in 1999, the CDC recommended that surveillance for HIV infection be performed in all states to track the AIDS epidemic more accurately. At that time (1999), 34 states had already implemented HIV surveillance by patients' names. By 2004, nine additional states (43 total) used name-based surveillance, New Hampshire being the lone state that only required reporting of new pediatric AIDS cases and had no requirement for the reporting of new adult cases. By 2010, all states offer confidential testing but not all offer anonymous testing. Twelve states offer only confidential testing.

Voluntary Testing

Experiences gained during the years of the AIDS epidemic have encouraged the U.S. Public Health Service (through the CDC) to recommend HIV testing and counseling for certain groups of individuals. Included in the recommendations are the principles that counseling should take place before and after testing when possible, that personal information will be held confidential, and that a person may decline testing without consequence, except where testing is required under mandate of law (e.g., blood donors, prisoners, and immigrants).

One group for which testing is recommended includes persons who may have sexually transmitted diseases. The CDC suggests that individuals who report for treatment at health clinics, offices of private physicians, or any other healthcare setting should routinely receive HIV testing and counseling, but only if they consent. Another group includes all injection drug users or persons seeking treatment for injection drug abuse. For these individuals, treatment programs should be sufficiently available to allow those seeking assistance to enter promptly, and each person should be counseled on how to modify his or her behavior to prevent the spread of HIV. Outreach programs are also encouraged to educate injection drug users on the risk of AIDS and to recommend treatment for substance abuse.

Testing is also recommended for any persons who consider themselves at risk, such as healthcare workers who have suffered needlestick injury with contaminated blood. Women of child-bearing age with identifiable risk are also singled out by the CDC. Such women include those who have used injection drugs, have engaged in prostitution, or have sexual partners who are bisexual, injection drug users, or hemophiliacs. In addition, testing is suggested for women of child-bearing age living in communities or born in countries where the HIV infection rate is high and for women who received blood transfusions after HIV entered the United States and before blood was being screened for HIV (i.e., between 1978 and 1985). The purpose of this recommendation is to counsel women to avoid a pregnancy that could transfer HIV to the newborn. For women already pregnant, testing can help identify the presence of HIV and ensure proper medical care for them and their newborns. Counseling on family planning and future pregnancies can also be provided. **Figure 7.10** illustrates the percentage of U.S. individuals aged 18 to 64 who reported being tested for HIV infection by race/ethnicity in 2011.

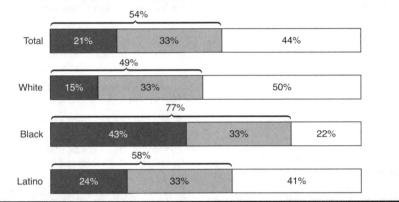

Percent of nonelderly, ages 18–64, who say they have been tested for HIV

■ Yes, in last 12 months   ▨ Yes, but not in last 12 months   ☐ No, never tested

Total — 54%
21% | 33% | 44%

White — 49%
15% | 33% | 50%

Black — 77%
43% | 33% | 22%

Latino — 58%
24% | 33% | 41%

FIGURE 7.10

Percentage of nonelderly who report being tested by HIV by race/ethnicity in 2011. *Source:* The Henry J. Kaiser Family Foundation HIV/AIDS Policy Fact Sheet: HIV Testing in the United States, July, 2012. This information was reprinted with permission from the Henry J. Kaiser Family Foundation. The Kaiser Family Foundation, a leader in health policy analysis, health journalism and communication, is dedicated to filling the need for trusted, independent information on the major health issues facing our nation and its people. The Foundation is a non-profit private operating foundation, based in Menlo Park, California.

Furthermore, voluntary testing is advised for persons undergoing medical evaluation or treatment because selected clinical signs may point to HIV infection. For example, individuals may have HIV infection if they display generalized lymphadenopathy, dementia, chronic fever or diarrhea, unexplained weight loss, or diseases such as herpes simplex, candidiasis, or tuberculosis. When medical evaluation is conducted in a hospital, the hospital staff may suggest routine testing on admission.

Crucial to the effort to widen the scope of HIV counseling and testing is the public's perception that patient information will be held confidential, and that persons found positive will not suffer discrimination. Confidentiality can be increased by improving the record-keeping practices of a health department, hospital, or other healthcare setting and by protecting the records within the parameters of state law. Inevitably, certain "need to know" situations will arise, and inappropriate disclosures and unauthorized releases of information will occur; however, public health policy should carefully consider ways to reduce the harmful impact of such disclosures.

There is also a difference between "counseling" and "effective counseling." In 2001, the CDC published revised guidelines for HIV counseling, which delineated an effective set of open-ended questions that promote client-centered HIV counseling and suggested certain risk-reduction steps that would more effectively change behavioral patterns. **Table 7.4** outlines these questions.

In the final analysis, the transmission of HIV can be reduced by an expanded program of counseling and testing, but the success of such a program depends on the level of participation. Individuals are more likely to participate when they believe they will not suffer discrimination in employment, school admission, housing, and medical services. No known medical evidence suggests that ordinary social situations permit the spread of HIV; thus, discrimination is not warranted. Diagnosis through various modes of testing can be a powerful ally in the effort to control AIDS; however, diagnosis must be used as it was intended, and that precludes discrimination.

TABLE 7.4 — Types of Questions that Interfere with and Promote HIV Counseling

Closed-Ended Questions that Might Interfere with Client-Centered HIV Prevention Counseling	Open-Ended Questions that Promote Client-Centered HIV Prevention Counseling
Have you ever injected drugs? OR	What are you doing that you think may be putting you at risk for HIV infection?
Have you (for a male client) ever had sex with a man? OR Have you (for a female client) ever had sex with a bisexual man?	What are the riskiest things that you are doing? If your test comes back positive, how do you think you may have become infected? When was the last time you put yourself at risk for HIV? What was happening then?
Have you ever had sex when you were under the influence of alcohol or drugs?	How often do you use drugs or alcohol? How do you think drugs or alcohol influence your HIV risk?
Do you (always) use condoms when you have sex? OR	How often do you use condoms when you have sex?
Can you always use condoms when you have sex?	When/with whom do you have sex without a condom? When with a condom? What are you currently doing to protect yourself from HIV? How is that working? What kinds of things do you do to protect your partner from getting infected with HIV? (for HIV-infected clients) Tell me about specific situations when you have reduced your HIV risk. What was going on that made that possible?
Can you always use clean works (i.e., needles, syringes, cottons, or cookers*) when you inject?	How risky are your sex/needle-sharing partners? For example, have they been recently tested for HIV?

*Cottons are filters used to draw up the drug solution. Cookers include bottle caps, spoons, or other containers used to dissolve the drugs.

Reproduced from CDC, *Morbidity and Mortality Weekly Report*, vol. 50/RR-19, November 9, 2001.

LOOKING BACK

Diagnosis has substantial importance in the effort to interrupt the AIDS epidemic because infected individuals can be identified, and measures can then be taken to counsel them on preventing the spread of HIV, while encouraging them to seek treatment. Among the most important laboratory procedures used to assist diagnosis are the HIV antibody tests. These are indirect tests based on the supposition that the body's immune system produces antibodies in response to exposure to an infectious agent such as HIV. The viral load test is another key diagnostic procedure. It is a direct test used to detect the RNA of HIV or the DNA associated with proviral HIV.

Two HIV antibody tests are the ELISA test and the Western blot analysis. In the ELISA test, patient's serum is added to HIV antigens, and a color change indicates that antibodies are present in the serum. For the Western blot analysis, HIV antigens are separated in a gel and the patient's serum is added. If HIV antibodies exist in the serum, they will react with the separated antigens and form bands detected by a staining reaction. Both tests have 99% sensitivity and specificity, but both can give false negatives and false positives. For example, an infected individual may not have had sufficient time to produce enough antibodies for the test to detect, and a false negative may result. Other serologic tests are also available for confirmatory testing, and some are available as FDA-approved home tests. For the latter, the individual provides a blood, saliva, or urine sample for testing. The first rapid home HIV test was licensed by the FDA in 2012.

The HIV viral load NAT is a direct test providing evidence of HIV RNA genomes in plasma. A gene probe composed of DNA seeks out and combines with DNA from HIV proviruses obtained from infected lymphocytes. Because too little viral DNA may be available for a reliable test, it is advantageous to amplify the DNA through a procedure called the PCR. In this technique, enzymes and primers are added to DNA from a patient's cells to increase the content of any viral DNA present. The gene probe is then more efficient. Contemporary tests detect HIV RNA and give results in terms of a number of copies of HIV RNA in a milliliter of blood. The viral load test has yielded new insights on the disease progression to an AIDS status of an individual as well as monitoring the patient for antiretroviral resistant HIV strains.

Considerable controversy exists about who should be tested for HIV and what should be done with the test results. When confidentiality is broken, various forms of discrimination can ensue. Currently, mandatory testing is required for blood, plasma, and organ donors, military applicants and active duty personnel, federal and state prisoners under certain circumstances, and newborns in some states. Voluntary testing is requested of those in high-risk groups and selected others to help interrupt the AIDS epidemic. High-risk groups encompass injection drug users, men who have unprotected sex with men, multiple partners or anonymous sex, and individuals who have been diagnosed with STDs like syphilis.

Healthline Q&A

Q1 After I took an AIDS test, the physician said that I had not seroconverted and that I was seronegative. What do those words mean?

A Serum is the clear fluid of the blood where antibodies are dissolved; it is used to test for HIV antibodies. When the serum gives a positive reaction (HIV antibodies are present), the person has seroconverted and is said to be seropositive. In contrast, if no reaction occurs (no HIV antibodies), the doctor uses the term "seronegative," as in your case, and says you have not seroconverted.

Q2 What does "anonymous testing" mean?

A Some states offer free testing for individuals who perform high-risk behaviors so that those persons can learn whether they have suffered HIV exposure. At the testing center, the person is given a number, and then the technician takes a small amount of blood for testing. Some days later, the test results are listed by number and given to the person having that number, along with counseling on the meaning of the results. No names or addresses are ever requested or used.

Q3 Why is a viral load test better than the AIDS antibody tests?

A The HIV antibody tests (ELISA, Western blot analysis, and others) search for indirect evidence of HIV by identifying antibodies produced in response to HIV's presence. In contrast, the viral load test searches for RNA fragments of HIV and thus is a direct test for HIV's presence. A direct test is considered more reliable than an indirect test.

Q4 Why is the viral load test better for detecting HIV in newborns than the AIDS antibody tests?

A Newborns receive antibodies from their mothers through the placenta and umbilical cord, and the AIDS antibody tests may be detecting the mother's antibodies rather than those of the child who is exposed to HIV. It is therefore advisable to use a direct test such as the viral load test to determine whether HIV is actually present in the newborn.

Q5 Who has to know about the results of my AIDS test?

A Different states have different regulations about the confidentiality of AIDS test results. The physician-patient relationship is one level of confidentiality that must be maintained. Many health departments offer confidential AIDS testing; others make anonymous testing available. Before you agree to be tested, you should be informed about what will be done with the results, and the information should be clearly spelled out.

REVIEW

Having completed this chapter on diagnosis and testing, you should be able to conceptualize the laboratory methods for detecting HIV and AIDS and summarize the uses, advantages, and shortcomings of the procedures. To test your knowledge, enter the word or words that best completes each of the following statements. The correct answers can be found elsewhere in the text.

1. _____ is the acronym used for the laboratory test most widely used to detect HIV antibodies.

2. _____ characterizes the result of an HIV antibody test when the person is infected but does not have sufficient antibodies to give a positive test.

3. _____ is the technique in which an electric current is used to separate HIV proteins (antigens) used in the Western blot analysis.

4. _____ are the cells that must be obtained from a patient to perform a gene probe.

5. _____ is the amount of HIV in the body when a test for HIV RNA is performed.

6. _____ is the name given to any test that determines the presence of antibodies in a patient.

7. _____ is the validation test most frequently used when the ELISA test is positive.

8. _____ is the result that can occur if an uninfected person has antibodies that react with HIV antigens in a laboratory test.

9. _____ are the main targets of HIV fragment tests.

10. _____ is the reaction used to amplify the amount of DNA in an HIV test using a gene probe.

11. _____ is the agency of the United States government that licenses HIV diagnostic tests.

12. _____ is the process wherein blood banks combine samples of donor bloods for HIV testing.

13. _____ is the material obtained from a patient to perform an ELISA test.

14. _____ is the source of the antibodies in a newborn's blood that makes an HIV antibody test inefficient.

15. _____ is a fragment of DNA that seeks out and binds with a complementary DNA fragment such as that from an HIV provirus.

16. _____ is the approximate time after exposure to HIV required for a person's immune system to make enough antibodies to give a positive ELISA test.

17. _____ is the level of specificity and sensitivity exceeded by both the ELISA test and Western blot analysis.

18. _____ is the material other than polymerase enzymes and nucleotides that must be added to DNA to carry out the PCR.

19. _____ is the serologic test in which a color change signals that HIV antibodies are present in the serum of the patient.

20. _____ is one group for which HIV testing is mandatory in the United States.

FOR ADDITIONAL READING

Branson, B. M., et al., 2006. "Recommendations for HIV testing of adults, adolescents, and pregnant women in healthcare settings." *MMWR* 55(RR14):1–17.

Florom-Smith, A. L., De Santis, J.P., 2012. "Exploring the concept of HIV-related stigma." *Nursing Forum* 47(3):153–175.

Ivers, L. C., et al., 2007. "Provider-initiated HIV testing in rural Haiti: Low rate of missed opportunities for diagnosis of HIV in a primary care clinic." *AIDS Research and Therapy* 4:28.

Korenromp, E. L., et al., 2009. "Clinical prognostic value of RNA viral load and CD4 cell counts during untreated HIV-1 infection—a quantitative review." *PLoS ONE* 4(6):e5950.

Lindholm, P.F. et al., 2011. "Approaches to minimize infection risk in blood banking and transfusion practice." *Infectious Disorders-Drug Targets* 11:45–56.

Luft, L. M., et al., 2011. "HIV-1 viral diversity and its implications for viral load testing: review of current platforms." *International Journal of Infectious Diseases* 15:e661–e670.

Mellors, J. W., 1998. "Viral load tests provide valuable answers." *Scientific American* 279:90–93.

Sullivan, P. S., et al., 2012. "Successes and challenges of HIV prevention in men who have sex with men." *The Lancet* 380:388–399.

The Henry J. Kaiser Family Foundation, 2012. *HIV/AIDS Policy Fact Sheet: HIV Testing in the United States, July 2012*. Retrieved November 27, 2012 from www.kff.org/hivaids/6094.cfm.

Treating HIV Infection and AIDS

LOOKING AHEAD

Development of therapeutic agents for treating HIV infection and AIDS has increased the length and quality of patients' lives and has strengthened optimism that additional therapies can be found. This chapter discusses some of the available approaches to treating patients. On completing the chapter, you should be able to. . .

- Explain the principles that apply to treatment of HIV infection and AIDS, including inhibitory mechanisms of AIDS drugs and unique problems associated with AIDS therapies.
- Identify the mode of action, benefits, and possible side effects of the drug azidothymidine (AZT). Name and discuss several other reverse transcriptase inhibitors used to treat AIDS.
- Explain the positive impact that protease inhibitors have had in fighting the AIDS epidemic. Summarize the biochemical activities of fusion inhibitors, entry inhibitors, coreceptor antagonists, and integrase inhibitors.
- Compare and contrast postexposure prophylaxis and preexposure prophylaxis and how they relate to HIV infection.
- List some drug therapies available for treating the opportunistic diseases associated with AIDS.
- Describe the methods for testing the effectiveness of experimental drugs before release for patient use, and discuss some of the ethical considerations involved in drug testing.

INTRODUCTION

Yellow fever stands out as one of the most savage diseases ever to strike the United States. During the 1700s, for example, historians chronicled 35 separate epidemics. The disease raged through the country like a firestorm, but no city was hit harder than Philadelphia.

In 1793, Philadelphia was the largest city and capital of the United States. When yellow fever broke out among the 40,000 residents, the panic rivaled that in Europe during the plague years. Thousands fled the city, and those who remained lived in fear. Officials posted warning notices where infected people lived and assigned guards to quarantine the sick. Uninfected individuals sought protection by wearing cloth masks soaked in garlic juice, vinegar, or camphor. For the sick, physicians prescribed blood-lettings, purges, and ice water baths. In the end, most of the 24,000 people remaining in the city contracted the disease. Almost 5,000 perished.

Yellow fever continues to occur in contemporary times, even though considerably more is known about it than 200 years ago. We know, for example, that yellow fever is caused by a virus that attacks the liver and causes bile to seep into the bloodstream. We understand that yellow fever is transmitted by mosquitoes and that epidemics can be interrupted by controlling mosquito populations. Unfortunately, we realize that for the person having yellow fever, the possibility of a cure is not much better than it was 200 years ago.

A similar pattern applies to many diseases. For instance, the agents of hepatitis, polio, rabies, mononucleosis, measles, herpes, and many other viral diseases are well known; the modes of transmission for these diseases have been identified, and epidemiologists can interrupt their spread. Drugs for treatment are relatively rare, however, and in many cases are unknown. Indeed, for viral diseases, therapeutic drugs are the exception rather than the rule.

For the first dozen years of the AIDS epidemic, physicians were generally pessimistic about the development of therapeutic drugs. In recent years, however, their pessimism has been replaced by optimism. Combinations of the drug **azidothymidine (AZT)** and **protease inhibitors** have proven helpful for lessening the symptoms of HIV infection and prolonging life. In addition, many other drugs are in various stages of development, and the search for therapeutic agents is well funded and has a high priority at several universities and pharmaceutical companies. Numerous novel approaches to therapy are also being pursued by researchers. We have reached a point in which new discoveries about HIV suggests a world without HIV/AIDS is within our reach. That being said, *Turning the Tide Together* was the theme for the XIX International AIDS Conference in 2012 held in Washington, D.C.

General Principles of AIDS Treatment

Through the early and mid-1980s, healthcare professionals viewed AIDS as an acute, almost immediately lethal crisis. Toward the end of the 1980s and into the 1990s, however, this viewpoint changed. During these years, new therapies for AIDS came into use, and with their acceptance, the length and quality of life of AIDS patients have improved. For those with access to reasonably good healthcare, AIDS has become a chronic disease, with an estimated 12-year course, on average. Greater physician expertise coupled with a higher level of patient motivation and improved therapies have changed the dim prospects once facing AIDS patients. HIV diagnosis is no longer a death sentence. The life expectancy of a male diagnosed with HIV at the age of 25 is only 0.4 years less compared to a healthy/uninfected male at the age of 25. If an individual has access to antiviral therapy and does their best to maintain their health through diet, exercise, and adhering to HIV treatment, the person can hold HIV in check and can live a relatively normal life.

At the forefront of this change are a number of therapeutic drugs. Before examining their nature and mode of action, here we explore some principles that apply to drug therapy, discuss the approaches taken by drugs that interfere with HIV activity, and identify some unique problems associated with AIDS therapy.

Objectives of Therapeutic Drugs

To be effective against a pathogen, a therapeutic drug must either kill the pathogen or prevent it from multiplying in the body; however, an effective therapeutic agent need not rid the body of all traces of the pathogen. By preventing the pathogen from multiplying, a drug can slow or halt the progress of disease. For example, people with tuberculosis have been able to lead fairly normal lives by taking therapeutic drugs to prevent the tubercle bacilli from proliferating. Unable to multiply, the bacilli do not cause symptoms or pass in significant numbers to the next individual.

To kill a pathogen or inhibit its multiplication, a therapeutic agent generally interferes with a sequence of biochemical reactions unique to that organism. Such interference is possible in bacteria because they are so different from humans; thus, interfering with bacterial biochemistry does not affect human biochemistry. Penicillin, for instance, interferes with the construction of the bacterial cell wall and prevents bacteria from proliferating (**Figure 8.1**). Because human cells have no cell walls, penicillin does not interfere with their normal activity. Viruses, in comparison, do not carry out independent biochemical reactions on which a drug can act, and thus, antibiotics do not affect viruses; however, drugs may interfere with viral replication, a biochemical process of high complexity.

To be effective, a therapeutic drug must do no damage or only minimal damage to the body; however, years of research and observation have made it clear that taking any therapeutic drug carries risk. Even penicillin, generally regarded as a "safe" drug, can induce severe and sometimes fatal allergic reactions. For many diseases, the risk associated with a drug treatment must be balanced against the benefit. When one has influenza, for instance, it may not be necessary to risk potentially toxic drugs because the disease is usually mild; when one has AIDS, in contrast, the illness is life threatening, and the risk of taking a potentially toxic drug may be warranted.

There is also the dilemma of when to institute AIDS therapy. Early intervention allows simpler therapeutic management in the long term because patients experience fewer side effects. Early intervention also provides the best chance for minimizing the development of **opportunistic infections** and reduces the possibility of immune system damage, especially in children in whom the large size of the thymus encourages a dramatic immune reconstitution.

FIGURE 8.1

A photomicrograph of a bacterium exploding on exposure to penicillin (×50,000). The penicillin has interfered with cell wall synthesis in the bacterium and left it with only a surrounding cell membrane. Internal pressures have led to membrane disruption. Penicillin is useful against bacteria but not viruses because the latter do not synthesize a cell wall. © CNRI/Science Source.

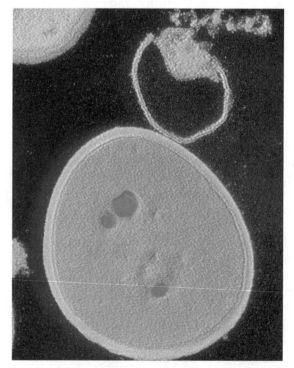

In March 2013, a Mississippi baby born with an HIV infection was reported in the news media to be "cured" and HIV-free. The child's mother was infected with HIV and had not received any antiviral treatment. The infant was started on an aggressive cocktail of antiviral treatment consisting of AZT, 3TC, and nevirapine 30 hours after birth. Initially the HIV viral load of the infant was high but became undetectable after one month. After a year, the infant stopped receiving antiviral medication. At 23 months the baby was tested and viral loads were still undetectable despite being without treatment for a year.

Also in March 2013, French researchers reported a study in the journal *PLoS Pathogens* in which 14 adult patients infected with HIV appeared to be in "HIV remission." All patients began HIV treatment early after diagnosis. They stayed on HIV antivirals for several years but then stopped taking anti-HIV drugs. This study argues that early initiation of HAART open up new therapeutic perspectives for HIV-infected individuals.

Those opposed to early therapy point out the reduced quality of life for patients who have no symptoms, the possible development of resistant HIV strains if patients do not adhere to the regimen of drugs, and the possible reduction in the useful lifespan of a drug if resistant strains of HIV arise. Ultimately, the treatment decision is made by the patient and his or her physician.

Approaches to Drug Therapy

Against this background, we can begin to appreciate some of the problems that must be considered in developing AIDS treatments. High among these problems is combating HIV in the body without damaging the body. To find HIV's "weak spot," researchers attempt to locate a step in viral replication that is dissimilar to events in the life of a normal host human cell. Locating and interfering with several steps, rather than one step, is clearly more advantageous, which is why a combination of drugs may be better than a single drug. By the middle of 2012, there were 24 Food and Drug Administration (FDA)-approved drugs available to treat HIV infections. These drugs are grouped into six distinct classes of drugs based on their molecular mechanism (**Table 8.1**).

TABLE 8.1	The Six Classes of HIV Drugs Licensed by the FDA
Drug Class	**Mechanism of Drug**
Nucleoside-analog reverse transcriptase inhibitors	Inhibits HIV reverse transcriptase by acting as a competing deoxynucleotide needed to create/reverse transcribe DNA from the HIV RNA. The HIV reverse transcriptase cannot add a natural nucleotide (A, T, C, or G) to the analog drug that has been incorporated into the growing strand of new DNA, thereby blocking the replication of HIV DNA, which is also referred to as DNA chain termination.
Non-nucleoside reverse transcriptase inhibitors	Inhibits HIV reverse transcriptase by binding to a different site on the HIV reverse transcriptase, directly inhibiting it from its enzymatic functions.
Integrase inhibitors	Blocks HIV integrase activity, abolishing the integration of a DNA copy of the HIV RNA genome into the host cell chromosome.
Protease inhibitors	Inhibit HIV maturation step by blocking an HIV-specific protease, resulting in a noninfectious virus.
Fusion inhibitors	Blocks the structural changes required for envelope of HIV to fuse with the host cell plasma membrane, blocking entry/infection of the CD4 containing T-cells.
Coreceptor antagonists	Block coreceptor function required for viral entry into host cells.

One approach to drug therapy might be to block the binding of HIV to its host cells by altering the gp120 molecules in the envelope spikes of HIV. Binding may also be inhibited by blocking CD4 or the CCR5 co-receptor molecules on the surface of helper T-lymphocytes. Another possibility might be to keep viral RNA and reverse transcriptase from escaping their protein coat after HIV has penetrated the cytoplasm. Preventing RNA-to-DNA synthesis via **reverse transcriptase** (the function performed by AZT) is also a desirable approach. Interference with reverse transcriptase activity is particularly attractive because the enzyme exists only in cells infected with retroviruses such as HIV. Interfering with the transcription of HIV mRNAs and HIV integrase activity can be targets. **Integrase inhibitors** are designed to block HIV integrase activity required for the integration of HIV proviral DNA into the host cell chromosome. Viral proteins must be modified before assembly with RNA to form new infectious HIV particles, and here lies another opportunity for possible drug action (this is where the **protease inhibitors** work). Finally, the budding of HIV from infected cells conceivably could be prevented. There are six distinct steps in the HIV-1 life cycle that are current or potential targets for antiretroviral drugs (**Figure 8.2**). Each step has its own time window of drug inhibition (Figure 8.2).

Although there appear to be many places where HIV seems vulnerable, there are also many unique features of HIV that require attention. As noted previously, HIV presents an elusive target because it can reside in cells as a provirus. Moreover, HIV infects a variety of cells and thus, any drug devised must take into account the possible toxic effect on different types of cells. A particularly difficult problem arises with HIV infection of brain cells: Brain cells are separated from the remainder of the body by the **blood–brain barrier**, a series of membranes that prevent substances from moving out of the blood into the brain cells. To be effective, a drug must cross this barrier. In addition, there is the problem of possible brain cell damage by the drug. Because adult brain cells are among the body cells least able to regenerate themselves, it may be impossible to reverse such brain damage.

Treating children with AIDS presents special challenges for physicians because HIV tends to behave more aggressively in children than in adults (indeed, children progress to AIDS much more rapidly than adults). Moreover, physicians have a relatively small body of research to consult when prescribing anti-HIV medications for children, and fewer drugs are available in acceptable formulations to treat young patients (swallowing pills, for example, can be difficult for young patients). In addition, treating children has greater urgency because HIV invades the brain early in the disease and the infection retards intellectual development, while impairing motor coordination and stifling physical growth. Families can also experience difficulty in following intensive and strict drug regimens (e.g., when a child is at day care). Despite these difficulties, great strides have been made in reducing the incidence of AIDS in children.

Treating **Kaposi's sarcoma** and opportunistic diseases while concurrently attempting to destroy HIV represents another challenge. Most infectious diseases are due to a single microorganism, and the physician can usually marshal available resources against that one pathogen. AIDS, in comparison, is a multifaceted disease, bringing numerous opportunistic diseases in addition to the HIV infection. Drug therapy must therefore be multifaceted, and physicians must battle on many fronts at one time. In addition, drugs may interfere with one another, and a patient may be weakened by drug toxicity. The cumulative stress on the body from several concurrent illnesses may make drug therapy very difficult.

Mutations and Other Considerations

Another key issue concerns HIV mutation and the drug resistance related to mutation. **Mutation** results in a permanent change in the HIV genome, a change that results in an altered protein. If the altered protein is a drug-targeted enzyme such as protease or

(a)

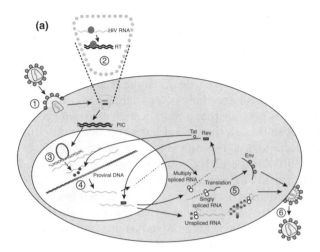

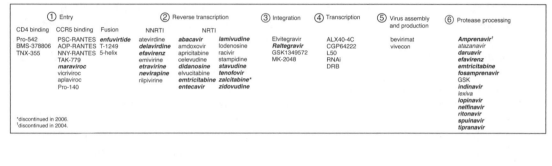

① Entry			② Reverse transcription			③ Integration	④ Transcription	⑤ Virus assembly and production	⑥ Protease processing
CD4 binding	CCR5 binding	Fusion	NNRTI	NRTI					
Pro-542	PSC-RANTES	*enfuvirtide*	atevirdine	*abacavir*	*lamivudine*	Elvitegravir	ALX40-4C	bevirimat	*Amprenavir[†]*
BMS-378806	AOP-RANTES	T-1249	*delavirdine*	amdoxovir	lodenosine	*Raltegravir*	CGP64222	vivecon	atazanavir
TNX-355	NNY-RANTES	5-helix	*efavirenz*	apricitabine	racivir	GSK1349572	L50		*daruavir*
	TAK-779		emivirine	celevudine	stampidine	MK-2048	RNAi		*efavirenz*
	maraviroc		*etravirine*	elvucitabine	*stavudine*		DRB		emtricitabine
	vicriviroc		*nevirapine*	*emtricitabine*	*tenofovir*				*fosamprenavir*
	aplaviroc		rilpivirine	*emtricitabine*	*zalcitabine**				GSK
	Pro-140			*entecavir*	*zidovudine*				*indinavir*
									lexiva
									lopinavir
									nelfinavir
									ritonavir
									spuinavir
									tipranavir

*discontinued in 2006.
[†]discontinued in 2004.

(b)

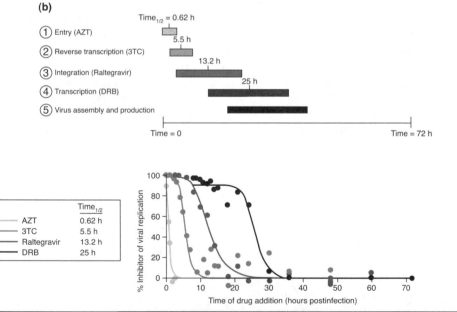

① Entry (AZT) — Time$_{1/2}$ = 0.62 h
② Reverse transcription (3TC) — 5.5 h
③ Integration (Raltegravir) — 13.2 h
④ Transcription (DRB) — 25 h
⑤ Virus assembly and production

Time = 0   Time = 72 h

	Time$_{1/2}$
AZT	0.62 h
3TC	5.5 h
Raltegravir	13.2 h
DRB	25 h

% Inhibitor of viral replication
Time of drug addition (hours postinfection)

FIGURE 8.2

(a) Six possible approaches for developing anti-HIV therapies by interfering with the life cyle of HIV-1. Note: any drugs not italicized were abandoned in pre-clinical trials. Drugs in *bold italicized* text have been FDA-approved. (b) *In vitro* time course showing the antiviral drug action HIV-1 inhibitors that target five different steps in the life cycle of HIV-1 in infected cell cultures. Some drugs inhibit early in the life cycle such as AZT while others like DRB, a transcription inhibitor, takes at least 25 hours to have an impact against HIV-1. A and B modified from Arts, E. J., Hazuda, D. J. 2012. "HIV-1 antiretroviral drug therapy." *Cold Spring Harbor Perspectives in Medicine* 2;2:a007161.

reverse transcriptase, then it will resist binding to the drug and the latter will become useless. For example, nevirapine and related drugs inhibit HIV particles by reacting with their reverse transcriptase molecules, but a single mutation in the enzyme-encoding gene changes the active site where the reaction occurs and binding fails. Scientists estimate that at least one mutation occurs each time an HIV particle undergoes replication; thus, in the approximate 10 billion HIV particles produced daily in a patient's body, a host of mutants is probably present. Moreover, it is possible that a mutant with resistance to an anti-HIV antiviral drug has developed even though the patient has never taken the drug.

The ability of HIV to mutate and become drug resistant is best exemplified by the "superstrain" of HIV first reported in 2005. At the 12th Conference on Retroviruses and Opportunistic Infections in Boston, Dr. David Ho, director of the Aaron Diamond AIDS Research Center in New York City and chairman of the conference, presented evidence of a new mutant "**superstrain**" of HIV that appeared to cause a rapid acceleration into AIDS. This strain was isolated from a gay man who acknowledged having anonymous unprotected anal sex with hundreds of men, often while under the influence of crystal methamphetamine. Dr. Ho believes the man was originally infected with HIV in October 2004 and by January 2005 his CD4 cell count had declined to 80, an eventuality that usually takes years. Dr. Ho's laboratory detected viral mutations in this strain that conferred resistance to 19 of the 20 drugs commonly used to treat HIV. The fact that this new superstrain is both resistant to most anti-HIV antivirals and exhibits a rapid onset to AIDS makes this virus of particular concern to health officials. Because the man from whom this virus was isolated had sex with hundreds of men, the potential spread of this virus will be watched very closely over the next several years to assess its real threat.

How, then, does a population of resistant viruses emerge in the body? The answer is based on the Darwinian process of selection. An anti-HIV antiviral drug, such as nevirapine, for example, will react with all of the HIV particles that contain susceptible reverse transcriptase and thereby neutralize 99.99% of them; however, the few resistant viruses will survive and continue to replicate, and soon they will comprise the dominant population (**Figure 8.3**). Their replication will result in more mutants, probably with increased resistance to the nevirapine, and soon the drug will become ineffective. As we discuss presently, physicians learned the implications of mutation-based resistance in the 1990s, and they switched from single-drug therapy (**monotherapy**) to combinations of drugs [multidrug or triple combination therapy also known as **highly active antiretroviral therapy (HAART)**] to reduce the likelihood of encountering drug-resistant

FIGURE 8.3

How drug resistance emerges in a population of HIV particles. An anti-HIV drug such as nevirapine reacts with susceptible HIV particles and neutralizes them. In the population, however, some resistant HIV particles exist as a result of mutation, and these resistant particles soon multiply and become the dominant members of the viral population. Nevirapine is no longer useful.

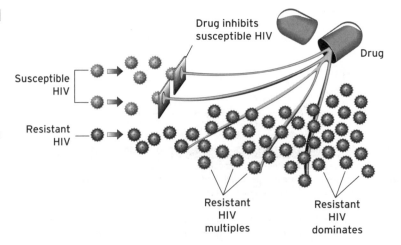

Treating HIV Infection and AIDS

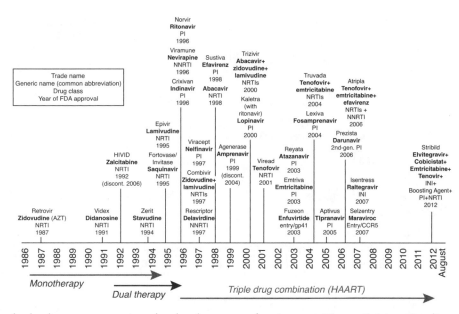

FIGURE 8.4

Timeline of antiretroviral drug development and therapy regimes. Data from Arts, E. J., Hazuda, D. J. 2012. "HIV-1 antiretroviral drug therapy." *Cold Spring Harbor Perspectives in Medicine* 2;2:a007161.

viruses in the body or encouraging the development of resistance. **Figure 8.4** is a timeline of FDA-approved antiretroviral drugs.

Because drug resistance is a widespread problem in anti-HIV drug therapy, methods of assessing the resistance are a focus of research efforts. In general terms, the presence of HIV in the blood four to six months after therapy has begun is a signature of HIV replication and potential drug resistance. A **nucleic acid test** (NAT) approved by the FDA in 2002 verified that resistance was occurring. The test, called Trugene, examines the genetic composition of the patient's HIV and checks for mutations, using a software program to compare the genome of the patient's HIV to a list of more than 70 mutations linked to resistance to specific drugs. This is called a **genotypic test**. A **phenotypic test**, in comparison, checks for resistance by measuring the ability of a specific drug to decrease replication of the patient's HIV by 50% under laboratory conditions. If resistance to several drugs is encountered, regaining control of viral multiplication requires salvage therapy. This therapy employs a fresh set of drugs, preferably new to the marketplace; a lower reduction in viral load is anticipated at the outset of the therapy. The FDA has approved five additional NATs that are also used for patient monitoring, especially detecting mutations in the HIV protease and reverse transcriptase genes of HIV that confer resistance to specific types of antiretroviral drugs.

The problem of general inexperience in treating viral diseases is also worthy of note. Through the 1980s, most pharmaceutical companies concentrated on drugs that target bacterial diseases. Past efforts had turned up few antiviral drugs, and most viral diseases were not viewed as life threatening (except, for example, rabies, smallpox, and polio). In addition, viral vaccines were effective in halting epidemics and preventing new ones. Pharmaceutical companies, therefore, were not inclined to conduct exhaustive searches for antiviral drugs; this meant that little experience in the development of such drugs was acquired. This inexperience may be the most significant problem that needs to be resolved if a successful therapeutic agent for HIV infection and AIDS is to be located (**Table 8.2**).

On the positive side of the ledger, a substantial effort is under way to develop effective therapies for HIV infection and AIDS. As early as 1989, a features reporter for the *American Society for Microbiology News* wrote that endeavors "to develop drugs for combating acquired immune deficiency syndrome (AIDS) make the U.S. 'War on Cancer' seem like a mere skirmish." In the next paragraph, the reporter wrote, "Unquestionably,

Focus on HIV

Combinations of antiretroviral drugs can safely and reliably suppress HIV replication in the body below the limits of detection in most HIV-infected people receiving this therapy.

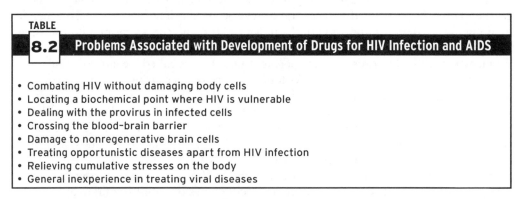

Mapping HIV

Canada

- An estimated 68,000 adults and children were living with HIV in Canada in 2009.

- An estimated 21,000 women (age 15+) in Canada were living with HIV in 2009.

- Approximately 26% of Canadians are unaware that they are infected with HIV.

- The estimated percentage of people living with HIV who are unaware of their HIV status varies by exposure category: 25% of them are injection drug users, 35% are heterosexuals, and 19% are men who have sex with men.

- HIV testing first became available in Canada in 1985.

- Anonymous HIV testing is only available in seven Canadian provinces.

- Given the new treatments available for HIV, it is more important than ever that all Canadians should be able to access HIV testing.

Data from: *UNAIDS Report on the Global AIDS Epidemic*, 2010 and *HIV/AIDS Epi Update*. Public Health Agency of Canada, Centre For Communicable Diseases and Infection Control–July, 2010.

TABLE 8.2 Problems Associated with Development of Drugs for HIV Infection and AIDS
• Combating HIV without damaging body cells
• Locating a biochemical point where HIV is vulnerable
• Dealing with the provirus in infected cells
• Crossing the blood-brain barrier
• Damage to nonregenerative brain cells
• Treating opportunistic diseases apart from HIV infection
• Relieving cumulative stresses on the body
• General inexperience in treating viral diseases

a massive AIDS drug development campaign is under way in the United States." To be sure, the search for treatments for HIV infection and AIDS is one of the most intensive research efforts of our time. Led by the National Institutes of Health (NIH), the search consumes hundreds of millions of dollars in federal funds annually. This commitment is viewed as a cause for optimism. The news headlines have changed. For example, in August 9, 2012 it was: "*HIV Prevention Drug OK for Heterosexuals, CDC Says.*" In the following sections, we examine some of the fruits of the search.

Reverse Transcriptase Inhibitors

Reverse transcriptase is the enzyme that catalyzes the synthesis of DNA using the RNA in the genome of HIV. This synthesis occurs after the HIV particle has penetrated a host cell, a T-lymphocyte. Reverse transcriptase is a valuable target for anti-HIV drugs because it does not occur in human cells (and thus the drug will not interfere with

T-lymphocyte metabolism) and because the gene that encodes reverse transcriptase is an essential element in viral replication. For these reasons, various reverse transcriptase inhibitors have been developed, as we explore next.

Azidothymidine

In 1964, Jerome P. Horwitz, an organic chemist at the Michigan Cancer Foundation, in Detroit, synthesized a new drug for treating cancer. The drug was similar to a building block of DNA but was counterfeit; that is, it was designed to confuse the cancer cells' genetic machinery and stop a tumor's growth. Unfortunately, when Horwitz and his colleagues injected the drug into mice with leukemia, the drug had little effect on the cancer. The drug was *azidothymidine (AZT)*. It would sit on the shelf with other failed anticancer drugs for 20 years.

During the 1970s, interest mounted in retroviruses, and a number of investigators tested AZT and similar drugs for activity against mouse retroviruses. The drugs were somewhat effective, but because there were no known retroviruses of humans, a practical benefit from the research was lacking. The drugs therefore remained in relative obscurity.

Then the AIDS epidemic broke out, and researchers began an active search for therapeutic drugs. Two of the researchers were Hiroaki Mitsuya and Samuel Broder of the NIH. In the summer of 1984, Mitsuya and Broder obtained a sample of HIV (then known as HTLV-III) from Robert Gallo, and they began testing various drugs against the virus. They took more than 300 drugs "off the shelf" and evaluated them. Fifteen of the drugs interfered with HIV replication in test tubes. One was AZT.

Broder and Mitsuya began an intensive effort to develop AZT as a therapy for AIDS patients. Working with the Burroughs Wellcome Company and researchers at Duke University, they found AZT to be a potent inhibitor of HIV in cultures of T-lymphocytes, and they worked out a concentration at which the toxic effects in human cells would be minimal. The first patients received AZT in July 1985.

A year later, Broder and Mitsuya announced the results of clinical trials conducted at 12 medical centers. Doctors gave the drug to 145 AIDS patients and an inert placebo to 137 AIDS patients. In the test group receiving AZT, the helper T-lymphocyte count rose, the immune response improved, resistance to *Pneumocystis jiroveci* pneumonia was enhanced, and the patients' lifespans increased over what was expected. In the spring of 1987, the FDA licensed AZT as a therapy for AIDS patients in the United States. In 1988, Burroughs Wellcome changed the drug's name from AZT to zidovudine and sold it by the trade name Retrovir; however, because most people know the drug as AZT, we continue to use that name.

Mode of Action

Since its introduction to general use in 1987, AZT has been a mainstay for treating AIDS. The chemical name of the drug is 3'azido-2',3'-deoxythymidine; it is a compound closely related to deoxythymidine (**Figure 8.5**). Deoxythymidine is an essential component of DNA. AZT acts by replacing deoxythymidine in the synthesis of DNA, a synthesis catalyzed by reverse transcriptase.

The key to AZT's activity lies in its similarity to deoxythymidine (Figure 8.5). Deoxythymidine is a nucleoside consisting of deoxyribose (the carbohydrate in DNA) and thymine (one of the four nitrogenous bases in DNA). In building a DNA molecule, molecules of deoxythymidine are linked by reverse transcriptase to other nucleoside building blocks via phosphate molecules. To form the link at the 3' (pronounced "3-prime") position, reverse transcriptase removes the —OH group from the deoxythymidine and attaches a phosphate molecule. The phosphate molecule then acts as a bridge to link the deoxythymidine to the next nucleoside in the DNA chain. The

FIGURE 8.5

The mode of action of AZT. (a) The normal DNA molecule consists of a series of nucleosides linked to one another by phosphate molecules. Four types of nucleosides are involved, each having a different nitrogenous base: A stands for adenine, C for cytosine, T for thymine, and G for guanine. Each nucleoside contains the carbohydrate deoxyribose (D). The phosphate molecule links at the 3' position of the deoxyribose. (b) AZT has a chemical structure similar to that of the nucleoside containing thymine. Thus, when AZT is present, it is erroneously taken up in place of the thymine-containing nucleoside as DNA is being formed; however, the 3' position of AZT contains an N_3 group, and a phosphate molecule cannot link here. Thus, DNA chain formation comes to an end with AZT's incorporation. In the absence of DNA, HIV replication comes to a halt because proviruses cannot form.

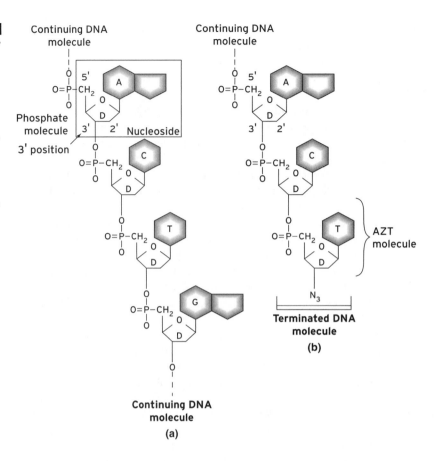

next nucleoside can be deoxycytidine, deoxyadenosine, deoxyguanosine, or another deoxythymidine molecule. The phosphate molecule is thus a linkage between the DNA building blocks, a "bridge" that joins the nucleosides together.

This is where AZT enters the picture. Because of its abundance in the cellular cytoplasm (the patient has taken a therapeutic dose) and its resemblance to deoxythymidine, AZT is taken up by reverse transcriptase and is slotted into the position where deoxythymidine should be placed, but AZT lacks the —OH group at the 3' position; it has a nitrogen group there instead. Reverse transcriptase cannot remove this nitrogen group to attach a phosphate molecule, and without the phosphate "bridge," a link cannot be forged to the next nucleoside. The effect is to abruptly halt the elongation of the building DNA chain. This mechanism, known as chain termination, thus prevents the production of **proviral HIV DNA**. Essentially, it interrupts the replication cycle of the virus.

AZT can also operate via a second mechanism. Once again, the key is the chemical similarity between AZT and deoxythymidine. Reverse transcriptase normally binds to deoxythymidine to incorporate it into the growing DNA chain, but when AZT molecules are abundant, the enzyme mistakenly binds to the AZT. This binding is irreversible. Reverse transcriptase becomes nonfunctional because it is unable to free itself from the AZT molecule. In effect, the AZT has successfully competed with the nucleoside for the active site on the enzyme molecule; in so doing, AZT has inhibited the enzyme. This mechanism is thus called **competitive inhibition**.

Theoretically, what AZT accomplishes could be carried out by other nucleoside substitutes, but this does not necessarily occur. A drug molecule, for example, must be able to enter affected cells, and not all nucleoside-like compounds can do so. Furthermore, a drug molecule must hook easily to phosphate at the "upper" end of the molecule (the 5' position), an ability not possessed by most nucleoside substitute. The question of enzyme preference also arises: Reverse transcriptase prefers AZT to deoxythymidine, whereas host cell enzymes prefer deoxythymidine to AZT for DNA synthesis. The result is that AZT can inhibit reverse transcriptase without affecting other cellular enzymes. Other compounds might not be preferred in this way. In addition, there may be the problem of decomposition: AZT escapes decomposition by cellular enzymes, but a substitute molecule might easily be broken down. A substitute molecule may be a potent inhibitor of HIV replication, but it will have little value if it is rapidly destroyed in cells.

Benefits and Uses

For persons with HIV infection or AIDS, AZT has been found beneficial for a number of reasons: It slows the spread of infection among the T-lymphocytes, and as the T-lymphocytes regenerate themselves, the immune system functions are partially restored (the count of helper T-lymphocytes, or CD4 cells, doubles or better in some patients); patients experience a weight gain. For many patients, the chronic fever leaves, and in some patients, the fungal infection caused by *Candida albicans* lessens or disappears. Perhaps the most significant benefit of AZT therapy is that the level of HIV declines demonstrably in the infected patient.

Several factors add to the value of AZT. It can be given orally because it is absorbed through the intestine; levels of the drug that inhibit viral replication in the laboratory can be achieved in the patient. Most of the drug remains active for at least an hour before being converted to an inactive compound by liver enzymes and excreted through the kidneys (the drug must be taken at regular intervals because of this elimination), and the drug can apparently penetrate the blood–brain barrier, enter the cerebrospinal fluid, and reach infected brain cells. Unfortunately, the toxicity of AZT somewhat tempers these positive factors, as we see here.

In 1988, researchers at the National Cancer Institute first reported that AZT could also alter the course of disease in children displaying the neurological symptoms of AIDS. Twenty-one children ranging in age from one to 12 years were given the drug by continuous intravenous infusion, and all showed improvement. In many children, the intelligence quotient (IQ) rose a significant number of points; in some cases, the youngsters' intelligence level returned to the level at which it was before they became ill. Scans of brain tissue showed that in some cases, shrunken tissue returned to its normal condition after treatment; younger children regained the ability to walk or talk, or they exhibited other developmental improvements. Because children appear to suffer more brain damage from HIV infection than adults do, the study was particularly encouraging (**Figure 8.6**).

One of the more remarkable success stories associated with AZT is its use in pregnant women to reduce the risk of HIV transmission to their unborn offspring. Dramatic results reported in 1994 summarized the effects of AZT in 748 HIV-infected women: 25% of children born to women administered placebos had HIV infection, whereas only 8% of children born to AZT-treated women were infected with HIV. In the years thereafter, the number of babies who were infected with HIV from their mothers declined substantially, largely because of AZT therapy. For example, in 1992, in the United States the number of newborns who developed pediatric AIDS was 907, whereas in 1997, the number was 297, a 67% decline. Some health officials were postulating that prenatal AZT treatment could reduce the risk of acquiring HIV to 3% (i.e., only 3% of HIV-infected women would give birth to infected babies).

Even limited AZT treatment in pregnant women is of significant value. In 1999, for example, researchers reported that administering AZT and another drug called 3TC during the last few weeks of pregnancy reduces transmission of HIV to the fetus

Studies using AZT in young children infected with HIV have shown that the drug can stem the progression to AIDS and can reverse some of the brain damage that often occurs in children. These babies being held by healthcare workers at the Birk Childcare Center in Brooklyn may be candidates for AZT therapy. © Hank Morgan/Science Source/Science Source.

by 50%. Furthermore, treating a woman intravenously with AZT at the beginning of labor and administering AZT to babies just after birth reduces HIV transmission by 37% (**Table 8.3**). Studies show that the risk of HIV transmission can be further reduced if a woman is treated with AZT and gives birth by cesarean delivery, as the infant has no opportunity to swallow blood or other fluid from the mother, nor can

TABLE 8.3	Efficacy in Newborns of HIV-Positive Pregnant Women Treated with AZT and 3TC			
Treatment	Number of Babies Tested	Number of Babies HIV+	Percentage Infected	Percentage Reduction in Risk
From 36 weeks to 1 week postpartum; baby treated for 1 week	359	31	8.6	50
From onset of labor to 1 week postpartum; baby treated for 1 week	343	37	10.8	37
From onset of labor to delivery; no treatment of baby	351	62	17.7	0
Placebo	273	47	17.2	0

Data from Cohen, J., *Science* 283 (1999): 916–917.

any of the mother's fluids contact the infant's mucous membranes or abrasions on its skin. Updated studies in developing countries conclude that without AZT intervention, 35% of HIV positive women will pass HIV on to their children: 5% of transmissions occur in utero, 15% during delivery and 15% during breastfeeding. Long-term studies indicate that AZT therapy poses no unusual risk for the newborn.

By the start of the twenty-first century, it was clear that AZT, when used in a combination of drugs, slows the progression to AIDS in persons who have HIV infection (as we discuss later here). The effects were far reaching: Many physicians and health officials who had opposed HIV testing because little could be done for those testing positive changed their minds and urged people with a history of high-risk behaviors to take the test. Their recommendations also dramatically increased the pool of potential AZT users to several hundred thousand. Scientists also found that a lower dose of AZT might be adequate for delaying the symptoms of AIDS. This was good news for two reasons: Lower doses would effectively reduce the cost of therapy and save the patient thousands of dollars annually, and lower doses would lessen side effects, as we discuss next.

Obstacles

No drug for microbial disease is taken without risk. Penicillin, for example, can lead to severe and sometimes fatal allergic reactions. Other drugs upset the microbial population of the intestine and permit fungal diseases such as **candidiasis** ("yeast disease") to emerge. Still others cause liver or kidney damage. In terms of risk, AZT is no different from such other drugs (**Table 8.4**). As long ago as 1985, the toxic side effects of AZT were known, and bone marrow suppression was a notable problem. The chief effect of this suppression is **anemia**, often so severe that a choice has to be made between continued use of AZT and the debilitating effects of anemia. Blood transfusions may be necessary to compensate for the anemia's effects.

Another side effect in many individuals who take AZT is **thrombocytopenia**. This condition develops when the body's blood platelets are progressively destroyed ("thrombocytes" are blood platelets; "penia" refers to reduction). Because platelets are required for blood clot formation, their reduction means that the blood fails to clot easily. Increased hemorrhaging, longer clotting times, and greater susceptibility to injury may result. Other side effects of AZT use include headaches, nausea, vomiting, seizures, and confusion. Symptoms such as these vary with individuals, with dosage, and with extent of the HIV infection or AIDS.

TABLE 8.4	Benefits and Obstacles to Use of AZT
Benefits	**Obstacles**
Increases patient life span	Suppresses activity in bone marrow
Interrupts viral replication	Induces anemia
Prevents progression to AIDS	Affects platelets and blood clotting
Penetrates into infected cells	Possible cause of cancer in animals
Remains active for 1 hour	Leads to drug resistance in viruses
Passes blood-brain barrier	May cause headache, nausea
Preferred by reverse transcriptase	Dose reduction causes symptom rise
Can be given orally	High cost

Another troublesome problem is the so-called **rebound effect**. Physicians have reported that when they recommend a lower dose of AZT to minimize anemia and other side effects, the dose reduction incites a dangerous and unexpected flare-up of symptoms. Neurologic problems were noted as a characteristic sign in one report, and a sharp increase in the level of bloodborne HIV was pointed out in a second report. Researchers speculated that AZT keeps the virus in check, but viral replication may occur at an increased rate when the drug is withdrawn.

Another problem is drug resistance. Drug resistance can develop when a drug destroys sensitive strains of a microorganism but allows more resistant mutants in the population to survive. As early as 1989, it became obvious that resistant strains of HIV were emerging in persons taking AZT. Of particular concern were people whose immune systems were so severely compromised that heavy doses of AZT were required to keep the HIV in check. In advanced cases of AIDS, increased viral multiplication yields increased possibilities for a strain with resistant traits. In a study at Canada's McGill University, for example, physicians gave AZT to 72 AIDS patients for 36 weeks and found that 20% of the patients harbored viruses with AZT resistance at the end of the study. Findings such as these also influence the use of AZT in HIV-infected patients in whom AIDS has not developed because treating people early and for long periods of time might encourage resistant strains of HIV to emerge and eventually make AZT useless.

Other Dideoxynucleosides

AZT belongs to a group of compounds known as **dideoxynucleosides**, that is, nucleosides missing oxygen groups at two positions, position 2' and position 3' ("dideoxy-"). Two other dideoxynucleosides that show anti-HIV value are dideoxycytidine (ddC) and dideoxyinosine (ddI). Both are inhibitors of the enzyme reverse transcriptase, working as chain terminators in a manner similar to AZT. Their side effects are less severe, however, and include skin rashes as well as peripheral neuropathy, such as headache, pain, and decreased touch, pinprick, temperature, and vibratory sensations. The drugs are often referred to in the technical literature as **nucleoside reverse transcriptase inhibitors** (**NRTIs**).

When ddI first became available in 1989, the FDA announced that it would allow wide distribution at the same time as tests were continuing to determine the effectiveness of the drug. This unusual landmark step was taken because the toxicity of AZT was too high for some AIDS patients and because ddI works in essentially the same way and could be expected to show benefits similar to those of AZT. Marketed as Videx, the drug could be used in rotation with AZT to minimize the respective toxicities of the drugs while maximizing their potency. In 1991, the FDA licensed ddI for use against HIV infection and AIDS. The drug is also known as didanosine.

The second dideoxynucleoside, ddC, was FDA approved in 1992. It was the first drug to be licensed through the Accelerated Approval Program established by the FDA. Another name for ddC is zalcitabine, and the trade name is Hivid. Like AZT and ddI, the drug is an inhibitor of reverse transcriptase, but it is thought to act at a different site. Varying combinations with AZT and ddI are currently recommended for patients. Another compound in the same class of dideoxynucleosides is d4T, which is also known as stavudine (Zerit).

Another dideoxynucleoside is a sulfur-containing derivative of deoxycytidine known as 2'-deoxy-3'-thiocytidine, or 3TC (also known as lamivudine or Epivir). In 1996, this drug became the first initial therapy drug approved to treat AIDS since AZT had been approved nine years previously. Since then, 3TC has also been prescribed in combination with AZT and a protease inhibitor to constitute the three-drug

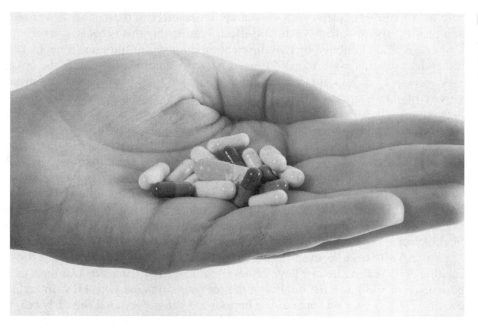

FIGURE 8.7
A patient who has had AIDS since 1990 shows the "cocktail" of drugs he takes daily. © Martin Allinger/ShutterStock, Inc.

regimen that has significantly reduced the viral load in AIDS patients. As we discuss here, the three-drug therapy has played a major role in the developing view that AIDS may one day be considered a chronic disease that can be maintained and controlled, rather than an acute disease bringing certainty of death (**Figure 8.7**).

In 1999, the FDA approved an AZT-like drug called abacavir (Ziagen). When used in combination with AZT and 3TC, this drug is significantly more effective in reducing the viral load in patients to "undetectable" levels (i.e., 400 or fewer HIV RNA copies per mL) than AZT and 3TC used together. The drug is particularly useful for treating children and young patients because it is formulated as a palatable liquid that is not bitter (most pills are difficult to swallow and are bitter).

The newest dideoxynucleoside is tenofovir (Viread), which achieved FDA approval in 2001. Tenofovir is called a nucleotide reverse transcriptase inhibitor because it contains an extra phosphate group. This extra group makes the drug a more active chain terminator. Taken once a day, tenofovir appears to circumvent HIV mutations that confer resistance to other drugs. Mild to moderate gastrointestinal disturbances are the only known side effects. The timeline of NRTIs approved are shown in Figure 8.4.

Nonnucleoside Analogs

Although the dideoxynucleoside drugs have achieved a measure of success, some patients cannot tolerate the side effects or are infected with HIV particles resistant to the drugs. For these individuals, nonnucleoside analogs offer a viable option. The drugs target reverse transcriptase and react directly with it (in contrast to the dideoxynucleosides, which trick reverse transcriptase into producing faulty DNA). Some researchers recommend that physicians prescribe nucleoside analogs in combination with the dideoxynucleosides, whereas others recommend them when the three-drug combination involving a protease inhibitor fails. In the technical literature, the drugs are called **nonnucleoside reverse transcriptase inhibitors** (**NNRTIs**).

The three most prominent nonnucleoside analogs are nevirapine, delavirdine, and efavirenz. Nevirapine (Viramune) and delavirdine (Rescriptor) elicit HIV resistance rapidly; thus, they have received FDA approval for use only with a dideoxynucleoside such as AZT or ddI. Such combinations are more effective than either drug used alone, as a result of the so-called **synergistic effect**. Presumably this is because reverse transcriptase is being inhibited by two different mechanisms at the same time. In 1999, researchers reported that, like AZT, nevirapine can significantly reduce the possibility of HIV transfer from HIV-infected pregnant women to their newborns. The substantially lower cost of nevirapine adds an economic advantage to its use, especially in developing countries; however, damage to the liver (hepatotoxicity) and skin reactions has been reported in patients taking the drug, and appropriate precautions have been issued.

Efavirenz (Sustiva) received FDA approval in 1998. This drug is taken once a day with normal doses of AZT and 3TC in a three-drug combination therapy. In one study, efavirenz used with ddI and an AZT-like drug called emtricitabine (FTC or Coviricil) reduced viral loads substantially in 98% of patients followed over a three-year period. Dizziness, insomnia, impaired concentration, and drowsiness are notable side effects of the drug, and women are strongly advised not to use it, as it could cause birth defects. The timeline of NNRTIs approved are shown in Figure 8.4.

Since AZT was approved for use in 1987, physicians have shown that the drug is capable of increasing helper T-lymphocyte counts, enhancing survival rates of AIDS patients, and adding to the quality of life for those infected with HIV. In addition, use of AZT has relieved some of the hopelessness that pervaded the early years of the AIDS epidemic, while providing impetus for additional efforts to develop treatments and therapies for AIDS. The work with other compounds has also been significant because it shows that scientists can use their understanding of the biochemistry of HIV to synthesize other specific drugs.

AZT and its related compounds, however, are only a few of the multitude of drugs currently in various stages of experimentation and testing. In this section, we survey some of the other drugs that are in use or hold promise for use against HIV.

Protease Inhibitors

Protease inhibitors (PIs) first made headlines in 1989. Development of these drugs resulted from determination of the three-dimensional structure of an HIV enzyme known as protease. The chemical determination was so detailed that it specified the arrangement of the enzyme's individual atoms. Proteases are essential to the final stages of HIV replication; it trims bulky, unprocessed viral proteins down to working size before they are assembled as protein coats for the new viruses. In other words, the proteases aid in the maturation of HIV particles. Once the maturation process is completed, the HIV particles are infectious.

As soon as the exact three-dimensional structure of protease was known, researchers designed drugs to fit precisely into the enzyme's griplike active site and jam its action. They were encouraged by the observation that a form of HIV with an altered protease enzyme could not trim proteins and assemble them to make protein coats. By 1995, 12 pharmaceutical companies were working on the development of an entire series of protease inhibitors that yielded lower levels of HIV in the bloodstream as well as higher counts of T-lymphocytes. The drugs were used alone and in combination with other drugs, and because they affected a relatively small enzyme, the emergence of resistant viruses was rare.

By 1996, the protease inhibitors had assumed their place in medicine as accepted therapies for HIV infection and AIDS. That year, three drugs received FDA approval: saquinavir (Invirase), indinavir (Crixivan), and ritonavir (Norvir). In clinical trials conducted in several parts of the world, all three drugs were shown to decrease viral concentrations in patients' blood and slow disease progression by up to 50%

in late-stage patients. Side effects were apparently limited to mild nausea, vomiting, and diarrhea as well as fat buildup on the torso and face (i.e., lipid-dystrophy). Soon researchers were recommending a three-drug combination (a "cocktail" of drugs) consisting of AZT, 3TC, and one of the protease inhibitors. The cocktail or triple therapy came to be known as *highly active antiretroviral therapy (HAART)*.

The impact of HAART on the AIDS epidemic was immediate; for example, in 1996, there were 38,025 AIDS-related deaths in the United States, whereas in 1997, the number dropped to 21,999 and continued to drop thereafter to 18,017 in 2003 (**Table 8.5**). Physicians described in glowing terms how ill patients responded to the therapy, and with the introduction of the viral load test, researchers could chart the disappearance of HIV from the blood, lymph nodes, and other tissues. Some scientists were so excited about the paradigm shift created by HAART (the $10,000 cost per year notwithstanding) that they forecast the imminent eradication of HIV from the population. It was not uncommon to hear the words "AIDS" and "hope" in the same sentence, and *Science* magazine, the preeminent journal of science in the United States, declared protease inhibitors to be the *1996 Breakthrough of the Year*.

The success of HAART led to numerous other studies involving drug cocktails. In one study, for instance, a combination of two protease inhibitors (ritonavir and saquinavir) showed a reduction in viral load to less than 400 copies per µL (considered undetectable) in 88% of patients. Other studies including or excluding protease inhibitors combined delavirdine, AZT, and 3TC; or efavirenz and indinavir; or nevirapine, AZT, and ddI. HAART's success also led to the view that maximum suppression of the virus would minimize damage to the immune system and possibly avert opportunistic diseases. This so-called **"hit-it-early, hit-it-hard" approach** was spearheaded by David Ho of New York's Aaron Diamond AIDS Research Center (**Figure 8.8**). Ho was among the first to show the efficacy of the protease inhibitors in three-drug combinations and to raise hope for new treatments. Ho was honored in 1996 as *Time Magazine's* Man of the Year.

Unfortunately, by 1997, reality was setting in, and HAART drugs were being seen as far from perfect. Many patients had already chosen to stop taking the drugs, complaining of difficulty in adhering to the drug regimen, a complicated affair that involved taking up to 15 pills a day—some alone and others in tandem, some on a full stomach and others on an empty one (in one study, only 60% of patients said that they adhered to the regimen). Others complained of the side effects of the drugs (**Table 8.6**), and still others expressed the feeling that the drugs were failing to work. That year, the FDA, in an unusual step, sent thousands of letters to physicians advising them that the protease inhibitors were linked to high blood pressure (related to raised cholesterol levels) and new or worsened cases of diabetes. Scientists also noted the emergence of multidrug-resistant HIV strains, often the result of patients' continual switching among the many available drugs.

By 2002, the newer policy advanced by many medical professionals was to back off the "hit-it-early, hit-it-hard" approach and consider deferring treatment until the patient showed signs that the immune system was weakening. Moreover, *Guidelines for the Use of Antiretroviral Agents* developed by the United States Department of Health and Human Services had been in place for a year. Among other things, the guidelines recommended beginning HAART drugs when the infected individual's T-lymphocyte count dropped below 350 cells/µL (as compared with the previous recommendation of 500 cells/µL). The guidelines are available at the HIV/AIDS Treatment Information Service website at http://www.hivatis.org.

Researchers have also discovered that as soon as patients stop taking their HAART drugs, HIV rebounds in the body and rises to high levels, possibly because of the

TABLE 8.5 Deaths of AIDS Patients in the United States Between 1999 and 2006, and Cumulative, 1981–2006

Exposure Category	Year of Death					Cumulative Through 2006
	1993	1996	1997	2003	2006	
Male adult/adolescent						
Men who have sex with men	23,956	16,854	8,666	6,015	4,930	266,272
Injection drug users	9,325	8,551	5,346	4,166	2,641	107,173
Men who have sex with men and inject drugs	3,188	2,591	1,447	1,233	1,064	40,993
Heterosexual contact	1,600	2,111	1,464	1,644	1,470	26,236
Other, including: hemophilia; receipt of blood transfusion, blood components, or tissue; risk not reported or identified	839	529	288	140	80	9,867
Male subtotal	38,908	30,636	17,212	13,198	10,184	450,541
Female adult/adolescent						
Injection drug users	3,152	3,289	2,137	2,056	1,456	43,143
Heterosexual contact	2,662	3,439	2,297	2,584	2,258	42,620
Other, including: hemophilia; receipt of blood transfusion, blood components, or tissue; risk not reported or identified	332	232	133	95	70	4,131
Female subtotal	6,146	6,960	4,567	4,736	3,784	89,895
Pediatric (<13 years old)	544	429	221	83	48	4,848
Total*	45,598	38,025	21,999	18,017	14,627	565,927

*Because column totals were calculated independently of the values for the subpopulations, the values in each column do not sum to the column total.

Data from *HIV/AIDS Surveillance Report*, Vol. 15, 2003, and Vol. 18, 2006.

influence of **chemokines** (**cytokines**) that stimulate the lymphoid cells to produce HIV particles; soon thereafter, the classic **opportunistic infections** surface. Scientists once believed that if viral replication could be suppressed for a few years, all the pools of HIV in the body would be exhausted; however, it now appears that HIV is able to find sanctuary in long-lived dormant ("resting") cells of the immune system, lurking in a latent state for many years. This reservoir of latent infection is apparently what prevents HAART from curing patients; instead, it puts patients into remission. The level of HIV is pushed way down, but it fails to hit zero (**Figure 8.9**).

Still, the reduction of the viral load is apparently very beneficial. In 2000, for example, European investigators reported that with HAART and other new drug therapies, about 80% of AIDS patients lived at least 10 years after becoming infected with HIV; before the advent of such drugs, only 55% lived 10 years or more. As a writer in the journal *Science* explained, "AIDS research ricochets from breathtaking optimism to stomach-wrenching disappointment and back again." Although the numbers continue to improve, it was still disappointing to see that the dramatic decrease in deaths due to AIDS between 1996 and 1997 had not been replicated each year (Table 8.5). The total number of deaths decreased by 42% between 1996 and 1997. The percentage decrease in deaths between 1997 and 2006 was 33%—which is good—but again, it took nearly a decade to accomplish this decrease (Table 8.4). Furthermore, the decrease in death rates leveled off, with only a 19% decrease between 2003 and 2006 (Table 8.5).

Therapies continue to evolve and a several more antiretrovirals were licensed since 2006. Cocktails of increased potency and higher barriers to drug resistance were also approved by the FDA. For example, in August 2012, Stribild was approved by the FDA as a single daily dose regimen for HIV infection. Stribild is a 4-in-1 combination drug. It contains elvitegravir (an HIV integrase inhibitor), emtricitabine (a protease inhibitor), tenovir (a nucleoside reverse transcriptase inhibitor), and cobicistat (a boosting agent which enhances the effects of elvitegravir). A warning came with the once-a-day treatment about side effects, which include severe liver problems, buildup of lactic acid, and common sides such as nausea and diarrhea.

TABLE	

8.6 Some Side Effects Associated with Antiretroviral Drugs

Antiretroviral Class/Agent	Primary Side Effects and Toxicities
Nucleoside reverse transcriptase inhibitors (NRTIs)	
Zidovudine (Retrovir; ZDV; AZT)	Anemia, neutropenia, nausea, headache, insomnia, muscle pain, and weakness
Lamivudine (Epivir; 3TC)	Abdominal pain, nausea, diarrhea, rash, and pancreatitis
Stavudine (Zerit; d4T)	Peripheral neuropathy, headache, diarrhea, nausea, insomnia, anorexia, pancreatitis, increased liver function tests (LFTs), anemia, and neutropenia
Didanosine (Videx; ddI)	Pancreatitis, lactic acidosis, neuropathy, diarrhea, abdominal pain, and nausea
Abacavir (Ziagen; ABC)	Nausea, diarrhea, anorexia, abdominal pain, fatigue, headache, insomnia, and hypersensitivity reactions
Nonnucleoside reverse transcriptase inhibitors (NNRTIs)	
Nevirapine (Viramune; NVP)	Rash (including cases of Stevens-Johnson syndrome), fever, nausea, headache, hepatitis, and increased LFTs
Delavirdine (Rescriptor; DLV)	Rash (including cases of Stevens-Johnson syndrome), nausea, diarrhea, headache, fatigue, and increased LFTs
Efavirenz (Sustiva; EFV)	Rash (including cases of Stevens-Johnson syndrome), insomnia, somnolence, dizziness, trouble concentrating, and abnormal dreaming
Protease inhibitors (PIs)	
Indinavir (Crixivan; IDV)	Nausea, abdominal pain, nephrolithiasis, and indirect hyperbilirubinemia
Nelfinavir (Viracept; NFV)	Diarrhea, nausea, abdominal pain, weakness, and rash
Ritonavir (Norvir; RTV)	Weakness, diarrhea, nausea, circumoral paresthesia, taste alteration, and increased cholesterol and triglycerides
Saquinavir (Fortovase; SQV)	Diarrhea, abdominal pain, nausea, hyperglycemia, and increased LFTs
Lopinavir/Ritonavir (Kaletra)	Diarrhea, fatigue, headache, nausea, and increased cholesterol and triglycerides
Entry inhibitors (EIs)	
Maraviroc	Rash (excluding cases of Stevens-Johnson syndrome), liver toxicity with or without rash
Integrase Strand Transfer Inhibitor (INSTI)	
Raltegravir	Hypersensitivy, muscle weakness and rhabdomyolysis, Stevens-Johnson syndrome

Data from: CDC, *Morbidity and Mortality Weekly Report*, vol. 50/no. 25, June 29, 2001 and vol. 58/no. 13, April 10, 2009.

Entry Inhibitors: Membrane Fusion, and CD4 and CCR5 Binding Inhibitors

A new class of drugs called **fusion inhibitors** represents an alternative approach to treating HIV infection and AIDS. Like protease inhibitors, the fusion inhibitors were derived from an understanding of the basic replication strategies of HIV, in this case,

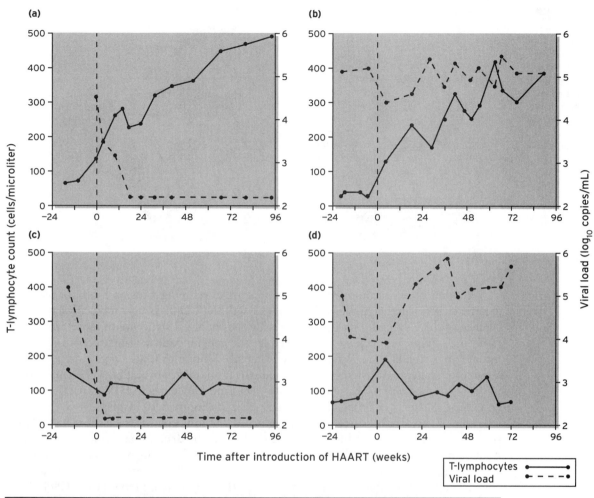

FIGURE 8.9

Four possible results of the anti-HIV therapy HAART and their interpretations. (a) Complete success is recognized as the viral load is suppressed and the T-lymphocyte count rises. In (b) and (c), the treatment is deemed partially successful, while in (d), it is deemed a failure. Modified from Perrin, L. and Telenti, A., *Science* 280 (1998): 1871–1873. Reprinted with permission from AAAS.

the method by which HIV fuses with its host T-lymphocyte. The drugs are designed to interfere with steps taking place before reverse transcriptase and protease act in the host cells.

Enfuvirtide was approved by the FDA in 2003. This is a synthetic peptide containing 36 amino acids. The peptide mirrors a key part of the gp41 molecule, the HIV glycoprotein that anchors the gp120 molecule after the latter has contacted the cell membrane of the T-lymphocyte. When enfuvirtide binds to the gp41 molecule, the gp120 molecule cannot contact the T-lymphocyte, and fusion is interrupted. It was shown effective for reducing the viral loads in patients for whom HAART therapy failed. Because it is a large molecule that must be injected, researchers were studying smaller peptides that can be taken orally and will survive the intense acidity of the stomach environment. These smaller molecules target a pocket on the gp41 molecule that is exposed when that molecule unites with the host T-lymphocyte. The research illustrates how the science of structural biology has application in the development of a possible new therapy.

FIGURE 8.10

The mode of action of soluble CD4. (a) An HIV particle unites with a susceptible host cell when its gp120 molecules bind with CD4 molecules at the host cell surface. (b) Genetically engineered soluble CD4 molecules bind to a virus and prevent it from binding with the natural CD4 molecules. Unable to bind to the host cell, the virus is rendered harmless.

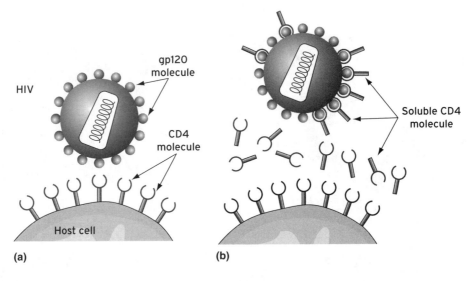

Another group of innovative researchers are developing a fusion inhibitor using genetic engineering technology. They begin with the **CD4 receptor molecules** at the surface of the T-lymphocytes and determine the chemical structure of these protein molecules. Working backward, they deduce the genetic code for the protein and formulate an artificial gene. When this gene is placed into a producer organism such as a bacterium, the organism synthesizes industrial quantities of soluble CD4 (Receptin) molecules. Injected into patients, soluble CD4 bind molecules to HIV particles and enhance the effects of other drugs such as AZT, as **Figure 8.10** displays.

One of the disadvantages of using soluble CD4 molecules is their short half-life in the bloodstream (the half-life is the time for half of the drug to be metabolized by the body and thus disappear). Researchers have partly resolved this problem by linking the CD4 molecules to the stem sections of anti-HIV antibodies to produce immunoadhesions. The immunoadhesions remain active in the bloodstream longer than CD4 molecules, and they bind more efficiently to HIV particles, thereby blocking fusion and inactivating the virus at the same time. The preparation called PRO542 was shown to enhance the activity of enfuvirtide, and vice versa. None of the CD4 binding drugs in development (Pro-542, BMS-378806, TNX-355) have been licensed by the FDA.

Before we leave the fusion inhibitors, we should note recent investigations of the **zinc fingers** in the capsid of HIV. Zinc fingers are sections of capsid proteins that contain large amounts of the amino acids cysteine and histidine combined with atoms of zinc in arrangements that extend out from the surface. A drug being developed by biochemists at the National Cancer Institute attacks the sulfur atoms in cysteine molecules; this attack leads to ejection of the zinc atoms and incapacitation of the capsid protein. The effect is to limit the ability of HIV to fuse with its host cell and to synthesize the protein it later needs for capsid formation. Because the zinc fingers resist mutation (i.e., they are "mutationally intolerant"), the drug is an attractive anti-HIV therapy.

Elucidation of the central role played by chemokines and chemokine receptors in HIV's attachment to T-lymphocytes has kindled hopes for therapeutic molecules that will block HIV infection. HIV slips into T-lymphocytes by commandeering CCR5 or CXCR4 and other **coreceptor** molecules on the cells' surface that normally bind to chemokines. *Chemokines* are a set of proteins that act as chemical messengers when produced by cells of the immune system. This discovery, made in 1996, encouraged scientists to develop treatments that would exploit the HIV/chemokine nexus. One research group developed a synthetic chemokine that works within the cytoplasm of the T-lymphocyte to inactivate the CXCR4 proteins as they are synthesized and before they can be dispatched to the cell surface, where they will provide docking sites for HIV. The synthetic chemokine has been called an *intrakine* because it works within the cell, rather than at its surface. In laboratory tests, this intrakine effectively blocked expression of the surface CXCR4 protein and thereby prevented formation of the CXCR4 coreceptor site. Essentially, the HIV particle found itself unable to enter its host cell.

Despite the encouraging results, researchers were quick to point out that the intrakine is not a new therapeutic agent because the gap between basic research and applied technology is usually vast and talk of bridging it is often premature. Nevertheless, the research has opened new possibilities for therapies. Indeed, in 1999, another group found that gene mutation causes some individuals to overproduce a chemokine identified as SDF-1. Individuals with this mutated gene apparently progress more slowly to AIDS than expected and, because SDF-1 molecules normally bind to the CXCR4 coreceptor sites, it is conceivable that these abundant molecules occupy the sites otherwise used by HIV particles. This finding correlates with the 1996 discovery that individuals lacking the CXCR4 coreceptors are able to resist HIV infection.

The promising discoveries relating to chemokine receptors are counterbalanced by the possible unanticipated developments. For example, using chemokines or chemokine analogs to block an entry site might encourage HIV to develop a preference for an alternate site such as the CCR5 coreceptor. Nevertheless, researchers are encouraged to continue their work with chemokines because they have already developed profitable drugs that target receptors involved in rheumatoid arthritis, psoriasis, and other inflammatory conditions.

Using an innovative approach described as "set a thief to catch a thief," scientists have formulated a recombined virus to attack HIV particles. Investigators at Yale University began with a well-researched virus called the *vesicular stomatitis virus (VSV)*, which infects cattle and other livestock and causes ulcers on their hooves and tongue. The researchers biochemically removed the genes that encode the capsid proteins of the VSV and inserted in its genome the genes that encode the CD4 and CXCR4 receptors of a T-lymphocyte. When the VSV replicated in cells, its new genome encoded a strain of the virus with CD4 and CXCR4 proteins on the surface instead of the usual capsid proteins. In laboratory tests, this virus bound itself to gp120 and gp41 molecules in the spikes of HIV, thereby neutralizing the HIV. It also united with and eliminated infected T-lymphocytes, as these cells are studded with gp120 and gp41 molecules on their surfaces (the molecules are later incorporated in new HIV particles during budding). The Trojan horse virus, as the recombined VSV came to be called, demonstrates that it is possible to use HIV's so-called grappling hooks (i.e., its gp120 and gp41 molecules) as a target for drug therapy. To date, maraviroc has been the only coreceptor antagonist approved by the FDA. It was licensed in 2007 and is a CCR5 binding drug.

Integrase Inhibitors

After successfully developing therapies based on two HIV enzymes—reverse transcriptase and protease—researchers are zeroing in on the third enzyme, integrase. Integrase incorporates proviral DNA into the DNA of the host T-lymphocyte. Raltegravir received accelerated approval by the FDA in 2007 as the first antiretroviral drug in the new class of inhibitors known as *integrase inhibitors*. Raltegravir is used in combination with other HIV drugs as treatment for HIV infection.

Postexposure Prophylaxis

Before leaving the therapeutic agents, we should note that none of these drugs is intended for use as a "morning after" pill; that is, none of the drugs is FDA approved for use in those who fear they may have been exposed to HIV and wish to receive so-called **postexposure prophylaxis (PEP)**. Unofficially, however, some health clinics and practitioners offer anti-HIV drugs to high-risk patients, such as those who may have been exposed to HIV during a sexual assault, and studies are ongoing to determine the feasibility of such treatment. Part of the reluctance to prescribe drugs is their often serious side effects and the long duration of the drug regimen.

The situation is somewhat different for healthcare workers who may have been exposed to HIV through a needlestick or other accident during the course of their work. Guidelines for PEP associated with occupational exposures to HIV were updated by the CDC in 2005. If a healthcare worker has been exposed to body fluids from an individual who is infected with HIV, there are several PEP regimens. All of the regiments contain AZT in combination with at least one other antiretroviral drug. PEP should be initiated within hours of exposure and administered for 4 weeks unless follow-up testing has determined that the source of contamination was HIV-negative. PEP should be discontinued immediately if that were the case. There were some concerns that a worker could be exposed to HIV-negative sources during the **window period** for **seroconversion**, but no case of infection involving an exposure source during the window period has ever been reported in the United States.

The CDC also made recommendations for PEP to prevent HIV infection if individuals were injured during a mass casualty event such as bombings during an act of terrorism. During 1980 to 2005, the Federal Bureau of Investigation (FBI) investigated 318 acts of terrorism in the United States and its neighboring territories. Of these, 208 were attempted bombings. The majority of these acts did not result in deaths or injuries, however, 19 bombings did result in 181 deaths and 1,967 injured individuals.

Military healthcare providers must respond to mass-casualty events such as those incurred during Operation Enduring Freedom in Afghanistan and Operation Iraqi Freedom in Iraq. During these events, explosive devices wounded 27,441 persons and of these, 24,433 survived. In 2001 and 2002, the Israeli Ministry of Health reported tissue from suicide bombers tested positive for hepatitis B virus (a bloodborne pathogen). One case reported bone fragments from a suicide bomber who had hepatitis B that were traumatically implanted into a bombing survivor. These observations prompted the development of a protocol to assess the risk for infection by bloodborne pathogens (including HIV) and postexposure management of wounded individuals. The CDC did not recommend routine PEP for treating persons injured in mass-casualty settings because the risk of HIV transmission is 0.3% after percutaneous exposure (a puncture wound) of HIV-infected blood.

Preexposure Prophylaxis

Preexposure prophylaxis (PrEP) has been used is used to prevent disease caused by an infectious agent before an individual is exposed to the disease-causing pathogen. For example, if there is an influenza epidemic in a community, physicians may recommend certain residents of a nursing home community be prescribed anti-influenza drugs as PrEP to avoid contracting influenza. Some nursing home residents may not be vaccinated or the influenza strain causing the epidemic could be a new strain not in the vaccine. Therefore, a group of nursing home residents could be susceptible to infection. Influenza can cause high mortalities in the elderly or cause complications such as secondary bacterial pneumonia, which may result in death.

Focus on HIV
Certain AIDS drug combinations given to pregnant women block 99% of HIV transmission to breastfed babies.

So why doesn't anyone at risk for HIV infection use PrEP to prevent HIV infection? Probably the main reason is that the antiretroviral drugs can have toxic side effects and shouldn't be used unless absolutely necessary. The FDA approved **Truvada** in 2012 for use by healthy heterosexuals or homosexual people who are at high risk for contracting HIV. It was especially targeted toward people who have HIV-infected partners. Truvada was recommended as a once-a-day pill. Truvada contains two nucleoside reverse transcriptase inhibitors: tenofovir and emtriva. When Truvada is used to treat HIV-infected individuals, it is used in combination with at least one other anti-HIV drug, usually a protease inhibitor or a non-nucleoside reverse transcriptase inhibitor.

Treating Opportunistic Diseases

After three decades of dealing with the AIDS epidemic, it has become clear that AIDS patients do not die of AIDS. Rather, they die of encephalopathy, wasting, or any of a series of opportunistic infections in the tissues after HIV has ravaged the immune system. The obvious way to deal with AIDS is to eliminate the virus or develop a vaccine against it. As long as neither of these is imminent, a reasonable tactic is to treat the opportunistic diseases that commonly cause death.

Pentamidine Isethionate

Before the HIV/AIDS epidemic, *Pneumocystis* pneumonia was uncommon. From 1967 to 1971, a total of 194 patients were diagnosed with *Pneumocystis* pneumonia and reported to the CDC, which was the sole supplier of **pentamidine isethionate**, the only treatment for *Pneumocystis* pneumonia at the time. *Pneumocystis jirovecii* pneumonia is a frequent AIDS-defining diagnosis. The responsible fungal pathogen multiplies without control in the lungs, fills up all the air spaces, and induces oxygen suffocation. Over half the patients who die from complications of AIDS die of this type of pneumonia. Indeed, about 70% of all AIDS patients suffer at least one bout of the disease.

As HIV spread among the U.S. population, pentamidine isethionate became available locally. Physicians had to administer it by injection, however; thus, its use was generally limited to those hospitalized. Then, in 1989, the FDA approved pentamidine isethionate in a much easier to use aerosol form and suggested that it be employed for prevention of *P. jirovecii* pneumonia as well as treatment. The procedure specifies use of the aerosolized drug in AIDS patients who have had one bout of the pneumonia or whose helper T-lymphocyte count drops below 200 cells per microliter (the threshold at which studies show that patients are at risk of developing the pneumonia). Most people infected with HIV visit a clinic monthly to inhale a dose of the drug. However, some physicians are skeptical, believing that the drug may protect the lungs, but it does not lend resistance to other organs that *P. jirovecii* may infect, such as the thyroid gland, kidney, spleen, and liver. These rare infections may require therapy with a different drug. A side effect of the aerosolized treatment is a cough, but there are few other side effects. In contrast, injected pentamidine isethionate often leads to damage to the pancreas, reduced white blood cell counts, and altered blood glucose levels.

In 2009, the CDC published recommendations for the treatment of opportunistic infections in HIV-infected adults and adolescents in the *Morbidity and Mortality Weekly Report*. Their treatment of choice for *P. jirovecii* pneumonia is **trimethoprim-sulfamethoxazole (TMP-SMX)** as a double-strength daily tablet. Trimethoprim-sulfamethoxazole, an FDA-approved combination of two drugs, is commonly used for bacterial infections. TMP-SMX also protects the AIDS patient by protecting him/her against toxoplasmosis (a parasitic infection) and common respiratory bacterial infections.

Other Drugs

Two opportunistic viruses, the herpes simplex virus and cytomegalovirus (CMV), can also be held in check with drug treatment. For herpes simplex, the preferred drug is **acyclovir** (commercially known as Zovirax). Like AZT, acyclovir is a nucleoside that inhibits replication of the herpes simplex virus by interfering with its production of DNA. Approved by the FDA in the mid-1980s, acyclovir is available in injectable, cream, and pill forms. In 1989, the FDA approved ganciclovir for the treatment of CMV-induced **retinitis**, a condition of the eye that affects 25% of AIDS patients and often leads to blindness. Because its side effects include anemia, ganciclovir cannot be taken with AZT, which is also associated with anemia.

Other opportunistic diseases can also be controlled with drugs. For tuberculosis, the drugs of choice include isoniazid and rifampin; for preventing tuberculosis caused by *Mycobacterium tuberculosis* or mycobacteriosis caused by *Mycobacterium avium-intracellulare* (MAC infection), rifabutin (Mycobutin) has been recommended, and for treating established cases, clarithromycin or azithromycin is suggested. **Table 8.7** is a list of drugs recommended by the CDC to treat opportunistic infections of AIDS patients.

Anemia remains one of the most worrisome side effects of many drugs used to treat AIDS patients. Physicians have estimated, for example, that half of the

TABLE 8.7 A Summary of Drugs Used to Treat Opportunistic Diseases

Opportunistic Disease	Available Drugs	Comment
Pneumocystis jirovecii pneumonia	Trimethoprim-sulfamethoxazole	Taken as daily single- or double-dose tablets/protects against other infections, e.g., toxoplasma and respiratory infections caused by bacteria
Toxoplasmosis	Combination of pyrimethamine + sulfadizine + leucovorin	Patient should be monitored for severe side effects/toxicities
Herpes simplex virus (oral or genital)	Oral valacyclovir, famciclovir, or acyclovir for 5–14 days	Treatment choice for acyclovir resistant strains is foscarnet
Cytomegalovirus infections	Oral valganciclovir, ganciclovir	
Oropharyngeal and esophageal Candidiasis	Fluconazole	Alternatives are itraconazole and posaconazole but they have more side effects
Bacterial respiratory infections (e.g., *Streptococcus pneumonia* and *Haemophilus* sp.)	Amoxicillin, amoxicillin-clavulanate	If allergic to penicillin, substitute moxifloxacin, levofloxacin, or gemifloxacin
Tuberculosis (*Mycobacterium tuberculosis*)	INH-rifapentine, rifabutin, pyrazinamide	Long therapy required, 6–12 month regimens
Malaria	Artemether-lumefantrine	Use same treatment for HIV-negative people
Influenza A	Zanamivir, peramivir	
Shingles (herpes zoster)	Valacyclovir or famciclovir	

Adapted from CDC, *Morbidity and Mortality Weekly Report*, vol. 58/no. 13, April 10, 2009.

individuals taking AZT suffer anemia. To lessen the possibility of this disorder, the FDA has approved the clinical use of erythropoietin, which is a protein hormone normally produced by kidney cells. The hormone promotes red blood cell production in the bone marrow. Biotechnologists have deciphered the genetic code for this protein and have synthesized abundant supplies. Physicians recommend it for use in patients taking AZT to lessen the need for blood transfusions.

The reasonably effective arsenal of drugs for fighting opportunistic diseases should encourage persons with the early signs of AIDS to place themselves under the care of a physician. Physicians with long experience in treating AIDS have noted that persons who delay therapy for opportunistic diseases are at greater risk for future bouts because of damage to their tissues. If, however, a person seeks medical help at the first sign of an AIDS-related illness, there is a good possibility that the effects can be lessened.

Drug Development and Testing

To receive the most promising drugs, a person who has HIV infection or AIDS must usually find a way to enter a sophisticated **clinical trial** at a major medical center. There the individual must meet strict criteria set by researchers and agree to use the test drug or a placebo, usually without knowing which is being administered. The fortunate ones emerge from the trial healthier; the unfortunate ones seek solace in the fact that they have helped develop the knowledge base of science. Essentially, the formal scientific evaluation of unproven therapies is performed more for the benefit of tomorrow's patients than for today's.

This approach to testing, however, has undergone a gradual change, stimulated in part by the AIDS epidemic. Driven by the intense need for new drugs, the medical establishment has been relaxing its procedures to bring promising drugs to patients more quickly. Neighborhood clinics were allowed to conduct clinical trials, and in 1987, the FDA changed its traditional rules to put certain drugs on a **fast track for approval** for individual patients under certain circumstances. In some situations, for example, an **investigational new drug (IND)** can be released to physicians, even though the drug is still undergoing trials and awaiting FDA approval. Under this procedure, an IND for treating "life-threatening" illnesses can be released during its Phase II trial while clinical trials continue. In 2009, the final rules for fast track approval were revised, expanding access to INDs for treatment use to be available to individual patients and the treating physicians, including in emergencies, to intermediate-size patient populations, and to larger populations under a treatment protocol or IND. This change provides a means to obtain INDs for people if their disease has no good treatment and their physicians are very interested in trying a new drug under development, especially if the early results of a clinical trial suggest that the drug shows promise.

In 1992, the FDA instituted the **accelerated approval regulation** for treating "serious" illnesses that fill an unmet medical need. No one quarrels with the compassionate motives that underlie this innovation; however, opponents of the fast-track procedure question whether AIDS patients would remain in a clinical trial (where they might be receiving a placebo) if they could leave the trial and purchase the experimental drug or have it prescribed by their physicians. Other issues include the reluctance of uninformed physicians to prescribe treatment with INDs and the refusal of certain insurance companies to cover treatment with unapproved drugs.

For the **parallel-track procedure**, which the FDA put into place in 1992, individuals were enrolled when they were ineligible for controlled clinical trials and in need of experimental drugs (**Figure 8.11**). Such individuals include those who cannot tolerate

The parallel- and fast-track procedures for testing an experimental drug. In the parallel-track procedure, experimental drugs are made available to certain patients while Phase I trials are taking place. Under the fast-track procedure, an IND is released during Phase II trial while the trial continues. Presence of a life-threatening illness could be the basis for a fast-track release.

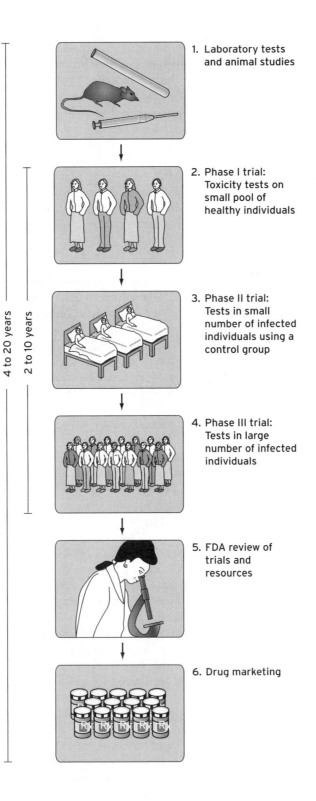

4 to 20 years

2 to 10 years

1. Laboratory tests and animal studies

2. Phase I trial: Toxicity tests on small pool of healthy individuals

3. Phase II trial: Tests in small number of infected individuals using a control group

4. Phase III trial: Tests in large number of infected individuals

5. FDA review of trials and resources

6. Drug marketing

TABLE 8.8 INDs with Expanded Access Enrollment Related to HIV/AIDS

Drug	Use	# Enrolled	Dates
AZT	HIV treatment	4,804	1986-1987
pentamidine isethionate	treatment of *Pneumocystis* pneumonia	728	1989
trimetrexate	treatment of *Pneumocystis* pneumonia	753	1988-1994
ddI	HIV treatment	>21,000	1989-1991
ddC	HIV treatment	6,705	1990-1992
atovaquone/proguanil	anti-malarial medication	1,054	1991-1993
rifabutin	used to treat tuberculosis	2,506	1992-1994
d4T	HIV treatment	12,551	1992-1994
3TC	HIV treatment	29,430	1993-1994
saquinavir	HIV treatment	2,200	1995
indinavir	HIV treatment	1,500	1995

Data from: FDA Information for Consumers: HIV Specific Resources.

a standard treatment or who have failed to improve. Those who live far from an institution-sponsored clinical trial are also eligible. The drugs to be used in this procedure are determined on a case-by-case basis. Opponents of the plan fear that people with AIDS may be diverted away from clinical trials where their participation is needed. The possibility also exists that some patients may stop using a good drug in favor of a poor one. Proponents point out that participants in the parallel-track procedure will believe they are receiving state-of-the-art, and possibly better, care.

Historical examples of drugs used to treat HIV or the complications of AIDS with expanded access are listed in **Table 8.8.** A recent example of emergency drug approval was President Barack Obama's action to declare a national state of emergency on October 24th, 2009 during a second wave of the H1N1 influenza pandemic. The national emergency waived certain FDA regulatory requirements, allowing an Emergency Use Authorization to treat hospitalized influenza patients who did not respond to treatment with current anti-influenza drugs (Tamiflu or Relenza). Peramivir was an unapproved drug in FDA Phase III clinical trials.

Drug Trials

A classical clinical trial for a new drug may take more than 10 years from laboratory synthesis to FDA approval. In preclinical testing, studies are conducted in laboratory animals to determine whether the drug is safe to use and is biologically active. Two years may be consumed by these studies. Then comes a three-part clinical trial (**Figure 8.12**). In Phase I, the drug is tested on fewer than 100 paid, healthy human volunteers, with a slow increase in dosage over a period of time to determine any toxic side effects and to see how the drug is handled by the body. This can be a frightening experience for both volunteers and researchers because although animal tests have been performed animals cannot express such symptoms as dizziness, nausea, or psychiatric symptoms. Serious side effects, not apparent in animals, could also develop in human tissues. Researchers

FIGURE 8.12

The process recommended by the U.S. FDA for gaining its approval of a drug.

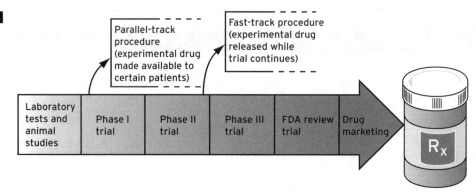

closely monitor the volunteers' blood pressure and temperature, and they carefully scrutinize drug distribution and drug elimination. A **Phase I trial** usually takes about one to two years.

Phase II trial is next. The drug is given to several hundred patients to pinpoint the conditions under which it will elicit the most effective response. The optimum **dose** must be established, and the endpoint must be set, that is, when can physicians say the drug "works" (for HIV infection, this may mean reduction of the **viral load** to less than 400 copies of HIV RNA per μL of blood, a level currently considered "undetectable"). A control group is used in a Phase II trial, and those participants receive either a **placebo** or the best therapy available. The **control group** allows scientists to gauge the effectiveness of the **candidate drug** in an experimental situation and observe any special side effects related to the drug. Ideally, the test should be double-blind—neither the participants nor the physicians know who is getting the drug and who is in the control group (ensuring that the placebo resembles the drug is helpful in this regard). A two-year period is commonly taken up by a Phase II trial.

In a **Phase III trial**, the drug is tested in thousands of volunteers over a three-year period or more. Statistics are compiled carefully to determine whether the drug is more effective than the control therapy. A successful trial is often followed by a second, confirmatory trial, or if the trial results are inconclusive, the data are examined to discover whether a Phase III trial with new parameters is justified. Finally, the results are submitted to either of two agencies for approval: In the United States, the FDA can grant or deny approval for sale or request additional information. In Europe, approval is sought from the **European Agency for the Evaluation of Medicinal Products**.

For every drug approved by this process, an estimated 5,000 drugs do not survive the rigorous tests. Supporters of the process suggest that it must be followed if potentially dangerous drugs are to be weeded out. Opponents point out, however, that thousands might die while a useful medication remains in lengthy testing phases. **Community-based clinical trials** may be a solution to this dilemma. Run by private organizations employing private doctors and clinics, the community-based trials allow patients to suggest how trials might be operated and to contribute their experience. They also permit patients to receive experimental drugs from their own physicians in a familiar setting, a less stressful situation than they would experience at a university center. Such community trials try to avoid placebos, opting instead to compare the experimental drug with an established drug such as AZT. In the 1990s, community-based groups in San Francisco and New York City were conducting trials on 15 potentially useful AIDS drugs. The San Francisco group was largely responsible for data leading to the approval of aerosolized pentamidine isethionate. The number of new drugs that the FDA approves each year has remained relatively stable over time. In 2011, the FDA approved 30 novel new drugs known as **new molecular entities** (**NMEs**). A NME is an active ingredient that has never been marketed in the U.S. in any form.

Key Questions About Drug Testing

Those who desire to participate in a drug trial may not find it easy because certain criteria must be met. The trial, for example, may be limited to persons in a specific stage of HIV infection. In addition, most trials place limits on other drugs the patient may take (e.g., those taking ganciclovir could not take AZT), and such limits may be difficult to enforce. Moreover, a study may be restricted to certain groups, such as injection-drug-using males. The possibility also exists that a participant may receive the placebo, a circumstance that may not be desirable to an infected individual, and a trial may be in a certain phase that is not personally acceptable (a Phase I toxicity trial, for instance, may require confinement to a hospital).

By 2000, approximately 75 antiviral agents and vaccines were undergoing investigation at university and government medical centers. Still unresolved were such questions as whether the small manufacturers of drugs would pay for the standard tests preceding approval, nor certain whether insurers would cover the costs of medical care, laboratory tests, and other expenses incurred during broad-based trials. Moreover, it had not yet been determined whether physicians, manufacturers, or the FDA would be held liable should unapproved drugs prove fatal.

There is the pressing question of whether scientists and doctors should prevent patients from having access to drugs that patients believe will help them. Should a patient be allowed to accept the risk that an experimental drug carries, or should the medical establishment have the final say? Who, essentially, has the ultimate responsibility? Which is the greater good?

Thinking Well

The idea that mental states can influence the body's susceptibility to and recovery from disease has a long history. The Greek physician Galen, for example, asserted that cancer struck more frequently in melancholy women than in cheerful women. During the past quarter-century, the concept of mental state and disease has been researched more thoroughly, and several links have been established between the nervous system and the immune system.

One such link occurs between the hypothalamus and the T-lymphocytes. The hypothalamus is a portion of the brain located beneath the cerebrum. It produces a chemical-releasing factor that induces the pituitary gland, positioned just below the hypothalamus, to secrete a **hormone** (ACTH) that targets the adrenal glands. The adrenal glands, in turn, secrete **steroid hormones (glucocorticoids)** that influence the activity of T-lymphocytes in the thymus gland.

Another link is established by branches of the autonomic nervous system that extend into the lymph nodes and spleen. The autonomic (or "automatic") nervous system typically operates on its own to regulate the involuntary actions of such organs as the heart, stomach, and lungs through myriad nerve fibers. The anatomical links between the nervous and immune systems permits direct two-way communication.

A third link between the nervous and immune systems involves **thymosins**, a family of substances originating in the thymus gland. When experimentally injected into brain tissue, thymosins stimulate the pituitary gland via the hypothalamus to release hormones, including the one that stimulates the adrenal glands. Although the precise functions of thymosins are still to be determined, it appears that they serve as specific molecular signals between the thymus and the pituitary gland. A circuit is apparently present that stimulates the brain to adjust immune responses and the immune system to alter nerve cell activity.

The outcome of these discoveries is the emergence of a strong correlation between a patient's mental attitude and the progress of disease (**Figure 8.13**). Rigorously controlled studies conducted in recent years have shown that the aggressive determination to conquer a disease can increase one's lifespan. Therapies can consist of relaxation techniques, including listening to music, as well as the use of mental imagery that HIV is being crushed by the body's stalwart defenses. Yoga exercises also support the body's health. Behavioral therapies of this nature can amplify the body's response to disease and accelerate the mobilization of its defenses.

FIGURE 8.13

Numerous studies have shown that a person's mental attitude can influence the course of an infectious disease such as AIDS. Such things as exercise (a) and the support of family members (b) can have a positive influence. (a) © Dynamic Graphics Group/Creatas/Alamy Images. (b) © LiquidLibrary.

Few reputable practitioners of behavioral therapies believe that such therapies should replace drug therapy; however, the psychological devastation associated with AIDS cannot be denied, and it is this intense stress that the "thinking well" movement attempts to address. Very often a person learning of a positive HIV test goes into severe depression, and because depression can adversely affect the immune system, a double dose of immune suppression ensues. Perhaps if the psychological trauma is relieved, the remaining body defenses can adequately handle the virus.

As with any emerging treatment method, behavioral therapies have numerous opponents. Some opponents argue that naïve patients might abandon conventional therapy; another argument is that therapists might cause enormous guilt to develop in patients whose will to live cannot overcome failing health. Proponents counter with the growing body of evidence showing that AIDS patients with strong commitments and a willingness to face challenges—signs of psychological hardiness—have relatively greater numbers of T-lymphocytes than passive, nonexpressive patients. No study has yet proven that mood or personality has a life-prolonging effect on immunity. Still, doctors and patients are generally inspired by the possibility that the mind can be used to help stave off the effects of AIDS. Although unsure of what it is, they generally agree that something is going on.

LOOKING BACK

Toward the end of the 1980s, new therapies for AIDS came into use, and with their acceptance, the length and quality of life for AIDS patients have improved. Drugs may not affect a cure, but they can control HIV infection to allow individuals to lead almost normal lives. Among the HIV activities that can be targets for drug inhibition are binding of HIV to host cells, the functioning of reverse transcriptase, and the budding of HIV from host cells. Toxicity, the variety of cells that can be involved in AIDS, and brain cell damage limit the use of drugs.

AZT has been approved by the FDA since 1987 for treating AIDS. The drug replaces deoxythymidine in the synthesis of proviral DNA mediated by reverse transcriptase. Additional nucleosides cannot attach to the DNA chain with AZT in place, and DNA chain development comes to an end. The major side effect is bone marrow suppression leading to anemia.

Other therapeutic agents used for therapy for HIV infection and AIDS include other dideoxynucleosides, which work in the same way as AZT but are less toxic. Examples include ddI, ddC, 3TC, abacavir, and tenofovir. Often one or more of these drugs are used in combination with AZT to overcome HIV resistance. Similar drugs that react directly with reverse transcriptase include nevirapine, delavirdine, and efavirenz. Easier dosage and limited side effects are notable characteristics of these drugs.

The protease inhibitors achieved prominence in 1996 as part of a combination drug therapy known as HAART. Combined with AZT and another drug (such as 3TC), the protease inhibitor can bring about a significant decline in AIDS-related deaths. The success of this therapy was tempered by its difficult regimens, side effects, and the HIV rebound effect when drug therapy was interrupted.

New classes of anti-HIV agents include the fusion, entry, protease, integrase inhibitors, and coreceptor antagonists. One fusion inhibitor seeks to react with gp41 molecules of the HIV envelope and thus prevent the HIV particle from fusing with its host T-lymphocyte; another uses synthetic CD4 molecules. Entry inhibitors attempt to block HIV entry into host cells by destroying the necessary receptors on the cells' surface.

For treating the opportunistic diseases associated with AIDS, aerosolized pentamidine isethionate has been found effective to preclude the development of *P. jirovecii* pneumonia, and ganciclovir is used for CMV-induced retinitis. Many other drugs can lessen the effects of *Mycobacterium*, *Candida*, *Toxoplasma*, and other opportunistic microbes or viruses.

Before drugs are approved by the FDA, they are subjected to a multiyear series of trials, including preclinical trials (in animals), Phase I trial (to determine dose and toxicity in healthy humans), Phase II trial (in small groups of infected humans to determine antimicrobial effects), and Phase III trial (in large groups of volunteers). Parallel-track and fast-track procedures have been used to reduce the time between drug development and FDA approval.

Healthline Q & A

Q1 Can any other drugs be used in HAART?

A Yes. A physician has a choice of numerous drugs approved by the FDA for HAART. Recommendations are constantly changing as a result of new research findings, and physicians are free to use whichever combination they believe will most benefit the patient.

Q2 Can I get unapproved drugs for personal use?

A Yes, even though the FDA may not give its approval for a drug to be sold commercially in the United States, it may permit importation from foreign countries so long as the drug is for "personal use." This approval implies that the drug is not for sale to others.

Q3 Suppose I choose to take an unapproved drug. Can I get a physician to help me?

A Physicians will generally not recommend that you take unapproved drugs, but there is no law that prevents a physician from helping you obtain and use an unapproved drug. A physician may also help you gain entry to a clinical trial where you can benefit from an experimental drug.

Q4 Is pentamidine isethionate the only drug available to prevent *Pneumocystis* pneumonia?

A Several drugs are now in use to prevent this pneumonia. If you experience toxic side effects from pentamidine isethionate, there are alternatives, including trimethoprim-sulfamethoxazole (Bactrim or Septra); however, pentamidine isethionate is the recommended drug.

REVIEW

After you have finished this chapter, you should be familiar with the drugs used for therapy in patients with HIV infection and AIDS. To check your knowledge, match the drug or therapeutic agent on the right with the characteristic listed on the left by placing the appropriate letter in the space (a letter may be used more than once). The correct answers can be found elsewhere in the text.

_____ 1. Preferred drug for treating herpes simplex infections

_____ 2. Combined with sulfamethoxazole for treating *Pneumocystis jiroveci* pneumonia

_____ 3. Approved by the FDA in 1987 but causes suppression of bone marrow and anemia

_____ 4. Used with AZT; also known as lamivudine _____

A. Alpha interferon

B. IL-2

C. Trojan horse virus

D. Pentamidine isethionate

E. Antisense molecule

F. AZT

G. NRTI

_____ 5. Binds to gp120 and gp41 proteins of HIV

_____ 6. Inhibits tumor necrosis factor

_____ 7. Primary drug for treating *Pneumocystis jirovecii* pneumonia

_____ 8. Acts on reverse transcriptase, but unlike AZT

_____ 9. Approved for treating HIV infected persons to forestall progression to AIDS

_____ 10. Possible alternative to AZT with fewer side effects; also called didanosine

_____ 11. Commonly used to treat tuberculosis

_____ 12. Aerosolized form used to prevent *Pneumocystis jirovecii* pneumonia

_____ 13. Induces T-lymphocytes to mature

_____ 14. Synthetic antiviral substance used against Kaposi's sarcoma

_____ 15. Marketed as Videx and used in rotation with AZT; a similar nucleoside

_____ 16. Synthetic molecule that combines with HIV-specific mRNA during viral replication

_____ 17. Also known as zidovudine and Retrovir

_____ 18. Structured to react with a protein-processing enzyme in viral reproduction

_____ 19. Has a synergistic effect when used with AZT

_____ 20. Recommended for CMV-induced retinitis

_____ 21. Protease inhibitor; also known as Norvir

H. Thalidomide

I. Protease inhibitor

J. Ganciclovir

K. 3TC

L. Acyclovir

M. Isoniazid

N. Ritonavir

O. ddI

P. T-120

FOR ADDITIONAL READING

Arts, E. J., Hazuda, D. J., 2012. "HIV-1 antiretroviral drug therapy." *Cold Spring Harbor Perspectives in Medicine* 2;2:a007161.

Asier, S.-C. et al., 2013. "Post-treatment HIV controllers with a long-term virological remission after interruption of early initiated antiretroviral therapy ANRS VISCONTI study." *PLoS Pathogens* 9:3:e1003211.

Chapman, L. E., et al., 2008. "Recommendations for postexposure interventions to prevent infection with hepatitis B virus, hepatitis C virus, or human immuno-deficiency virus, and tetanus in persons wounded during bombings and other mass-casualty events—United States, 2008. *MMWR* 57(RR06):1–19.

Dieffenbach, C. W., Fauci, A. S., 2011. "Thirty years of HIV and AIDS: Future challenges and opportunities." *Annals of Internal Medicine* 154:766–771.

Dockrell, D. H., et al., 2011. "Evolving controversies and challenges in the management of opportunistic infections in HIV-seropositive individuals." *Journal of Infection* 63:177–186.

Essex, M., et al., 2011. "AIDS at 30: Hard lessons and hope." Harvard Public Health Review: Spring/Summer. Retrieved November 28, 2012 from http://www.hsph. harvard.edu/news/magazine/files/spring_11_aids.pdf.

June, C., Levine, B., 2012. "Blocking HIV's attack." *Scientific American*, March:54–59.

Kaplan, J. E., et al., 2009. "Guidelines for prevention and treatment of opportunistic infections in HIV-infected adults and adolescents." *MMWR* 58(RR04):1–198.

Munawwar, A., Singh, S., 2012. "AIDS-associated tuberculosis: A catastrophic collision to evade the host immune system." *Tuberculosis* 92(5):284–387.

Panlilio, A. L., et al., 2005. "Updated U.S. public health service guidelines for the management of occupational exposures to HIV and recommendations for postexposure prophylaxis." *MMWR* 54(RR-9):1–17.

Pendick, D. A., 1993. "Structure-based drug designers eye HIV protease." *ASM News* 59:382–387.

Skerlj, R., et al., 2011. "Design of novel CXCR4 antagonists that are potent inhibitors of T-tropic (X4) HIV-1 replication." *Bioorganic & Medicinal Chemistry Letters* 21:1414–1418.

Van Sighem, A. I., et al., 2010. "Life expectancy of recently diagnosed asymptomatic HIV-infected patients approaches that of uninfected individuals." *AIDS* 24:1527–1535.

The Kesho Bora Study Group, 2011. "Safety and effectiveness of antiretroviral drugs during pregnancy, delivery and breastfeeding for prevention of mother-to-child transmission of HIV-1: The Kesho Bora Multicentre Collaborative Study rationale, design, and implementation challenges." *Contemporary Clinical Trials* 32:74–85.

Zivin, J. A., 2000. "Understanding clinical trials." *Scientific American* 282:69–75.

Hopes and Hurdles Towards an HIV Cure or Vaccine

LOOKING AHEAD

More than 60 million individuals have been infected with HIV and more than 25 million have died of AIDS since 1981. Behavior modification, drug therapy, and screening can reduce HIV-1 infection rates. Nearly 30 antiretroviral drugs have been approved to treat HIV patients. Antiretroviral therapy has extended the lifespan of HIV-infected people but it does not eradicate the infection and HIV multi-drug resistance emerges. Novel antiretroviral therapies are being developed to eradicate HIV from infected patients, if they are even possible, but it may take many years if not decades for them to come into routine use.

Developing a vaccine for AIDS is among the highest priorities of today's medical researchers. HIV-associated co-infections with pathogens that cause tuberculosis (TB), toxoplasmosis, hepatitis C, and malaria are major concerns of AIDS patients. Since 1987, many candidate HIV-1 vaccines have been developed and tested in monkey studies as well as human clinical trials. Scientific advances suggest that for the first time in the history of HIV/AIDS, the possibility of controlling and ending the pandemic is within reach. This chapter surveys many aspects of accomplishments toward HIV prevention, a possible cure, and vaccines. On completing the chapter, you should be able to . . .

- Recognize the limitations of current HIV therapies.
- Identify interventions that can close the gap in controlling the HIV pandemic.
- Discuss the challenges of bone marrow transplants as a cure for HIV infection.
- Discuss anti-HIV gene therapy approaches.
- Explain how a strategy to activate reservoirs of HIV in infected cells could lead to a cure.
- Identify how vaccines work in the body and understand the various forms a microbial vaccine can take.
- Appreciate the various strategies for developing AIDS vaccines, and differentiate between sterilizing and therapeutic immunities.

- Describe the various components that can be used in an AIDS subunit vaccine, and delineate the advantages of each.
- Explain the composition of a viable-vector vaccine, and specify how it works to elicit immunity to HIV.
- Identify the whole virus vaccines that are being developed, and indicate some advantages and disadvantages of each.
- Discuss the various problems associated with development of an AIDS vaccine, and describe how some of these may be overcome.
- Describe how clinical trials for AIDS vaccines are conducted and anticipate some of the technical, ethical, and social difficulties that such trials will encounter.

INTRODUCTION

The theme of the 2012 International AIDS conference was "Turning the Tide Together." Secretary of State Hillary Clinton remarked at the conference that, "an AIDS-free generation will be achieved when virtually no children are born with HIV; when teenagers and adults are at a far lower risk of becoming infected with HIV than they are today, because of the availability of a range of prevention tools; and when people who do become infected with HIV have access to treatment that helps prevent them from developing AIDS or passing the virus on to others (**Figure 9.1**)."

FIGURE 9.1

Secretary of State Hillary Rodham Clinton speaking at the Plenary Session of the XIX International AIDS Conference 2012. Used with permission. © UPI/Kevin Dietsch/Landov.

Hopes and Hurdles Towards an HIV Cure or Vaccine

Geographically, wherever HIV screening is done accurately and combination antiretroviral therapy is accessible, the rates of HIV-associated deaths have declined sharply. Studies indicate that HIV-infected people treated with combination antiretroviral therapy early during the course of disease, and have other health benefits available, can now anticipate living a near-normal lifespan. By the end of 2010, 6.6 million HIV-infected people were receiving antiretroviral medication. In the absence of a cure, it is unlikely the world will ever have the money, healthcare workers, or infrastructure to treat all HIV-infected patients indefinitely. Untreated HIV-infected individuals are the main source of ongoing HIV transmission and will cause the infected population to increase. Therefore, the importance of prevention continues to be a top priority.

Reducing HIV Rates of Infection

Several methods or approaches can reduce HIV-1 infection rates in people at risk of exposure. Screening the blood supply and testing blood-derived products for human use, counseling programs, monogamy and the use of condoms, male circumcision, and the immediate use of postexposure antiretroviral therapy are known to dramatically reduce HIV-1 infection rates.

Sexually active South African women who used **tenofovir** vaginal gel before and after sexual intercourse lowered their chances of HIV infection by 39%. Another study reported that women who used the tenofovir microbicide 80% of the time had 54% fewer infections than those women who used a placebo microbiocide. A study conducted by the National Institutes of Health demonstrated that a daily dose of **Truvada** (combined tenofovir and **emtricitabine**) decreased chances of HIV infection by 44% among men who have sex with men.

Treating the infected mother with antiretroviral drugs can prevent transmission of HIV-1 from mother-to-child resulting in 2% or fewer children born to HIV-infected mothers. Even though these preventative measures have the potential to strongly reduce the trajectory of the HIV/AIDS pandemic, logistically it is difficult to scale up these prevention efforts because of financial obstacles and the resistance to behavioral change. An effective and safe vaccine would prevent HIV infection and/or AIDS.

Eliminating Viral Reservoirs

During HIV replication the viral genome may be integrated into the host cell chromosomal DNA. These cells are referred to as **latent reservoirs of HIV** or "sleeper cells" and are the main obstacles to HIV elimination from the body. The HIV genome, called a **provirus**, can silently ride along with the host chromosomal DNA for decades. The provirus is invisible to immune system defenses and effective antiretroviral therapy. Research led by David Margolis, a molecular virologist at the University of North Carolina at Chapel Hill found that a lymphoma cancer drug, **vorinostat**, activates or kick-starts the proviral DNA into transcription of the HIV genes, resulting in the production of new HIV particles.

In Margolis' small, elegant clinical trial, the CD4 cells of six HIV patients were removed and treated with vorinostat to break HIV latency. Within 8 hours of vorino-stat treatment, HIV RNA transcription was induced within the sleeper cells, forcing HIV out of its hiding place and making it vulnerable to the body's immune response and antiretroviral therapy. This study suggests that drugs targeting viral latency could build a path that leads to a cure for HIV. It provides hope that HIV can be eradicated from the body completely.

Vorinostat shocks HIV into a productive life cycle, allowing the virus particles to be seen and attacked by the body's immune system and antiviral therapy. Adapted from Deeks, S. G., 2012. "Shock and kill." *Nature* 487:439–440.

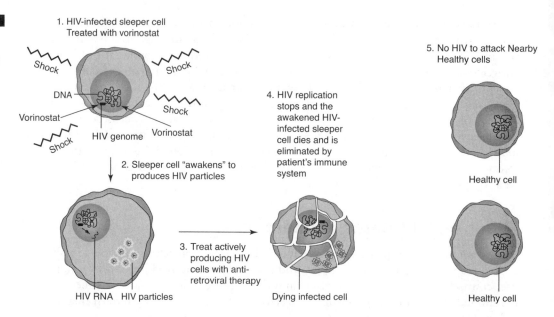

Vorinostat is one of a class of anti-cancer drugs known as **histone deacetylase** (**HDAC**) inhibitors. Vorinostat blocks the unregulated growth of cancer cells by regulating HDAC activity with little or no toxicity to normal cells. HDACs in normal cells acetylate histones, altering the structure of chromatin and regulate the transcription of cellular genes. In contrast, HDAC inhibitors close the chromatin structure such that unregulated gene transcription is repressed in cancer cells. Ironically, vorinostat activates transcription (instead of repressing) the LTR promoter region of the HIV proviral DNA present in sleeper cells. In doing so, the drug shocks HIV into a productive life cycle, new HIV particles become visible to immune system and antiretroviral attack (**Figure 9.2**).

Other drugs that also break HIV latency, such as **disulfiram** (Antabuse), have not resulted in significant HIV production or a reduction in the pool of HIV sleeper cells in a small clinical trial led by Liang Shan, a researcher at Johns Hopkins University in Baltimore, Maryland. Overall, a combination of approaches will be needed to eradicate HIV from the body. HDAC inhibitors have opened the door to a new strategy in which a drug can safely purge HIV into production from dormant cells containing the HIV integrated genome in patients. Flushing HIV (shock) allows the virus to be detected by the immune system and thus vulnerable to its attack (kill) and to antiretroviral therapy (Figure 9.1). It offers an alternative to drug therapy and hope for a cure.

Bone Marrow Transplants

One alternative therapy that may hold promise for the future is the **bone marrow transplant**. In this procedure, bone marrow cells from a **donor** are transplanted into the bone marrow of the AIDS patient (**recipient**) in the hope that they will restore the immune system cells that have been destroyed. The body normally rejects foreign tissue and, thus, the donor and recipient must be closely matched genetically (the ideal pairing is identical twins, a perfect match).

In the past, researchers have performed bone marrow transplants among sets of identical twins, each set consisting of a healthy twin and a twin with HIV infection or AIDS. Although some partial restoration of the immune system was observed, HIV eventually attacked the incoming cells and destroyed them. More encouraging reports appeared in the 1990s when AZT treatment accompanied the transplant; HIV appeared to be absent in the tissues while the marrow cells were establishing themselves.

Inherent in a bone marrow transplant are several problems that must be circumvented. The process is very dangerous because the remnants of a patient's own bone marrow must be completely destroyed with radiation or toxic chemicals; otherwise, patient cells will immediately attack the incoming cells. Finding suitable donors is also difficult, and the transplant procedure is extremely expensive.

In 2007, a breakthrough bone marrow transplant *cured* the **Berlin patient** (Californian Timothy Ray Brown living in Germany) of HIV. His case was well documented by researchers at Charite Universitatsmedizin Berlin and the Robert Koch Institute. It entailed transplanting stem cells from a donor who had two copies of a defective *CCR5* gene known as *CCR5del32* into Timothy Ray Brown who suffered from acute myeloid leukemia and HIV-1 infection. Persons with two gene copies (making them homozygous) of this defective *CCR5* gene are highly resistant to HIV-1 infection (**Figure 9.3**). CCR5 is a chemokine receptor. It is a coreceptor for HIV-entry into target cells (e.g., CD4 positive T-helper cells). The *CCR5del32* gene codes for a shorter gene product resulting in a protein that is missing 32 amino acids. HIV-1 cannot gain entry into cells because it is unable to use the truncated (shorter) CCR5 coreceptor to enter host cells.

The procedure was brutal, involving chemotherapy and radiation treatments prior to receiving the stem cell transplantation in an attempt to re-populate Brown's immune system with HIV-1 resistant T-cells. Brown's viral loads were undetectable and antiretroviral therapy was discontinued. Unfortunately, the leukemia returned and the procedure was repeated. His cancer is in remission at the time of this writing and he remains HIV-free. To prove that Brown did not harbor any sleeper cells or viral reservoirs in body tissues, he underwent repeated blood draws, and colon, rectal, and liver biopsies to screen for the presence of HIV. He underwent brain biopsies

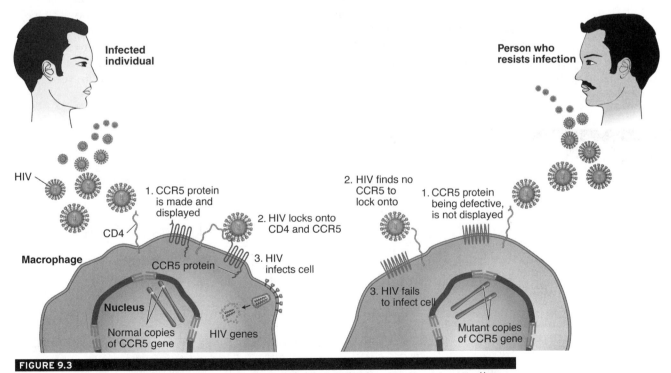

FIGURE 9.3

One mechanism of HIV resistance in humans. Adapted from Coleman, F. "HIV," Chem101: Chemistry, http://www.wellesley.edu/Chemistry/Chem101/hiv/resist.gif.

after suffering from chemotherapy complications that caused neurological problems resulting in coordination problems, vision, speech, and memory problems as well as personality changes. By July 2012, a frail-looking Brown told news reporters he felt great and believed he was cured of HIV.

Traces of HIV genes were found in Brown's tissues in 2012. Researchers argue that this HIV genetic material is unable to replicate and others suggest that the remnants of HIV genetic material could be contaminating HIV nucleic acids in the labs doing the assays instead of HIV from Brown's tissue samples. Yet others remain skeptical that Brown is cured.

Daniel Kurtzkes, infectious disease expert and Professor of Medicine at Harvard Medical School, described two additional HIV-positive American men "cured" through stem cell transplantation at the 2012 International AIDS Conference held in Washington, D.C. Both of the men underwent stem cell transplantation treatment for lymphoma. Their chemotherapy was milder than Brown's, allowing the men to remain on antiretroviral treatment throughout the transplant period. Neither patient had detectable levels of HIV and their antibody levels against HIV decreased, indicating a lack of ongoing exposure to HIV. In contrast to the "Berlin patient," these two men received donor stem cells that had two copies of a normal *CCR5* gene. The doctors believe that the antiretroviral therapy effectively cleared HIV from the patient, protecting the donor cells from HIV infection. At the time of this writing, this study was not published and the men's tissues were being sampled. They remain on antiretroviral therapy. Antiretroviral therapy must be discontinued and the patients carefully monitored for HIV infection to determine if these men are cured. **Figure 9.4** is a general overview of how this type of therapy is carried out.

Although theoretically possible, the short-term prospects remain dim that bone marrow transplants will become a commonplace therapy for treating AIDS. This approach is arduous and expensive. Roughly only 1% of the Caucasian population and zero percent of the African-American population have the homozygous *CCR5* *CCR5del32* in their genetic makeup. These studies have unraveled a path to a potential cure. Researchers are exploring the possibility of removing the T-cells of HIV patients and genetically engineering them (a *gene therapy* approach) to resist HIV infection.

FIGURE 9.4

Overview of bone marrow transplant procedure.

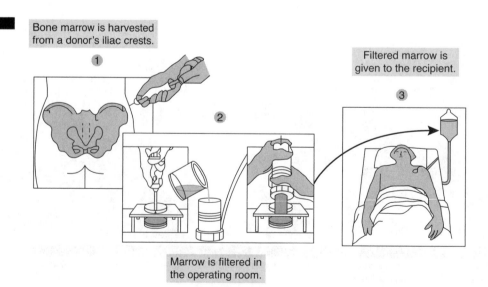

Bone marrow is harvested from a donor's iliac crests.

Filtered marrow is given to the recipient.

Marrow is filtered in the operating room.

Gene Therapy

Gene therapy is one of the most innovative and imaginative uses of the DNA technology and genetic engineering that emerged in the 1970s. For gene therapy, cells are taken from the patient, altered in the laboratory by adding genes, and then placed back into the patient. It is anticipated that the new genes will provide the genetic codes for certain proteins not normally produced by the cells.

Gene therapy may be used one day to relieve the effects of HIV infection and AIDS. In 1993, the Recombinant Advisory Committee of the U.S. government approved the first clinical test involving patients with HIV infection. The test used a mutant strain of HIV having defective forms of the *rev* and *env* genes. This strain, produced by researchers at the University of California at San Diego, cannot replicate with the mutated genes (*rev* is a regulatory gene and *env* encodes envelope proteins). Researchers removed T-lymphocytes from HIV-infected patients and inserted mutated viruses into the cells. Then they cultivated large numbers of the cells and injected them back into the patient. The virus-containing T-lymphocytes could not produce viruses, but they did stimulate the body to produce cells called killer T-lymphocytes. These killer cells are specifically manufactured to react with HIV-infected cells. Tests with patients are ongoing.

In another form of gene therapy, genes that encode HIV proteins are attached to the DNA of mouse viruses, which are harmless for humans. The re-engineered viruses are then injected into HIV-infected individuals, where they enter receptive cells. Researchers hope that the HIV genes will stimulate normal body cells to produce HIV proteins. The proteins will then act as a vaccine and stimulate the immune system to produce anti-HIV antibodies. These antibodies may help prevent HIV expression in the patient. Along these lines, research led by David Baltimore at the California Institute of Technology is using genetically engineered adenoviruses carrying genes that encode anti-HIV antibodies isolated from people infected with HIV. The genetically engineered adenoviruses were injected into the muscle cells of mice. The DNA encoding the antibodies was incorporated into muscle cells where it programed the cells to manufacture anti-HIV antibodies that were secreted into the bloodstream of the mice. In 2011, Baltimore announced to the press that two of the antibodies, called b12 and VRC01, completely neutralized HIV in mice infected with larger numbers of HIV-1 than would occur in a natural infection. This type of procedure is called **vectored immunoprophylaxis (VIP)**. The next step is to determine if VIP will be safe and effective in clinical trials.

A very promising development in gene therapy is the application of gene editing enzymes to permanently alter or inactivate host or viral genes necessary for HIV infection and replication, ultimately creating resistance to HIV infection. One of the earliest gene editing experiments entailed the use of genes that encode site-specific DNA editing proteins known as **zinc finger nucleases. Zinc finger proteins** are naturally occurring proteins that bind DNA during transcription. In human cells there are more than 2,500 different types of zinc finger proteins. Each type binds to a different set of nucleotides present on chromosomal DNA. **Nucleases** are proteins that cleave DNA. Scientists have engineered designer hybrid zinc-finger nuclease genes that encode proteins that bind to a section of the *CCR5* gene and to cut the DNA like a scissors, snipping out the *CCR5* gene or enough of the gene such that the CCR5 HIV coreceptor is defective and resistant to HIV-1 infection (**Figure 9.5**). The DNA is also repaired or rejoined together by the host cell's machinery. During the repair process, repair enzymes chew the ends of the snipped DNA, ensuring that the cut DNA cannot recombine or be glued back into the cellular *CCR5* gene.

Intensive research effort into understanding the mechanisms of HIV-1 replication has led to the identification of several new therapeutic targets for antiviral gene

FIGURE 9.5

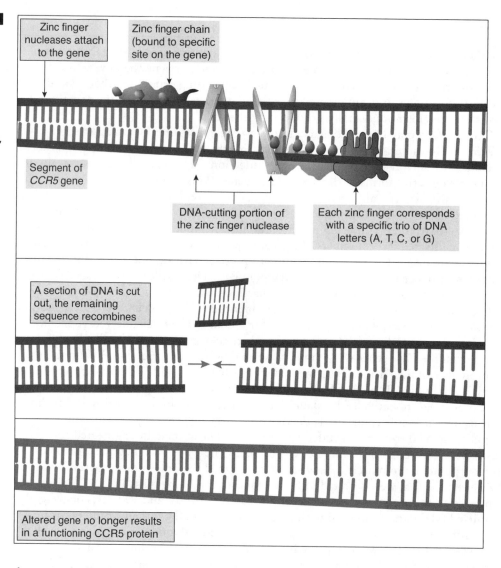

FIGURE 9.5
Zinc finger nuclease technology used to cleave CCR5. The body's repair enzymes will glue the DNA strands together, resulting in a nonfunctional CCR5 gene that HIV cannot use to gain entry to host cells. Adapted from June, C. and Levine, B., 2012. "Blocking HIV's attack." *Scientific American*, 306:54–59.

Zinc finger nucleases attach to the gene

Zinc finger chain (bound to specific site on the gene)

Segment of *CCR5* gene

DNA-cutting portion of the zinc finger nuclease

Each zinc finger corresponds with a specific trio of DNA letters (A, T, C, or G)

A section of DNA is cut out, the remaining sequence recombines

Altered gene no longer results in a functioning CCR5 protein

therapy. Ideally gene therapy against HIV will involve gene transfer in hematopoietic stem cells. Experiments such as these illustrate the novel approaches gene engineers are taking to fight HIV. Practical use of these methods remains in the future, but it is comforting to know that alternatives to drug therapy are both possible and feasible.

Introduction to Vaccines

Smallpox historically has been among the worst killers of humanity. In the 1600s and 1700s, for example, smallpox epidemics were so severe in England that one third of all children died before reaching the age of three years. Many smallpox victims were blinded by the disease, and those who recovered were pock marked for life.

For those who survived smallpox, however, the disease did not strike again. People therefore sought a way to contract the disease during a "mild" year, hoping to acquire immunity. At one point, the custom of "buying the pox" arose. Unfortunately, this method of acquiring immunity was haphazard and quite risky. Thus, excitement surfaced in the late 1700s when the method for immunization was improved considerably by an English country surgeon named Edward Jenner. Jenner observed that people who had suffered from a similar disease—cowpox—were apparently immune to the more deadly smallpox. Although cowpox (technically known as vaccinia) primarily occurs in the udders of cows, it also occurs as a mild skin disease in farmers, herders, and those who tend cows.

Jenner pondered whether intentionally giving cowpox to people would protect them against smallpox. In 1796, he tested his theory, enlisting the help of Sarah Nelmes, a dairy maid who had cowpox, and a young boy named James Phipps (**Figure 9.6**). Jenner took pus from lesions on Nelmes' arm and injected the boy's skin with the material. Phipps developed a swelling at the inoculation site but little else. Several weeks later, Jenner carefully injected Phipps once again, this time with pus from the lesion of a smallpox patient. Within days, the boy developed an uncomfortable feeling, but he did not develop smallpox. Phipps had been successfully immunized. Prominent physicians confirmed his work, and word of Jenner's method of "vaccination" spread quickly throughout the world. By 1801, an estimated 100,000 people in England had been vaccinated.

A hundred years would pass before scientists understood that the cowpox virus stimulates the immune system to produce antibodies, which neutralize the deadly smallpox viruses as well as the mild cowpox viruses. Hailed as one of the great medical–social advances in history, Jenner's immunization method was among the first attempts to control disease on a national scale. It was also the first effort to protect the community as a whole, rather than the individual.

Modern vaccines are as effective as Jenner's but more refined. They are composed of weakened or killed microorganisms, or in some cases, they contain molecular fragments of microorganisms or their chemically altered toxins. Such vaccines work by

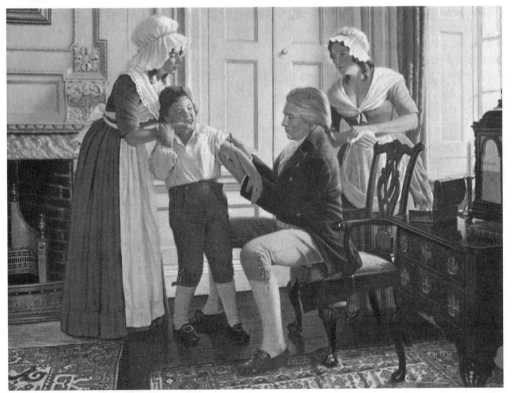

FIGURE 9.6

A painting by Robert Thom depicting Edward Jenner vaccinating the young James Phipps. The dairy maid at the right, Sarah Nelmes, is holding the lesion on her wrist where the cowpox material was taken from. Collection of the University of Michigan Health System, Gift of Pfizer, Inc. UMHS.23.

exploiting the immune system's ability to recognize specific antigens and respond with antibodies and other defenses that later attack and destroy invading microorganisms and their toxins. A good vaccine prevents infection or disease and protects against different strains of the pathogenic virus. It gives long-lasting protection against both free viruses and cells infected with viruses, and it induces an immune response not only within the blood and tissues, but also at body surfaces, as in the gastrointestinal, respiratory, and urinary tracts.

In this chapter, we first discuss some general principles of immunization and then focus on the intensive effort to develop a vaccine for AIDS. In a sense, AIDS is the smallpox of our era. Indeed, the fear about contracting AIDS rivals the fear people expressed about contracting smallpox in past centuries. Developing an AIDS vaccine carries the priority once given to developing a smallpox vaccine, because over 60 million people of the world have been infected with HIV over the past few decades and more than 25 million have died of AIDS. AIDS remains the number one killer of individuals in the 25- to 44-year-old age bracket. AIDS affects not only those who engage in risky behavior (i.e., injection drug users) but also the countless numbers of heterosexuals who cannot be reached by educational resources, and the masses of illiterate individuals who simply do not understand that certain behaviors are irresponsible. Preventing AIDS with a vaccine is vital to the economic health of nations. Caring for and treating an infected individual in the United States with drugs costs over $600,000 over his or her lifetime, whereas immunizing with a vaccine costs only $200 per person. For these reasons and numerous others, the effort to develop an AIDS vaccine is proceeding at full speed in laboratories throughout the world.

General Vaccine Strategies

One of the significant developments of modern medicine was devising a tissue culture technique for the propagation of viruses. In the 1940s, the technique was refined for culturing animal cells by Nobel laureates John Enders, Thomas Weller, and Frederick Robbins. Their technique led to the development of vaccines against polio, measles, mumps, and other diseases because it allowed laboratories to grow sufficient viruses for vaccine production.

Types of Vaccines

By the 1960s, vaccine manufacturers were able to produce two types of vaccines: killed or **inactivated vaccines** contained inactivated viruses, that is, viruses unable to multiply in the body because of chemical or physical treatment, and a second type called **attenuated** or **live vaccines** contained attenuated (weakened) viruses, that is, viruses able to multiply in the body but at a rate so low that disease does not occur. The **Salk injectable polio vaccine** typifies a vaccine containing inactivated viruses; the **Sabin oral polio vaccine** has attenuated viruses. Both are referred to as first-generation vaccines because they contain entire microorganisms (**Figure 9.7**).

During the 1970s, researchers developed another type of vaccine, the **subunit or protein vaccine**. Manufacturers found that they could isolate protein subunits from the infectious agent (virus or bacterium) and use these individual viral protein(s) as a vaccine to illicit an immune response to the whole virus. Such a vaccine is safer than a whole-virus vaccine because subunits cannot possibly infect cells. If the microbe invades the body, antibodies bind with the portion of the microbe corresponding to the subunit and neutralize it.

The vaccine currently used against bacterial pneumonia typifies a subunit vaccine. Licensed in 1983, the vaccine contains polysaccharides obtained from the capsule

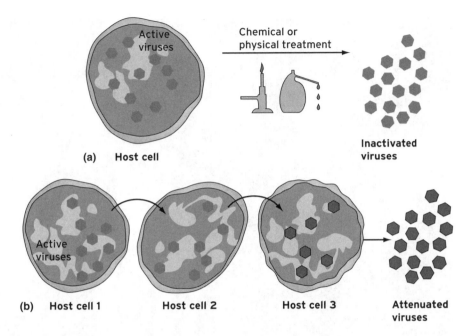

FIGURE 9.7

Methods for producing two types of viral vaccines. (a) Viruses are cultivated in cells in tissue culture and then treated with a chemical or physical agent to yield inactivated viruses. (b) Viruses are transferred from cells in tissue culture to new cells in tissue culture until a variant with reduced virulence emerges. These are attenuated viruses. The Salk polio vaccine contains inactivated viruses, while the Sabin polio vaccine has attenuated viruses.

(outer layer) of the bacterium *Streptococcus pneumoniae*. Specific antibodies form against the polysaccharides and circulate in the blood, binding to the capsule when the bacterium enters the body at a later date. The antibodies cripple the bacterium, preventing its proliferation and aiding its destruction by other cell defenses. Such a vaccine composed of subunits is known as a second-generation vaccine.

Another form of vaccine is the **recombinant vaccine**. In this case, a gene from the infectious agent is isolated and put in a culture of microorganisms, such as a bacterium or fungus. The viral gene is then expressed into proteins, and the protein can be isolated and used as a subunit vaccine as described previously here. Alternatively, the recombinant gene can be introduced into the person so that the viral protein is produced, which leads to a vaccination without having to isolate the protein subunit. The current vaccine for hepatitis B is an example of a recombinant vaccine. Licensed since 1987 and available as Recombivax, the synthetic vaccine is considered safer than the previous hepatitis B vaccine, which used proteins isolated from human blood.

A type of recombinant vaccine known as a **DNA vaccine** consists of plasmid DNAs modified to carry one or more protein-encoding genes. Plasmids are submicroscopic loops of DNA found in bacterial cells (**Figure 9.8**). DNA vaccines are safer than vaccines with whole microorganisms because the plasmids are not infectious or able to replicate, nor do they encode any proteins other than those specified by the genes. When injected into muscle tissues, the plasmids enter the muscle cells and stimulate them to synthesize the gene-encoded proteins and place them at the cell surface; here the proteins are recognized as foreign and stimulate an immune response. These viral proteins also tend to hold the native structure that they are expected to have in the virus. One problem with DNA vaccines is that they are difficult to produce because the plasmids must contain various high-technology genes that will ensure their survival and activity in the host cell. Another problem is the concern that this mutagenesis may be oncogenic.

Another vaccine approach is the **vector vaccine**. In this case, a mild or virulent virus to humans is used as a vector to carry HIV genes into host cells. Example vector viruses for HIV vaccines include the vaccinia virus, adenovirus, and the canarypox virus. These vectors will have some of its genome removed but retain the genes

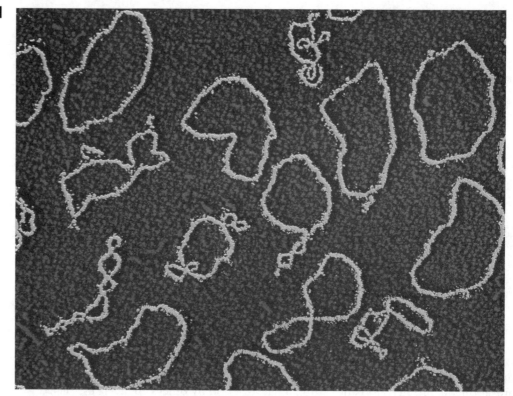

FIGURE 9.9

Safety and efficacy of the
different types of vaccines.

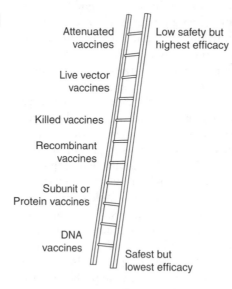

Attenuated vaccines — Low safety but highest efficacy

Live vector vaccines

Killed vaccines

Recombinant vaccines

Subunit or Protein vaccines

DNA vaccines — Safest but lowest efficacy

involved in the ability to infect cells, and then HIV genes will be introduced into it. This virus can then infect the host and lead to the expression of HIV proteins that can serve as the vaccine by eliciting an immune response. This vector vaccine approach also leads to the production of native HIV proteins, but the modified vector virus does not cause disease or have the potential for reactivation as described with attenuated live HIV vaccines. **Figure 9.9** compares the safety and efficacy of the different types of vaccines. Vaccine efficacy refers to how well the type of vaccine prevents disease for which it was developed.

AIDS Vaccine Strategies

Because of the high mortality associated with AIDS, developing a vaccine presents one of the major medical challenges of modern times. In the United States alone, tens of millions would be candidates for the vaccine, among them almost three million men at risk because of their homosexual practices with infected partners and one to two million men who are bisexual. Also at high risk are several million Americans who regularly inject illicit drugs, as well as their sexual partners. In the medium-risk group are persons likely to come in contact with contaminated blood, such as millions of allied health science professionals. These include doctors, nurses, dental hygienists, pathologists, paramedics, and funeral service employees. Police, correction officers, and members of the military are also vaccine candidates, as are highly sexually active, nonmonogamous heterosexuals.

Developing a vaccine for AIDS is far from easy, however: HIV has an extraordinarily high mutation rate. It takes several weeks or more to mount an effective immune response against a specific target. Within this time, HIV is able to change its "look" many times by mutating. These gene mutations cause the protein structures of the virus to change. Because the specific recognition of a virus by the immune system is dependent on recognition of the three-dimensional structure of the target proteins, the rapid mutation of the HIV virus makes it impossible for the immune system to consistently recognize it. For example, by the time the immune system makes an effective antibody against the HIV, the virus has already mutated, so the antibodies produced are ineffective. Now the immune system begins to mount a response to the new mutant, and again, by the time those antibodies are made, the virus has mutated and changed it look. In this regard, the HIV virus is very stealthy and able to evade the immune system. The same problem obviously exists for HIV vaccines. Indeed, many HIV vaccines have shown promise in that they elicited a strong immune response to HIV; however, once again, after the recipient receives the vaccine, the virus in the person mutates, and the immune response to the vaccine is ineffective against the new viral forms. In the case of a vaccination of an HIV-negative person, there is a great deal of probability that the virus they may become exposed to will "look" very different than the vaccine, again through mutations within the carrier.

Another dilemma is deciding which immune response to trigger. Most viral vaccines work by calling forth a strong antibody response. The antibodies bind to the virus and prevent it from infecting its host cell, but the other arm of the immune system depends on activity of the cytotoxic T-lymphocytes. These cells interact with and destroy microbe-infected body cells, such as those with indwelling HIV in its proviral state. Researchers must study how best to exploit the capabilities of both arms of the immune system.

Furthermore, researchers have found that even if pathogenic HIV could be modified to a nonpathogenic attenuated variant, retroviruses in general do not elicit sufficient amounts of antibodies to prevent subsequent infection. For example, the retrovirus that causes feline leukemia elicits a low level of protection when used in the inactivated form as a vaccine. Furthermore, researchers have also reported that non-pathogenic attenuated viruses could revert, or reactivate, to a pathogenic form. Thus, there is concern that an inactivated nonpathogenic variant of HIV could revert to its pathogenic form. It is also possible that gene segments of viruses in a vaccine could integrate into human chromosomes and promote tumor formation, as certain retroviruses are known to be tumor-inducing.

To counter this pessimism, virologists point out that in contrast to the feline leukemia virus, HIV apparently calls forth a strong antibody response when it enters the body. Studies have also shown that HIV antibodies can hold the virus in check for considerable periods of time (months or years). One objective of vaccine research is to locate the part of HIV (e.g., capsid protein, reverse transcriptase, or envelope protein) that elicits the strongest immune response and amplify that part for a possible vaccine. This may be a considerable task because no individuals completely immune to AIDS have yet been found, and thus, no one is really sure about the nature of the immune state. Nevertheless, major interest has focused on various subunits of HIV, as we see next.

Sterilizing and Therapeutic Immunities

Since HIV was first isolated, scientists have been hunting for a vaccine that could prevent it from infecting humans, similar to the vaccines for measles, mumps, rubella, hepatitis B, yellow fever, smallpox, and polio (**Figure 9.10**). The key word here is "prevent" because the definition of a vaccine implies a preparation that brings about immunologic surveillance and stops microorganisms before they establish themselves in the body. In modern terminology, this form of immunity has been termed *sterilizing immunity* to distinguish it from a newer concept of immunity known as *therapeutic immunity*.

In the early days of AIDS vaccine research, the preventive vaccine was the gold standard. Then, in the early 1990s, tests performed on monkeys yielded dispiriting results, indicating that a preventive vaccine might be an impossible goal to attain. Some frustrated scientists turned to therapeutic vaccines as a method of preventing "disease" rather than "infection." They accepted the fact that a vaccine could not stop HIV from establishing itself in body cells and turned their efforts to preventing the symptoms of

FIGURE 9.10

An adult being immunized against a number of "childhood" diseases. Courtesy of Barbara Rice/CDC.

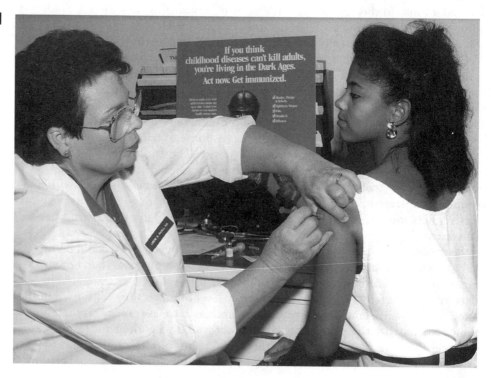

Hopes and Hurdles Towards an HIV Cure or Vaccine

disease from developing. With that assumption, their definition of success changed, and many researchers set modified goals for themselves.

Therapeutic vaccines are used to prolong life, promote a disease-free period, prevent progression of the disease, and possibly eliminate the virus. In addition, a therapeutic vaccine could be used to protect the offspring of a pregnant woman who is HIV positive. For example, a therapeutic vaccine could induce an immune response to reduce the amount of virus present in her bloodstream and reduce the possibility of passage across the placenta to the fetus.

The advantages of therapeutic immunity are numerous: Antibodies induced by the vaccine reduce the viral load in the body and make the patient less likely to spread HIV to the next individual; the delay of symptoms is clearly beneficial to the patient; and the economic pressure on the healthcare system is reduced. Moreover, as the concept of sterilizing immunity is increasingly rejected and that of therapeutic immunity grows in acceptance, scientists will have a better selection of vaccines for clinical trials; also, reduction of viral load can be used to compare vaccines' efficacies much better than the current approach of determining how many immunized persons fail to contract HIV infection. The current thinking is therefore dualistic: Many researchers still contemplate a preventive vaccine, but others have begun to consider the advantages of a therapeutic vaccine. Both paradigms are represented in the vaccines currently being developed, as the next sections survey.

AIDS Subunit Vaccines

Using viral subunits such as proteins or glycoproteins is a viable alternative to using whole-virus vaccines. In the 1990s, technologic advances using recombinant DNA made it possible to produce an unlimited supply of subunits employing various types of cells as "factory organisms," or vectors (**Figure 9.11**). The genes that provide the genetic code for a particular subunit are first identified, and then these genes are copied in huge numbers. Next, the researcher selects a vector—an organism such as a bacterium, yeast, or insect cell—and biochemically inserts the genes into the nuclear material of the vector. When the vector's genes express themselves during protein synthesis, the subunit's genes also express themselves, and large quantities of the subunit proteins are synthesized. These proteins can then be isolated and purified from the cellular material.

Among the major subunits being considered for an AIDS vaccine are glycoprotein 120 (gp120) and glycoprotein 41 (gp41). Both glycoproteins exist in the HIV envelope spikes; gp120 juts out and contacts the CD4 receptors and coreceptors, and then gp41 penetrates the membrane and serves as an anchor for gp120 molecules. Both gp120 and gp41 are essential for the binding of HIV to host T-lymphocytes and brain cells. Antibodies produced against these glycoproteins presumably would unite with the glycoproteins and prevent HIV's interaction with its host cells.

Additional HIV subunit or protein vaccines include the HIV glycoprotein known as gp160. This glycoprotein is a precursor molecule (a forerunner) to both gp120 and gp41. Researchers have shown that gp160 can stimulate antibody production when used in a vaccine. Still another vaccine targets the HIV protein known as p17. This protein is located in the capsid surrounding the HIV genome. The protein is believed to protrude into the envelope of HIV and come near enough to the viral surface to provoke an immune response. Biotechnology companies are using recombinant DNA technologies to produce the protein in bulk and then chemically cut the molecule into smaller and smaller subunits to determine which subunit most efficiently elicits antibody production by the immune system. The commercial vaccine made by this

Procedure for preparing viral protein subunits for vaccine use. (a) The viral RNA that codes for a specific protein (viral protein 1) is identified and isolated. (b) Reverse transcriptase uses the viral RNA as a template to form viral DNA containing the gene for viral protein 1. (c) A plasmid is isolated from a bacterium such as *Escherichia coli* and cleaved using a restriction enzyme. (d) The viral DNA is inserted into the cleaved plasmid, and (e) the latter is inserted into a fresh *E. coli* cell, thereby transforming the bacterium. (f) When the recombined bacterium multiplies and metabolizes nutrients, the bacteria produce proteins, one of which is viral protein 1. (g) The viral protein is then isolated for use in vaccine production.

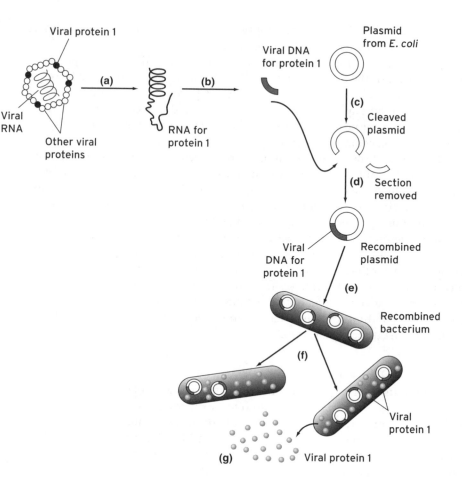

process is called HGP-30. It contains a 30-amino acid region of the p17 protein. One distinct advantage to using p17 is that this protein does not display the propensity to mutate as do envelope proteins such as gp120 and gp41. This is because the *gag* genes for p17 are apparently more stable than the envelope (*env*) genes for the envelope proteins.

Unfortunately, isolated protein subunits—such as gp120, gp41, and p17—generally do not elicit large amounts of antibodies. One reason is that they are not easily engulfed by the body's macrophages or efficiently delivered to the immune system during the immune process. Subunits are therefore made more attractive to macrophages by chemically binding them to "carrier" particles or to chemical substances referred to as adjuvants. Adjuvants include molecules of aluminum sulfate (alum), globules of peanut or mineral oil, artificial membranes called liposomes, and certain noninfectious viruses, as we discuss here. A potentially affective adjuvant now in testing stages is granulocyte-macrophage colony stimulating factor. This chemical substance is often used to boost blood counts in transplant and cancer patients.

Even the best adjuvants are not always able to encourage a strong antibody response. For example, in the early 1990s, optimism was high that a gp120 vaccine might become the weapon of choice against HIV (**Figure 9.12**). However, two gp120 vaccines formulated independently by two different companies failed to elicit a strong antibody response to freshly isolated HIV, and in 1994, the National Institutes of

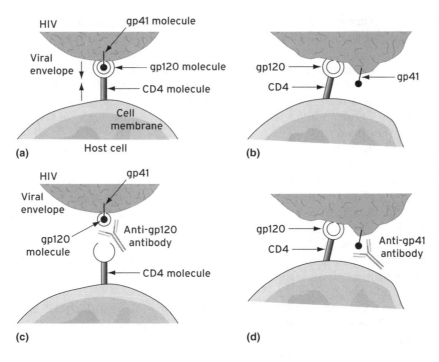

FIGURE 9.12
How antibodies prevent binding of HIV to host cells. (a) When HIV associates with a host cell, its gp120 protein binds to the CD4 receptor site on the host cell. (b) After the initial binding has taken place, the gp41 protein anchors HIV to the host cell membrane and encourages penetration. (c) Antibodies against gp120 will combine with gp120 and prevent the latter's binding with CD4, and (d) antibodies against gp41 will inhibit gp41's anchorage to the host cell. By interfering with both gp120 and gp41 activities, the attachment of HIV to host cells can be prevented.

Health (NIH) was prompted to abandon human trials using gp120 molecules. That decision dampened enthusiasm for vaccine development for several years.

The gp120 vaccine was back in the news by the end of the decade, however. In 1998, VaxGen, Inc. of San Francisco was given Food and Drug Administration approval to launch a trial of its newly formulated alum-adjuvant gp120 vaccine called AIDSVAX. The so-called bivalent vaccine contains two types of gp120 molecules, one type from a laboratory strain used in previous vaccines and one from currently circulating HIV. In the United States, about 5,000 uninfected gay men volunteered for the trial (two thirds received the vaccine and one third a placebo); in Thailand, about 2,500 uninfected injection drug users offered their services. Effectiveness of the vaccine was determined by recording how many in each group became infected with HIV and the viral load in those infected. A total of seven inoculations over a 30-month period were administered. Early results generated optimism that the vaccine could induce at least some immunity. The benchmark for effectiveness of the vaccine was that 30% of those receiving it should develop cytotoxic T-lymphocytes for protection against HIV. In 2002, plans were drawn up to use the gp120 vaccine in the "boost" phase of the so called prime-and-boost approach that we discuss later here. The canarypox vaccine (explored presently) was used in the "prime" phase.

Another approach seeks to uncover the vulnerable parts of the gp120 molecule (the V3 loops discussed in that are crucial to infection and that elicit the strongest possible antibody response). These hidden portions of gp120 are not normally available to react with antibodies, but they become exposed when HIV interacts with the host cell's CD4 receptors and coreceptors. Indeed, the camouflaging of these sites may be a reason why they are poor targets for antibodies and why infected individuals mount limited antibody responses.

In 1999, to expose the V3 loops, scientists at the University of Montana added HIV genes to cultured cells and induced the cells to produce gp120 molecules at their surface. Then they engineered neuroblastoma cells to produce the targets for the V3 loops, and CD4 and CCR5 molecules at their surface. Next, they combined the two cell types, and just as the cells fused, they added formaldehyde to "freeze" the cells during fusion, a time when critical V3 regions of the gp120 molecules are exposed. Injections of the fused cells into laboratory animals resulted in anti-gp120 antibodies that thwarted HIV infection but did not protect against simian immunodeficiency virus (SIV) infection (thus implying that the antibodies were very specific for the structures they neutralized).

Finally, some vaccine work has been done using the tat protein as an immunizing agent. The *tat* gene is one of the nine HIV genes. It encodes a regulatory protein (tat) that appears to enhance the expression of HIV genes, although this observation is not universally accepted. Tests have shown that in humans some delay in progression to AIDS can be achieved when antibodies that neutralize the tat protein are present in the blood. However, working with the tat protein is difficult because it occurs in two forms (encoded by two regions of the *tat* gene), and the two forms appear to trigger different actions. Nevertheless, the experiments continue, and in 1999, five of seven primates immunized with tat protein and challenged with viruses appeared to resist infection.

As of October 2012, there were 37 ongoing HIV vaccine trials in humans. Of these 37, only nine involve HIV protein subunit vaccines. The reasons these vaccines are not as popular as other methods are, again, that the proteins can denature, or fail hold the structure that they have in the live virus, and because the immune response to these vaccines can be weak. However, the reason they continue to be used is that they do not have the risks associated with attenuated viral and DNA vaccines.

Viable-Vector Vaccines

A high priority of vaccine researchers is to complex an HIV subunit with a nonpathogenic microbe. To form this combination, a poxvirus such as the **vaccinia** (the virus used as a live vaccine to prevent smallpox) is genetically engineered to incorporate HIV genes into its genome. In the laboratory, this is accomplished by infecting a culture of lymphocytes with HIV and permitting the RNA of HIV to be transcribed into DNA. Then the DNA is extracted and chemically recombined into the DNA of the vaccinia virus. Next, vaccinia is allowed to multiply in host cells to produce enough for a vaccine. Animals inoculated with the recombined vaccinia develop antibodies to both the virus and the HIV protein subunits whose genes are carried by the virus. The vaccine so produced is known as either a **viable-vector vaccine** or an **infectious recombinant virus vaccine**. All ongoing clinical trial vaccine candidates in 2012 were using a **modified vaccinia ankara** virus (MVA), which is a highly attenuated vaccinia strain missing 10% of the vaccinia genome and unable to replicate in primate/human cells, increasing its safety for vaccine delivery.

One viable-vector vaccine that has entered clinical trials begins with a weakened virus used to immunize canaries and other birds against **canarypox**. The large genome of the canarypox virus is recombined by adding the HIV *env* gene for surface protein, the HIV *gag* gene for core protein, and the HIV gene for protease, also a protein. The virus does not replicate in human cells, but when it enters them, they package genome-encoded proteins into empty lipid shells called **pseudovirions**. The latter are not infectious, but they trigger an immune response consisting of antibodies against the three HIV proteins and against the HIV-infected cells. Moreover, the infected cells display gp120 molecules and other HIV proteins on their surfaces, and these marked cells elicit an immune response from cytotoxic T-lymphocytes in cell-mediated immunity. As noted earlier, the vaccine was prepared in 2002 for use in the "prime" phase of the **prime and boost trials** slated to take place in the United States, Canada, the Netherlands, and Thailand. Over 25,000 volunteers were expected to take part in the trial.

One unique feature of the canarypox vaccine is its possible use by application to the mucosal surfaces of the respiratory, urinary, and reproductive tracts. These applications are novel routes of immunization. Laboratory tests are seeking to determine whether the vaccine elicits a special class of antibodies (called IgA) that are known to act at body surfaces and attack free-floating viral particles. Enhanced mucosal immunity is of obvious benefit when trying to develop protection from an infectious agent that more often that not enters humans through the vaginal mucosa. There were 29 completed candidate vaccine trials using an ALVAC (canarypox-based) recombinant vector vaccine in 2012.

The adenovirus is also a vector used to carry HIV viral proteins into the host. Adenoviruses are a common cause of respiratory infection in people and can also lead to gastrointestinal infections. The vector, however, has had a gene removed such that the virulence of the intact adenovirus is lost from the vector. The advantage to the adenovirus vector is the wide range of cells it is able to infect, thereby distributing the expression of the HIV genes into a wider range of cells in the immunized individual.

Another viable-vector vaccine being studied consists of cells of the bacterium *Salmonella* engineered with genes for the gp120 glycoprotein (**Figure 9.13**). The *Salmonella* used in the vaccine has been modified to prevent its reproduction in human tissues and thereby prevent infection. The theory is that macrophages will pick up the bacteria more efficiently than a chemical adjuvant would and transport them to the immune tissues, where antibodies against gp120 molecules will form.

Of the 37 ongoing HIV vaccine trials underway during the third quarter of 2012, 26 involve the use of viral vectors, often in combination with plasmid DNA. Thirteen of these use the adenovirus vector, whereas the modified vaccinia ankara is currently being used as a vector in eleven HIV vaccine trials and two trials are using more than one vector.

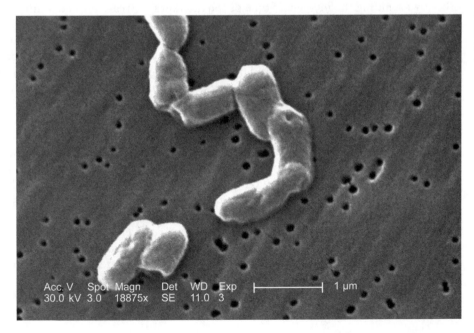

FIGURE 9.13
Scanning electron micrograph of *Salmonella* cells. Courtesy of Janice Carr/CDC.

A Whole-Virus Vaccine

Despite the drawbacks cited earlier, some researchers have continued their quest for a whole-virus vaccine in part because a whole virus would present immune stimulating proteins in their natural form and provide a better target for antibodies. Among the groups seeking such a vaccine is one formerly led by Jonas Salk, developer of the polio vaccine containing inactivated viruses (Salk died in 1995). In 1989, Salk presented evidence that HIV inactivated by chemical treatment and gamma radiation could be safely used as a vaccine. For vaccine production, the envelope of HIV was completely disintegrated by the inactivation treatment, thereby preventing any possible union with host cells. In the late 1990s, plans were under way to test the whole-virus preparation as a therapeutic vaccine in a 3-year study involving 3,000 people. Proponents pointed out that the vaccine was not intended to provoke an antibody response (as it had no envelope proteins) but rather to enhance cell-mediated immunity centered in cytotoxic T-lymphocytes. It was hoped that this immune process would keep the viral burden under control while the patient's immune system underwent repair.

Proponents of the whole-virus vaccine are encouraged by a 1995 discovery that an HIV strain lacking most of its *nef* gene apparently discourages the progression from HIV infection to AIDS. The discovery culminated a long study in which seven Australian recipients of HIV-contaminated blood did not progress to AIDS for 14 years (neither did the person who donated the blood). The virus lacking *nef* seemingly infects T-lymphocytes but encodes so few new viruses that the patient's viral load remains within limits that the body can control. Although the *nef* gene was seen as a possible target for drug or vaccine development, the hopes of researchers were dampened by the 1999 report that two of the seven Australian blood recipients had begun the progression to AIDS.

An alternative whole-virus vaccine is one containing SIV, the virus that causes AIDS in monkeys. In 1998, Ronald Desrosiers and his colleagues at Harvard University reported mildly encouraging results three years after vaccinating rhesus monkeys with two forms of SIV, one weakened by excluding three of HIV's nine genes and the other by excluding the *nef* gene. Some of the monkeys so treated resisted infection when given lethal dose injections of SIV. Moreover, very little SIV could be found in the tissues, and the T-lymphocyte count remained normal in some experimental animals. Because the possibility exists that weakened viruses can mutate to a virulent form or acquire deleted genes, Desrosiers and his colleagues shifted their attention to mutated SIV particles unable to synthesize parts of their gp120 molecules. Work on deleting the so-called V2 region of gp120 was apparently productive, as the group reported in 2001, and the research turned to producing a vaccine with whole, inactivated SIV having the deletion.

Evidence also exists that individuals infected with the milder (but nevertheless infectious) HIV-2 seem to be protected against infection by the more virulent HIV-1. A controversial, decade-long study of 756 African women was concluded in 1995, and researchers found that those initially testing positive for HIV-2 were 70% less likely to become infected with HIV-1 than women not previously infected with HIV-2. Studies such as these give hope that alternative milder viruses can protect against virulent viruses in much the same way cowpox viruses protect against smallpox viruses.

Proponents of the whole-virus approach to a vaccine suggest that using the traditional method would reduce the time necessary for vaccine development by several years. They argue that the subunit approach is basically a hit-or-miss technique; that is, it requires the development of costly vaccines that use every part of the virus until the most desirable part is found. Using the whole virus without taking it apart is the sensible approach, the proponents maintain. Opponents continue to point to the safety risks in using whole viruses as they continue to develop subunit vaccines.

In the final analysis, the ideal AIDS vaccine should stimulate two forms of immunity: **cell-mediated immunity (CMI)**, based on T-lymphocyte activity, would result in cytotoxic T-lymphocytes that attack and destroy HIV-infected cells, thereby ending HIV infection in the body, and **antibody-mediate immunity (AMI)**, a second form that is based on B-lymphocyte activity, which would result in antibodies that bind to HIV and prevent it from attaching to host cells. This viewpoint is now gaining favor as well as practical application: The prime and boost approach we noted earlier uses a whole-virus vaccine followed by a purified subunit vaccine. The first vaccine stimulates CMI, whereas the second elicits AMI.

As it stands in first quarter of 2008, there are no current human HIV vaccine trials that involve the use of attenuated or inactivated whole HIV vaccines.

Virus-Like Particle Vaccines

Virus-like particles (VLPs) are self-assembled viral proteins that make up the outer shell or coat of the virus. VLPs are like "empty viruses." They do not contain any genetic material, therefore they are unable to replicate inside of host cells and are unable to infect host cells. VLPs are not infectious but they mimic an infection in the body and can induce both CMI and AMI. VLPs are a recent alternative to killed or inactivated vaccines. In 2012, there was one completed vaccine trial using VLPs in the United States. There were 36 volunteers enrolled in the VLP trial. The VLP was composed of p17 and p24 HIV-1 antigens.

DNA Vaccines

Although enthusiasm for the new DNA vaccines ran high in the 1980s, tests performed in humans in the early 1990s did not yield strong immune responses. Nevertheless, researchers continued their work, and by the end of the century, the future looked more promising. For example, scientists at Chiron Corporation attached DNA molecules to microparticles of a new **adjuvant** called **polylactide coglycolide** (which is used in surgical sutures), and they found that the combination boosted both CMI and AMI in laboratory animals.

Another test performed by Harvard researchers was reported in 2000. In this case, rhesus macaque monkeys were immunized with plasmids containing the genes that encode proteins in both HIV and SIV, that is, the SIV *gag* gene and the HIV *env* gene. As an adjuvant, some monkeys were also administered human interleukin-2, a chemokine that encourages the proliferation of cytotoxic T-lymphocytes; alternately, some monkeys received the genes that encode interleukin-2. The monkeys were then challenged with HSIV, a hybrid virus containing a core of SIV and envelope proteins of HIV. They remained healthy. Although only modest antibody production was noted, there was a substantial increase in the number of cytotoxic T-lymphocytes specifically programmed to lock onto HIV particles. Moreover, the animals displayed suppressed replication of SIV and had low or undetectable viral loads when tested (control animals given no vaccine fared poorly in the test). One researcher suggested that a therapeutic vaccine combined with may be a long-range solution to keeping HIV under control in the body.

Between 1986 and 1992, Merck & Company was a leader in AIDS vaccine research, specializing in HIV vaccine components that elicit antibody response. When these vaccines proved questionable, the company changed its focus to vaccines that stimulate

Mapping HIV

Central and South America

- At the end of 2009, there were 1.4 million adults and children living with HIV.

- By the end of 2010, there were 1.1 million adults and children living with HIV.

- HIV epidemics in South and Central America have changed very little.

- The number of people living with HIV continues to grow due largely to the availability of antiretroviral therapy.

- About one third of all people living with HIV live in Brazil.

- The adult prevalence in Brazil has been under 1% for the past decade where early and ongoing treatment are available.

- Most of the HIV epidemics are in regions in and around networks of men who have sex with men.

- Social stigma has kept the epidemics among men who have sex with men hidden and unacknowledged.

- Fear of being stigmatized can compel many men who have sex with men to also have relationships with women. In Central America, more than 1 in 5 men who have sex with other men reported having sex with at least one woman in the previous six months.

- Injecting drug use has been the other main route of transmission, especially in the southern cone of South America.

- Two million people in Central and South America inject drugs and more than 25% of them are living with HIV.

Data from: *UNAIDS Report on the Global AIDS Epidemic*, 2010 and *HIV/AIDS Epi Update*, Public Health Agency of Canada, Centre for Communicable Diseases and Infection Control–July, 2010.

CMI. Ten years later, in 2002, Merck was ready to reenter the field with a number of new vaccines, among them three vaccines containing plasmids with the SIV *gag* gene, one with the HIV *gag* gene spliced into the genome of a cowpox virus, and one with the HIV *gag* gene spliced into the genome of a common cold virus known as the adenovirus. Tests with these new vaccines and a number of novel adjuvants are ongoing.

Currently, DNA vaccines are quite commonly used in combination with vector vaccines, particularly the adenovirus vector. There are currently six HIV vaccines trials being conducted that combine HIV DNA vaccines with the adenovirus vector with HIV genes. In the third quarter of 2012 there are at least 16 HIV ongoing vaccine trials that are solely DNA vaccines.

Problems in Vaccine Development

The optimism generated by continuing work on an AIDS vaccine is overshadowed by problems encountered during vaccine development and testing. Among the major problems that confront vaccine researchers are the complexity of the virus and its persistence in the body, the lack of animal models for testing vaccines, the tendency of HIV to mutate, and the nagging possibility that traditional vaccine approaches will not work with HIV. There are other problems to be addressed as well (**Table 9.1**). A synthetic vaccine, for example, may contain traces of carrier protein (e.g., bacterial, insect cell, or mammalian cell), and this protein may induce an allergic response in

TABLE	
9.1	**Problems Attending AIDS Vaccine Development: HIV Escapes Antibodies When Integrated to Host Genome**

- Antibodies are ineffective against proviral DNA.
- Animal models are not available for testing vaccines.
- Antibodies may encourage host cell penetration.
- Mutations occur in HIV envelope proteins.
- Volunteers can participate in only a single trial.
- Volunteers will test positive for HIV antibody.
- Challenge doses of HIV cannot be administered.
- Product liability insurance will be high.

the recipient. Another challenge is developing improved adjuvants to ensure uptake by macrophages. As mentioned previously, safety remains a key concern in vaccine development. Problems like these have made vaccine development a Herculean task.

Complexity of HIV

High on the list of concerns is the finding that HIV is much more complex than other viruses, such as those causing measles, mumps, rubella, or polio. In these cases, no integration of the viral genome into the host cells' DNA takes place, and antibodies elicited by the vaccines can neutralize the viruses in the extracellular fluids. HIV, by comparison, is an integrating retrovirus that installs its genes as a provirus, and because antibodies do not enter body cells, the virus can escape neutralization by remaining inside cells for months or years. It is therefore unlikely that antibodies elicited by an AIDS vaccine would be able to locate every viral particle or completely rid the body of HIV. Moreover, antibodies generally react with capsid proteins, and because HIV exists in its proviral form, the antibodies would be ineffective against its DNA (indeed, the viral protein does not even exist at the proviral stage). For these reasons, attention is focusing on prime-and-boost vaccines that stimulate CMI as well as AMI.

Ridding the body of every single viral particle may not be necessary, however. To be sure, controlling the virus in an infected individual may be a valuable function performed by a whole-virus vaccine such as that developed by Salk's group. The vaccine could be used to boost the level of HIV antibodies in the blood and prevent the viral replication that leads to host cell destruction and development of symptoms. Even if the vaccine contained some active viruses, there would be little effect on the person receiving the vaccine because infection has already occurred. Such a "postexposure vaccine" has traditionally worked for people infected with rabies viruses (rabies has an incubation period measured in months or years), and the approach could possibly be useful for those previously exposed to HIV.

The ability of HIV to mutate remains a lingering problem in the quest for a vaccine. Like the influenza virus, HIV has a tendency to undergo frequent mutations, especially in its *env* gene. These mutations are expressed as altered envelope proteins gp120 and gp41. In one patient, California researchers isolated dozens of variants of HIV over a period of a year. The possibility of the existence of multiple variants requires a vaccine to elicit different antibodies to be effective or to be directed at a single protein essential to viral replication or infection. In this context, it is also noteworthy that HIV exists as three main groups (M, N, and O), and many subgroups (or clades) as well.

Economic Issues

Another problem facing vaccine researchers is the overall "deflation factor." After years and years of trying, the pharmaceutical companies are feeling the pressures of defeat. The original estimate of candidates for a vaccine in developed countries has dwindled, partly because fewer individuals consider themselves at risk for HIV infection and because HAART drugs have lifted the cloud of doom over those infected. Thus, the market for vaccines will mainly come from developing countries, a fact that means reduced profits.

Fears of regulation by governments and lawsuits from people claiming injury from the vaccine have further let the steam out of the vaccine development effort (sick individuals can accept that a drug does not work or has side effects, but healthy individuals take exception to a vaccine that makes them ill). The investment community has become somewhat disillusioned by the failures, and that pessimism is fueled by the deflated profits anticipated from reduced use of a vaccine. Indeed, investing in a vaccine normally brings less profit than investing in a drug because a vaccine is used only a few times, whereas a drug is used innumerable times over the course of an illness. Moreover, cost of development remains high because the science continues to be tough, as demonstrated by NIH's 1994 decision to halt trials of two vaccines because of disappointing results.

The economic consideration is also related to vaccine effectiveness, especially because a flawed vaccine may be preferable to no vaccine. An NIH statistical analysis made in 1992 emphasizes this view: The analysis indicated that over an entire decade, a 60% effective vaccine introduced in 1992 would prevent nearly twice as many HIV infections as a 90% effective vaccine introduced in 1997.

Lack of Animal Models

Another major problem attending vaccine development is the fact that HIV has a restricted host range (humans); animals do not display the symptoms of AIDS, nor do most animal species harbor the virus. For vaccine developers, this problem has a substantial practical implication because a useful animal must be found for vaccine testing. Without an animal model, researchers cannot predict whether an experimental vaccine might work in humans.

Through the early 1990s, the only animals available for vaccine testing was the chimpanzee. These animals can be infected with HIV, but they do not display significant symptoms of AIDS, even after years of infection; however, they do mount an antibody response within three to nine months after infection with experimental AIDS vaccines. Chimpanzees are an endangered species, and by federal law they may not be imported (**Figure 9.14**). In the United States, only about 1,000 chimpanzees are available for all biomedical research and most are committed to other projects. About half of them are owned by the government. The National Institutes of Health (NIH) spends an average of $34 per day to take care of one chimpanzee in a research facility and $44 per day in a federal sanctuary while the federally-owned Alamogordo Primate Facility in New Mexico spends $66 per day/chimpanzee maintenance in 2012. In September 2012, NIH Director Francis Collins announced that 110 chimpanzees would retire from biomedical research funded by the NIH. Collins made a statement to *The Washington Post:* "*This is a significant step in winding down NIH's investment in chimpanzee research based on the way science has evolved and our great sensitivity to the special nature of these remarkable animals, our closest relatives.*"

FIGURE 9.14
Chimpanzees, which are considered an endangered species, are among the few animals available for testing of AIDS vaccines.
© Timothy E. Goodwin/ ShutterStock, Inc.

A possible solution to the animal model problem may lie in the Indian rhesus macaque monkey, a much more common species of primate. Rhesus macaque monkeys cannot be infected with HIV, but they develop an AIDS-like illness when infected with SIV or with the hybrid virus HSIV.

Still another proposed solution to the animal model problem was the so-called **humanized mouse model.** This animal is a special breed of mouse born with a **severe combined immune deficiency (SCID)** and genetically altered to include elements of the human immune system by implanting thymus and lymph node tissue from human fetuses. Many different humanized mice models were developed using SCID mice. Researchers are using genetic engineering techniques to produce mice and rats whose T-lymphocytes have human CD4 receptors and coreceptors and whose cells produce some of the unique proteins and other factors that human cells supply when HIV replicates itself (**Box 9.1**).

BOX
9.1

The Human Mouse

Investigators seeking an HIV/AIDS vaccine and wishing to determine drug effectiveness are seriously hampered by the lack of a good animal model. Vaccine developers need to know whether an experimental vaccine can elicit antibody production in a test animal, and drug producers require a small-animal model to bridge the gap between laboratory cultures and human trials.

The search for a good animal model frustrated researchers until 1988, when two laboratory teams from California transplanted components of the human immune system into mice and produced what the media immediately labeled "the human mouse." The mice involved had been identified five years previously as having severe combined immunodeficiency (SCID), an immune disorder in which no B- or T-lymphocytes are produced. The animals had essentially no immune system. One team performing the research was led by Michael McCane, an immunologist at Stanford University. McCane's group took thymus, lymph node, and liver tissue from an aborted human fetus and placed it under the capsule-like membrane surrounding the mouse's kidney. A week later, they injected fetal human immune system cells into the experimental mice, hoping that the immature cells would home in on the thymus, develop into mature T-lymphocytes, and circulate to the lymph nodes. Two weeks after the transplant, the mice could fight off infection when inoculated with *Pneumocystis jerovecii*. They had apparently acquired an immune system, and experiments showed that it was an immune system of human cells. The researchers even located antibody-producing B-lymphocytes in their prized mice.

The second team, working independently, successfully transplanted white blood cells from adult human tissues into SCID mice. They demonstrated the animals' immune functions by injecting tetanus toxoid into the animals and showing that the mice could produce antibodies against the toxin. The implanted white cells would conceivably protect the mice for life. The researchers hope that the lessons learned from the mice will protect human lives, as well.

Even if researchers succeed in building or locating an animal that supports HIV replication, there remains the problem of locating an HIV strain whose replication in animals mimics its replication in humans. As noted previously, HIV does not replicate well in chimpanzees and other animals, and thus, inoculating an animal with a candidate vaccine and then challenging the animal with HIV has not provided a rigorous test of whether the vaccine is likely to help humans. That concept appeared to change in 1999 when scientists at Emory University reported that an HIV-infected chimpanzee developed an AIDS-like illness in which the T-lymphocyte population was virtually depleted in six months as the viral load skyrocketed. Although this potentially lethal strain was equally devastating in other chimpanzees, some investigators pointed out that the strain is excessively virulent for testing purposes, as the same T-lymphocyte depletion in humans takes years to accomplish. Meanwhile, proponents of the strain voiced their opinion that it would make the chimpanzee challenge model more persuasive. Other researchers point out that the ability to study human immune system responses in a mouse model cuts costs and time in research needed to test candidate vaccines.

Vaccine Trials

At the beginning of the twenty-first century, over 30 groups were actively working in laboratories around the world to develop an AIDS vaccine (11 were in the United States). Each group consists of industrial, university, and government scientists combining their resources and talents.

To determine the efficacy of candidate vaccines, the Food and Drug Administration requires a procedure similar to that used for drug testing. First, there is preclinical testing in animals, intended to assess whether a promising vaccine is biologically active and safe enough to be tested in people. Next comes the Phase I trial, in which the vaccine is tested in relatively few individuals (perhaps 25) to determine safety in humans. In the Phase II trial, several hundred persons are enlisted to assess side effects and effectiveness as an immunizing agent. Finally, thousands of volunteers participate in the Phase III trial to certify safety and effectiveness.

After vaccine trials get to Phase III stage and involve thousands of volunteers, four questions about the candidate vaccine must be answered: How strong is the evidence that the vaccine offers any protection? How likely is it that the trials can prove whether the vaccine will work in the general public? Is it ethical to spend millions of dollars for such trials rather than using the money to encourage behavioral changes known to protect people? Is the government marching ahead with the trial for the sake of appearing to make progress rather than because the products look promising?

China, which does not have the same system of phases in vaccine development, reported in March 2005 that eight volunteers received an AIDS vaccine after signing waivers and receiving medical checkups. Chen Jie, then vice-director of the disease prevention and control center in China, reported that 49 volunteers had signed up for the trial. By 2006, China reported that the vaccine did produce a strong immune response in immunized individuals and that the vaccine was safe. In 2012, researchers led by Dr. Sidong Xiong of Fudan University (Shanghai, China) reported the development of a vaccine containing both *Mycobacterium tuberculosis* antigens and HIV p24 antigens that induced a strong cell-mediated response against both types of antigens in mice. In individuals who are infected with *Mycobacterium tuberculosis* and HIV, the two pathogens accelerate the depletion of T-cells, resulting in premature death if untreated. About 14 million people are co-infected with these two pathogens worldwide. An integrated approach to both pathogens may result in a vaccination to prevent both in the future.

In 2005, Switzerland and the United Kingdom enrolled 40 volunteers in the EuroVacc trial. This vaccine is composed of portions of the gp120 molecule and is ongoing. China also initiated a vaccine trial in 2005, which enrolled 49 people and is now complete. Although the results from this trial were seen as promising early on, it now appears that no benefit was associated with this vaccine. On September 21, 2007, the NIH, in collaboration with Merck & Co. Inc., announced that their STEP vaccine trials with HVTN 502 (the Merck V520-023 study) would be discontinued. This decision was based on the observation that this vaccine neither prevented HIV nor reduced the amount of virus in already infected individuals. This vaccine included the B clade gag-pol and nef subunits. Russia, Switzerland, Germany, and the United Kingdom all have ongoing vaccine trials that were initiated in 2006 and these trials continue.

According to *The Publication on AIDS Vaccine Research Report* (*IAVIReport*) trials network database updated on October 11, 2012, there were 12 vaccine trials completed that were located in the United States, South Africa, Uganda, Zambia, Thailand, Kenya, the United Kingdom, Canada, the Netherlands, France, and Puerto Rico. There were four clinical trials that were either closed to follow-up, withdrawn, or terminated and one new trial was scheduled in 2012. There were three vaccine candidate trials ongoing in Phase II or III efficacy clinical trials (**Table 9.2**).

TABLE 9.2	Ongoing HIV-1 Vaccine Phase II/III Vaccine Trials in 2012*			
Type of Vaccine/ Strategy	Number of Volunteers	Name of Vaccine Candidate	Phase	Location
Viral Vector-Pox Protein	N/A	ALVAC-HIV vCP1521 AIDSVAX B/E	II	Thailand
DNA/Viral Vector	225	PoxpGA2/JS7 DNA/MVA/HIV62	IIa	Peru, U.S.
DNA/Viral Vector	2,200	VRC-HIVDBA016-00-VP/VRC-HIVADV014-00-VP	IIb	U.S.

*Updated October 11, 2012. Data from: IAVI Report Trials Database http://www.iavireport.org/Trials-Database.

Issues to Consider

During vaccine trials, a number of technical, social, and ethical concerns can be expected to surface. For example, volunteers can only participate in a single trial because once they are inoculated with a particular vaccine, they have been "used," and injections with other vaccines would not give reliable results. Thousands of volunteers are needed for trials, and the requirement for one-time participation shrinks the pool of available participants considerably. Attracting volunteers may also be difficult because few animal studies will have been performed before human studies begin; therefore, volunteers might be reluctant to participate in a vaccine trial.

Another dilemma confronting volunteers in a vaccine trial is the possibility of improved vaccines. It may happen, for instance, that a volunteer vaccinated with a candidate vaccine may be unable to mount an immune response at some time in the future when injected with an improved vaccine. Moreover, after volunteers have been immunized, they will test positive whenever an HIV antibody test is performed. This positive result may suggest falsely that they have been infected by HIV, and they may suffer the discrimination associated with HIV carriers. (To prevent this possibility, volunteers are given notarized documents explaining their participation in the vaccine trial.) Also, for the volunteers, an HIV antibody test will be useless as a diagnostic measure should infection occur at a later time.

An ethical problem is how to test a vaccine's effectiveness while counseling persons on measures to avoid HIV (**Figure 9.15**). When a person volunteers to be in a vaccine trial, the doctor is ethically obliged to counsel the person on HIV-prevention measures. In the event that such measures are followed, how could researchers determine whether failure to contract HIV was due to the vaccine or the counseling? In the past, for trials of a measles vaccine, in contrast, it was assumed that a certain percentage of volunteers would be unavoidably exposed to measles over the course of the trial; however, that assumption could not apply to an AIDS vaccine trial because changes in high-risk behaviors might also limit exposure to the virus and account for failure to develop the disease.

Ethical questions also abound when vaccine trials are conducted in poor countries. For example, how can it be determined that participants are fully aware of what the trials involve and whether they truly give their consent? Also, should vaccines tested in a particular country use the same subtypes (or clades) of viruses present in that country? Another thorny issue is what to do if volunteers in poor countries

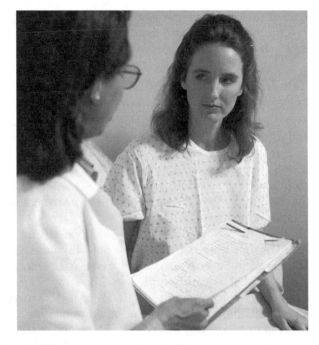

FIGURE 9.15

A potential test patient being counseled before entering a vaccine's test trials. © Photodisc.

become infected during the vaccine trials. Should they be given state-of-the-art drugs even if the country cannot afford them, or should they be administered the highest level of care attainable in that country? One obvious solution is to conduct initial trials in developed countries, as a 1993 document from an international medical group recommended, but officials in developing countries then wonder why they are being treated paternalistically. To help resolve the issue, the UNAIDS has decided to let each country answer the question for itself.

A further consideration is the excessively long incubation period before AIDS develops in an HIV-infected individual. This period can be 10 years or longer in some persons. Thus, with disease as the clinical endpoint, the value of a vaccine cannot be statistically measured until an extraordinary amount of time has passed (during which many volunteers would probably leave the study and the costs would become astronomical). Time is less of a consideration if a therapeutic vaccine is tested. In this case, the reduced amount of virus in the body (the viral load) can be measured as a test of the vaccine's effectiveness. That analysis can be accomplished in three years according to public health estimates, and far fewer people would be required since fewer would be lost. With less dangerous diseases, a vaccine's effectiveness can be tested by inoculating the vaccinated volunteers with doses of the pathogenic agent. It is unlikely that this practice will be followed in AIDS vaccine development because HIV is too deadly. Thus, another measure of a vaccine's success will be unavailable to researchers.

There are also a series of commercial product liability concerns that must be considered by vaccine manufacturers as well as by potential vaccinees. To minimize liability concerns, a researcher needs to practice good science, good medicine, and good ethics. For instance, the reason for undertaking a clinical study of an AIDS vaccine with human volunteers should be scientifically sound, especially as good animal models are currently lacking. The clinical trial should be designed and carried out to yield useful scientific information consistent with the protection of human subjects. Moreover, the selection and screening of potential vaccinees should be scientifically appropriate. The plan for the vaccine trial should be submitted for rigorous scientific review by an institutional review board that includes medical practitioners and ethicists, and the basic parameters of informed consent and medical ethics must be followed throughout the trial.

The Race to Vaccine

On May 25, 1961, President John F. Kennedy stood before a special joint session of Congress and challenged the United States to send an astronaut to the Moon by the end of the decade. Scientists and engineers were astonished by Kennedy's audacity, for at that time, the United States had not even carried out a manned spaceflight, let alone seriously considered a moonshot. Nevertheless, the American scientific community shifted its resources into high gear the next day as the campaign for the Moon began, and on July 20, 1969, astronaut Neil Armstrong made the historic first footprints of a human being on the Moon.

Now we fast-forward to May 18, 1997, and the commencement speech at Baltimore's Morgan State University. President Bill Clinton is at the podium addressing 850 graduates and proclaiming a new national goal: an AIDS vaccine by 2007. "If the twenty-first century is to be the Century of Biology," he says, "let us make an AIDS vaccine its first great triumph." He indicates that to spearhead the initiative, the U.S. government will create a new $30 million AIDS vaccine laboratory at the NIH; other countries will be enlisted in the effort to develop a vaccine, and the pharmaceutical industry will be challenged to make an AIDS vaccine a major priority. "There are no guarantees," President Clinton adds, but he expresses confidence in the collective wisdom of the world's scientists. Thus, with great fanfare, the race for an AIDS vaccine began in earnest.

By 2002, the Vaccine Research Center at the NIH was in full operation, and several candidate vaccines have been developed and brought to clinical trials (a primary focus of the center is to move basic research results more aggressively into clinical trials). With the influx of money from The Bill and Melinda Gates Foundation, the International AIDS Vaccine Initiative, a New York-based nonprofit organization that sponsors trials of AIDS vaccines has also joined in with clinical AIDS vaccine clinical trials (the World Bank and the British government also supplied funds). This renewed excitement in research laboratories has lead to new thinking and investigation of a vaccine that will provide both antibody- and cell-mediated immunities. Indeed, in 2007 several vaccines were brought to phase II clinical trials involving hundreds of people in China and other parts of the world. Unfortunately, much of the excitement engendered in the early part of this decade has been lost as these latest set of phase II vaccine trials failed to demonstrate a protective response against HIV.

Nevertheless, there is continued optimism as these vaccines were successful in generating an immune response to HIV in the absence of secondary complications. At the XIX International AIDS conference in Washington D.C. in July 2012, there were 23,000 participants from more than 170 countries attending the world's largest HIV/AIDS conference. It was the first time since 1990, thanks to action by Presidents Barack Obama and George W. Bush and Congress that the ban on people living with HIV was lifted, allowing HIV-positive conference delegates to enter the U.S. There were 142 abstracts/presentations on new ideas surrounding AIDS vaccine strategies and other prevention methods. The theme of the International AIDS conference was "Turning the Tide."

There is also a new spirit of cooperation among academic and industrial scientists, a cooperation fostered by new leadership. The Vaccine Research Center (VRC) at the National Institute of Allergies and Infectious Diseases is divided into 16 sections of vaccine research, including vaccine production (which is targeted to produce vaccines against HIV, Ebola, West Nile virus, and the virus that causes severe acute respiratory syndrome [SARS]) has been headed by Dr. Richard Schwartz since 2008. Schwartz is a molecular biologist with extensive experience in vaccine development. The AIDS Vaccine Research Subcommittee (AVRS) assists NIH in developing a research program aimed to develop an HIV/AIDS vaccine. The chairperson of the AVRS is Dr. Nancy L. Haigwood. She is also Director of the Oregon National Primate Research Center.

Among the notable cooperative efforts are the Waterford Project and EuroVac. The Waterford Project is a privately funded venture linking Robert Gallo's Institute of Human Virology with researchers at Harvard University and the University of California to seek a vaccine and combining the work of various researchers. For example, the group is working on a linking to a DNA molecule the genes that encode gp120 molecules and CD4 receptors and packing the new DNA into *Salmonella* cells for delivery as a vaccine to the body tissues. EuroVac is a similar effort, but it combines the talents of AIDS research groups in Europe. Moreover, several pharmaceutical firms have reentered the vaccine race with vigor—Merck & Company with its multiple gp120 vaccines is an example. Government guarantees of vaccine purchases have created incentives for these giant corporations.

Since 1997, the community of AIDS vaccine researchers has been heartened by these advances as well as by the increase in federal funding for vaccine development, which is now over $250 million per year. It might also be possible for the public to be more involved: In his 2001 book entitled *Shots in the Dark*, science writer Jon Cohen recounts the March of Dimes campaign that supported polio vaccine development and calls for a "March of Dollars" to spark the AIDS vaccine race. Engaging ideas like this reflect the spirit of scientists and spur the search for an AIDS vaccine. Indeed, on November 29th, 2012, World AIDS Day, President Barack Obama's proclamation included the following words, ". . . *Creating an AIDS-free generation at home and abroad—a goal that, while ambitious, is within sight. Through the President's Emergency Plan for AIDS Relief (PEPFAR), we are on track to meet the HIV prevention and treatment targets I set last year. We are working with partners at home and abroad to reduce new infections in adults, help people with HIV/AIDS live longer, prevent mother-to-child transmission, and support the global effort to eliminate new infections in children by 2015. . .*"

LOOKING BACK

Before drugs are approved by the FDA, they are subjected to a multiyear series of trials, including preclinical trials (in animals), Phase I trial (to determine dose and toxicity in healthy humans), Phase II trial (in small groups of infected humans to determine antimicrobial effects), and Phase III trial (in large groups of volunteers). Parallel-track and fast-track procedures have been used to reduce the time between drug development and FDA approval.

Certain alternative therapies for AIDS do not use drugs. Bone marrow transplants, for example, seek to replace damaged cells in the body, and gene therapy experiments are being attempted to encourage body cells to produce anti-HIV antibodies. Attitudinal and behavioral changes can also affect the outcome of the disease, and physiologic and anatomical links have been established between the immune and nervous systems.

The development of vaccines for viral diseases was an outgrowth of the cultivation of viruses in tissue cultures. Early vaccines, such as the first polio vaccine, were composed of whole viruses, but subsequent vaccines contained only fragments of viruses or microorganisms capable of antibody production. A sterilizing vaccine induces an immune response that will later provide protection against establishment of a pathogenic organism in the body. A therapeutic vaccine, in contrast, prevents infection from progressing to disease—for example, HIV infection to AIDS.

Most early efforts to develop an AIDS vaccine have focused on chemical sub-units of the virus such as the glycoproteins of the viral envelope or the proteins of the capsid. To increase the immune response, protein subunits such as gp120 are bound to adjuvants such as alum or to other viruses such as vaccinia, adeno, or canarypox viruses. Research is also being performed on the V3 loops of the gp120 molecules, where antibodies are believed to bind; however, most current vaccines are virus-like particles, DNA vaccines, viral vector vaccines, or combinations of these two.

Procedures similar to those for drug testing are employed with vaccine trials, and periods of many years are anticipated before final vaccine approval. Researchers must address the problems of locating volunteers and conducting scientifically valid trials while counseling volunteers on AIDS-prevention techniques. In addition, it is unethi-cal to inoculate vaccines with doses of infectious HIV to test a vaccine's effectiveness. Commercial product liability concerns also need to be considered. Despite these draw-backs, the worldwide community has 37 active vaccine trials as of the first quarter of 2012, and both the scientific community and the public remain optimistic that an AIDS vaccine will be developed during the next 10 years.

Healthline Q & A

Q1 Why can't an AIDS vaccine be made like a measles or other viral vaccine?

A Most viral vaccines are produced by taking whole viruses, inactivating (killing) them, and using them as vaccines. Using HIV in this way would be very dangerous and unethical, because if any HIV particles remained active (alive) in the vaccine they might cause HIV infection and AIDS. Also, HIV is far more complex than other viruses for which vaccines have been made.

Q2 Could I receive an injection of AIDS vaccine even if I had been previously exposed to HIV?

A The basic principle of a vaccine is to build up the body's immunity so that infection cannot take place at all; however, if infection has already occurred, a vaccine could be used to marshal the body's defenses and keep HIV under control, thereby preventing the development of symptoms. Drugs could be combined with the vaccine to further limit the spread of HIV in the body.

Q3 I've heard that if I participate in an AIDS vaccine trial I'll show up positive for the HIV antibody test. Is that true?

A Yes, it is. The object of a vaccine is to induce your immune system to produce HIV antibodies and mount other immune defenses. These same antibodies are the ones that show up in the AIDS antibody test. To prevent possible discrimination because of a positive antibody test, you will be given a notarized document explaining your participation in the test.

REVIEW

This chapter has explored the development of a vaccine to prevent AIDS. To test your knowledge of this topic, select the phrase that best completes each of the following statements. The correct answers are listed elsewhere in the text.

_____ 1. All the following statements are true except
  a. Certain vaccines contain whole viruses.
  b. A subunit vaccine is known as a second-generation vaccine.

    c. Developing a vaccine against HIV infection is an unrealistic expectation.

    d. All vaccines currently in use stimulate the human immune system.

_____ **2.** Among the major subunits being considered for use in an AIDS vaccine are
    a. fragments of the RNA of HIV.
    b. gp120 and gp41.
    c. the *nef* and *rev* genes.
    d. segments of proviral DNA.

_____ **3.** In subunit vaccines, the immune-stimulating agents are made more attractive to macrophages by
    a. binding the agents to adjuvant.
    b. converting the agents to DNA molecules.
    c. binding the agents to glucose molecules.
    d. converting the agents to RNA molecules.

_____ **4.** A viable-vector vaccine uses HIV genes linked to
    a. viral capsids.
    b. p17 protein associated with reverse transcriptase.
    c. envelope proteins of HIV and SIV.
    d. canarypox viruses cultivated in the laboratory.

_____ **5.** A therapeutic vaccine is one that
    a. prevents an infected person from progressing to disease.
    b. contains whole active viruses.
    c. cannot be taken by infected individuals.
    d. is used only in animals.

_____ **6.** Possible solutions to the problem of the lack of a suitable animal model for vaccine testing may involve any of the following except
    a. using antihuman antibodies.
    b. developing a mouse with a human immune system.
    c. using rhesus macaque monkeys.
    d. using test tube cultures of human cells.

_____ **7.** Liposomes have found value in vaccine technology as
    a. immune-stimulating reagents for viral subunits.
    b. substitutes for reverse transcriptase molecules.
    c. enzymes to produce subunits from macromolecules.
    d. adjuvants to enhance immune system stimulation.

_____ **8.** The whole-virus vaccine developed by Salk's group contains
    a. HIV particles without envelopes.
    b. envelope subunits rearranged to form whole viruses.
    c. normal HIV particles minus reverse transcriptase.
    d. normal HIV particles minus the protein capsid.

_____ **9.** The propagation of viruses for vaccine use depends heavily on the
    a. ability to cultivate viruses in tissue culture.
    b. identification of growth genes in the virus.

    c. action of enzymes similar to reverse transcriptase.

    d. formation of ISCOMs (immunostimulating complexes) by the viruses during cultivation.

_____ **10.** Among those who would probably be candidates for an AIDS vaccine are all the following except

    a. children between the ages of 5 and 15 years.

    b. persons who use injection drugs.

    c. individuals who practice anal intercourse.

    d. allied health science professionals.

_____ **11.** A vaccine containing subunits is considered safer than one containing whole viruses because

    a. subunits do not stimulate an antibody response.

    b. subunits do not infect cells or replicate in the body.

    c. whole viruses may induce antibody production.

    d. whole viruses may lack the protein necessary for immune system stimulation.

_____ **12.** A vaccine composed of attenuated viruses

    a. has viruses that multiply in the body.

    b. contains virus inactivated with chemicals.

    c. cannot be used to induce immunity.

    d. is in the developmental stage for AIDS.

_____ **13.** Aluminum sulfate and mineral oil have both been used in vaccines to

    a. produce segments of DNA from proviral DNA.

    b. inhibit reverse transcriptase activity.

    c. bind viruses to the CD4 receptor sites of cells.

    d. make subunits more attractive to macrophages.

_____ **14.** Mutations that occur in a virus tend to

    a. enhance the effectiveness of the immune response.

    b. limit the usefulness of a vaccine.

    c. enhance the binding activity of envelope proteins.

    d. limit the type of adjuvant that can be employed.

_____ **15.** All the following are ethical or social concerns that may surface during vaccine trials except

    a. Volunteers can only participate in a single trial.

    b. Multiplying viruses will be used in the AIDS vaccine.

    c. Counseling must be given to all recipients of the vaccine.

    d. Participating in a vaccine trial will prompt a positive test for HIV antibodies.

FOR ADDITIONAL READING

Abdool, K. Q., et al., 2010. "Effectiveness and safety of tenofovir gel, an antiretroviral microbiocide, for the prevention of HIV in women." *Science* 329:1168–1174.

Archin, N. M., et al., 2012. "Administration of vorinostat disrupts HIV-1 latency in patients on antiretroviral therapy." *Nature* 482–485.

Balazs, A. B., et al., 2011. "Antibody-based protection against HIV infection by vectored immunoprophylaxis." *Nature* 481:81–84.

Chatbar, C., et al., 2011. "HIV vaccine; hopes and hurdles." *Drug Discovery Today* 16:948–956.

Deeks, S. G., 2012. "Shock and kill." *Nature* 487:439–440.

Deeks, S. G., et al., 2012. "Towards an HIV cure: A global scientific strategy." *Nature Reviews Immunology* 12:607–613.

Fauci, A. S., Folkers, G. K., 2012. "The world must build on three decades of scientific advances to enable a new generation to live free of HIV/AIDS." *Health Affairs* 31:1529–1536.

Hatziioannou, T., Evans, D. T., 2012. "Animal models for HIV/AIDS research." *Nature Reviews Microbiology* 10:852–867.

Girad, M. P., et al., 2011. "Human immunodeficiency virus (HIV) immunopathogenesis and vaccine development: A review." *Vaccine* 29:6191–6218.

Hutter, G., et al., 2009. "Long-term control of HIV by CCR5 Delta32/Delta32 stem-cell transplantation." *The New England Journal of Medicine* 360:692–698.

June, C., Levine, B., 2012 "Blocking HIV's attack." *Scientific American* 306:54–59.

Kitchen, S. G., et al., 2011. "Stem cell-based anti-HIV gene therapy." *Virology* 411:260–272.

Perez, E. E., et al., 2008. "Establishment of HIV-1 resistance in CD4+ T cells by genome editing using zinc-finger nucleases." *Nature Biotechnology* 26:808–816.

Rerks-Ngarm, S., et al., 2009. "Vaccination with ALVAC and AIDSVAX to prevent HIV-1 infection in Thailand." *The New England Journal of Medicine* 361: 2209–2220.

Richon, V. M., 2006. "Cancer biology: mechanism of antitumour action of vorinostat (suberoylanilide hydroxamic acid), a novel histone deacetylase inhibitor." *British Journal of Cancer* 95, S2–S6.

Rosenberg, T., 2011. "The man who had HIV and now does not." *New York Magazine* November. Retrieved November 29, 2012 from nymag.com/health/features/aids-cure-2011-6/.

Shan, L., et al., 2012. "Stimulation of HIV-1-specific cytolytic T lymphocytes facilitates elimination of latent viral reservoir after virus reactivation." *Immunity* 36:491–501.

Shors, T., 2012. "HIV and bone marrow transplantation," in *McGraw-Hill Yearbook of Science & Technology*. New York: McGraw Hill, pp. 111–114.

Valdiserri, R. O., 2011. "Thirty years of AIDS in America: A story of infinite hope." *AIDS Education and Prevention* 23:470–494.

Walker, B. D., 2012. "Secrets of the HIV controllers: A rare group of HIV-positive individuals need no medicine to keep the virus in check. Their good fortune could point the way to more powerful treatments—perhaps a vaccine." *Scientific American* July:44–51.

Young, K. R., 2006. "Virus-like particles: Designing an effective AIDS vaccine." *Methods* 40:98–117.

AIDS in Perspective

LOOKING AHEAD

The epidemic of HIV infection and AIDS has many implications of a nonbiological nature that affect society in numerous ways. This chapter places the AIDS epidemic in the perspective of society and explores certain implications of the epidemic. On completing the chapter, you should be able to . . .

- Describe how social institutions are touched by the AIDS epidemic, and explain some of the effects of the epidemic on society.
- Summarize why HIV infection and AIDS are unique problems in the workplace, and indicate the steps for developing an AIDS policy in business settings.
- Discuss some legal issues affecting AIDS patients, including discrimination and confidentiality, and identify some economic consequences of the AIDS epidemic.
- Understand the economic impact of the AIDS epidemic on the United States and the world, especially in developing countries like Sub-Saharan Africa.
- Discuss the funding challenges for HIV-related interventions.
- Identify sources of treatment/health assistance programs for people living with HIV in the United States.
- List the goals of the U.S. HIV/AIDS policy priorities known as the **National HIV/ AIDS Strategy.**
- Appreciate the special problems that the AIDS epidemic poses to healthcare and medical professionals.
- Identify some special needs of the AIDS patient during care in the hospital and at home.
- Discuss the status of people living with HIV worldwide after three decades of the HIV pandemic; for example, are the numbers of infections stabilizing? increasing? decreasing?
- Describe some future expectations for the development of the AIDS epidemic, and specify certain lessons learned from the epidemic and why there is a clear path toward an "AIDS-free generation."

INTRODUCTION

In the 1990s, the *New York Times* reported the story of Ellen Ahlgren, a 71-year-old resident of Northwood, New Hampshire. Mrs. Ahlgren learned that thousands of babies infected with HIV were living out their short lives in cold and sterile hospital cribs and decided to do something to help them—she would make

FIGURE 10.1

AIDS quilt displayed on the Washington, D.C. Mall. Courtesy of Library of Congress, Prints & Photographs Division [LC-HS503-2457].

quilts for them. Together with several other women in her New England town, Mrs. Ahlgren sewed about a dozen quilts and sent them to Boston City Hospital.

The response from hospital administrators was so gratifying that the circle of quilters, begun in 1987, soon expanded. A series of fliers distributed around New England spread the word, and within weeks, Mrs. Ahlgren was even receiving quilts in the mail. An article in *Quilters Newsletter* magazine brought hundreds of letters and many offers of help. Today, Mrs. Ahlgren's modest group has become a nation-wide organization called ABC Quilts. More than 1000 persons aged nine to 90 years are sewing quilts in 44 states, all doing their share to help the children with AIDS. As of September 2002, ABC Quilts had provided 450,000 quilts. Parts of the AIDS quilt have been on display at the Washington D.C. Mall (**Figure 10.1**). In 2012, the AIDS Quilt had more than 48,000 panels and 94,000 names. It was roughly 1.3 million square feet and so large that it couldn't be displayed in its entirety in one piece. In 2012, The AIDS Quilt went digital and was put online using Bing mapping technology. The AIDS Quilt has raised more than $4 million dollars to fight AIDS.

If future historians were to judge the worth of our society by our response to the AIDS epidemic, they would doubtlessly find many positives as well as many negatives. Certainly, Mrs. Ahlgren's work would be among the positives, as would the efforts of tens of thousands of other volunteers who have come forward to give of themselves in a time of crisis. On the negative side is the discrimination shown to AIDS patients in employment, housing, and medical care; the deferral of services to AIDS patients; and the moral judgments verbalized by some who pontificate on what people do rather than who they are.

In this closing chapter, we place the AIDS epidemic in perspective and examine how it has influenced society. The epidemic has affected the way people live, where they work, how they interact with other people, where their money is spent, and how they receive health care. As a disease of troubling social complexity, AIDS managed to touch all people in one way or another, and all people are called on to react to it either directly or indirectly. For decades, the idea of controlling the HIV/AIDS pandemic was distant because we lacked the scientific evidence and key discoveries in HIV prevention. The past three decades in learning how to treat and prevent HIV infection brings hope that a world free of HIV/AIDS is achievable.

Focus on HIV

The digital map of the AIDS quilt can be found at: http://research.microsoft.com/en-us/um/redmond/projects/aidsquilt/.

AIDS and Society

When the bubonic plague swept across Europe in the 1300s, it profoundly changed the way people lived. The massive death toll contributed to the end of feudalism because landowners had to pay high wages for work from the decimated peasantry. Moreover, the authority of the clergy, already in decline, deteriorated further because the Church was helpless in the face of disaster, and with many priests and monks dying, a new order of reformers arose. Medical practices became more sophisticated. New standards of sanitation were imposed, and a 40-day period of detention (a **quarantine**) before vessels could dock at a port was instituted.

In many ways, the AIDS epidemic has also changed modern society. Reaching into virtually all social institutions (e.g., family, community, business, government, school), the AIDS epidemic has stimulated changes that have endured (**Figure 10.2**). A generation ago, for instance, it would have been unthinkable to exercise free expression about homosexuality, condom use, and sexual practices such as anal intercourse. Today, these subjects are discussed openly. Indeed, social taboos have lifted to the point that in 1991, the New York City Board of Education voted to distribute free condoms in the public schools to help prevent the spread of HIV. The New York Department of Health and Mental Hygiene gave out over 37.2 million free condoms in 2011 and continues to make free condoms available at hospitals, clinics, and homeless shelters. This averages to 70 condoms per minute! The NYC subways contain ads promoting free condoms (**Figure 10.3**). On Valentine's Day, 2011, (which was also **National Condom Awareness Day**) the New York Health Department launched a free smartphone application, **NYC Condom Finder**, designed to locate the five nearest New York City venues that distribute free NYC condoms.

Because HIV is transmitted directly from person to person, the AIDS epidemic is a universal problem. As such, it touches highly industrialized as well as developing countries and affects the life of virtually every human being. Who, for instance, in urban business centers has not known an HIV-infected coworker? Which educational

FIGURE 10.2

One of the social effects of the AIDS epidemic has been free expression regarding condom use. Posters and advertisements like these were much less visible a generation ago.

Brooklyn subway ad for free NYC condoms. Courtesy of the New York City Department of Health and Mental Hygiene.

institution has not been required to develop a policy on AIDS? How many times has the average healthcare worker needed to use gloves, masks, or other protective devices in the course of a day's employment? Which funeral directors have been reluctant to provide a dignified burial to a person recently deceased from AIDS?

The AIDS epidemic has also compelled individuals to make new judgments or has served to reinforce old outlooks on many practices taking place in society. It has heightened the awareness of homosexuality, developing tolerance and understanding in some individuals but aversion and repulsion in others. It has engendered a fresh look at the epidemic of drug addiction and stimulated compassion in the form of needle exchange and drug treatment programs, while hardening the stance of others that drug addiction is a punishable crime. It has increased attention on the plight of the poor and homeless, but it has also encouraged others to castigate those who take resources from social programs seen as more worthy of support.

Debates stimulated by the AIDS epidemic continue unabated in the media as well as in social gatherings. For instance, the issue of mandatory premarital testing has not yet been settled, and needle exchange programs still stimulate controversy. A lively debate continues on the extent of sex education and AIDS education in the schools, not to mention whether condoms should be freely distributed in these settings. Public health and healthcare professionals are not united on the nature and course of HIV therapies, nor do they agree on the usefulness of vaccines.

Debate also continues among ethicists who address the thorny issues surrounding therapy and vaccine use. For instance, ethicists argue about the high cost of using drugs in underdeveloped nations and whether this obstacle to equitable care is tolerable. They point out that the complexity of the **HAART regimen** often precludes its use, and they question the ethics of physicians who prescribe less complicated but also less effective regimens. They wonder aloud whether rich countries should be permitted to test new drugs in poor countries that could not afford the drugs if they were found to be effective. They debate the use of placebos in controlled experiments (when some patients are purposefully denied therapy), and they press for new statistical designs that gauge a study's success while it is in progress, rather than at its conclusion.

In the inner city, the stigma of AIDS continues without any foreseeable end. Treatments are generally too little and too late for the poor, and infected individuals may not even be inclined to ask for assistance. Poor women, for example, may not wish to seek help for fear of losing their children to foster care. In the inner city, the AIDS epidemic has superimposed itself on the epidemics of drug addiction, welfare dependency, crime, teenage pregnancy, and homelessness, and on the general social decay. None of these epidemics can be interrupted without attention to all.

The AIDS epidemic, however, has also served as a catalyst to bring the positive side of the human spirit to the fore. The theme of the 2012 International AIDS conference—titled "Turning the Tide Together"—stresses the actions needed to collectively change the course of HIV/AIDS and achieve these goals. Volunteers have emerged to service AIDS hotlines, develop educational materials, and give comfort to AIDS patients. In the hospitals, many doctors, nurses, and other professionals take pride in their work with HIV-infected persons (**Figure 10.4**). Business organizations have generously supplied funds to assist education, treatment, and

FIGURE 10.4

In many cases, the AIDS epidemic has served to bring the positive side of the human spirit to the fore. Many physicians and nurses take pride in giving comfort to AIDS patients. At a center for children in Albany, New York, a nurse cares for a young patient infected with HIV. © Photos.com.

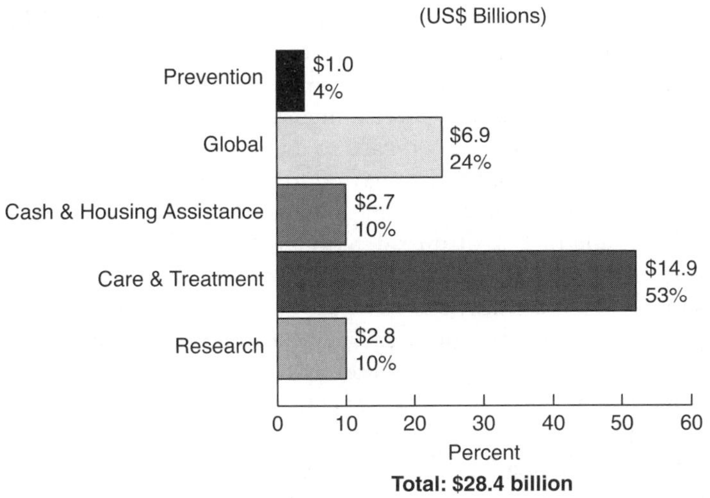

FIGURE 10.5

(US$ Billions)

Prevention $1.0 4%

Global $6.9 24%

Cash & Housing Assistance $2.7 10%

Care & Treatment $14.9 53%

Research $2.8 10%

Percent

Total: $28.4 billion

U.S. federal funding request for domestic HIV/AIDS by category in 2012. Data from: The Henry J. Kaiser Family Foundation.

prevention campaigns; a number of grassroots organizations have sprung up to act as advocates of persons with AIDS. President Barack Obama's 2012 fiscal year federal budget request for domestic HIV/AIDS for domestic AIDS-related programs and services was 28.4 billion dollars (**Figure 10.5**).

AIDS in the Workplace

Until well into the twenty-first century, successful management of the consequences of the AIDS epidemic in the workplace will be an ongoing challenge for the business community. Included under the general heading "business" are both managers and employees in public and private sectors as well as members of state agencies, labor unions, educational institutions, law enforcement agencies, and healthcare establishments.

HIV infection and AIDS pose a particular problem in the workplace because the disease strikes the majority of people in their economically most productive years, and although individuals with HIV infection may be asymptomatic for 20 or more years, their blood and body fluids are infectious. This presents special risks for workers in occupations where routine exposure to body fluids occurs.

Public health epidemiologists point out that in the United States most future cases of AIDS will occur in persons currently employed in the workforce. As new treatments come into use, these workers will remain on the job longer. Indeed, as the years unfold, having employees and coworkers with HIV infection or AIDS will become a more common experience in both large and small businesses.

The challenge of managing AIDS in the workplace should elicit compassion for those with HIV infection and AIDS, and prompt education and training to eliminate fear among coworkers. Aspects of a business's policy should address transmission methods for HIV, safe work practices, unique aspects of the work environment, protection for workers in these environments, and laws against discrimination. Basically, the major way to allay workers' fears is to educate workers through a consistent, well-thought-out approach for dealing with HIV infection and AIDS.

Developing an AIDS policy need not be a difficult proposition if established methods are followed. A person or task force should determine the organization's point of view, use medical facts to substantiate recommendations, take legal and cost issues into account, and consider organizational values and civic responsibilities. Existing policies can be factored in, and consistent style should be used. The policy may be AIDS-specific or more general in approach, as we explore next.

An AIDS-specific policy should seek to inform all employees about AIDS and recognize that persons with HIV infection and AIDS pose significant and sensitive issues in the workplace. Guidelines for handling such issues should include a commitment to a healthy and safe work environment, a statement that AIDS is to be treated the same as any other illness with regard to employee policies and benefits, and the assurance that employees with AIDS may continue to work with reasonable accommodation, as long as they are medically able. The guidelines should also call on coworkers to extend compassion or risk dismissal. Statements could reinforce the belief that all workers should be educated about AIDS and include an invitation to employees to contact their supervisor in confidence if they are concerned about or affected by AIDS.

A more general policy can be developed on the supposition that there is no medical reason to treat AIDS differently from other major illnesses. To this end, a policy statement addressing AIDS, cancer, heart disease, and other serious health problems should be prepared. Such a policy should indicate that employees with life-threatening illnesses will be allowed to continue their normal pursuits, as long as they can meet performance standards and do not pose a health threat to others. The policy should request managers to protect the confidentiality of employees with HIV and encourage employees to work with personnel relations experts on their concerns. Reasonable accommodations should be made for affected employees, and coworkers should be requested to exercise sensitivity and be responsive to those employees' physical and mental needs. A statement attesting to the therapeutic value of continued employment should be included in the policy.

Each of the two policy approaches has its advantages. The general approach projects the notion that AIDS is no different from other major illnesses and avoids possible discrimination in favor of AIDS. Also, it is likely to be read carefully by a broader range of people because more individuals will be affected. In comparison, the AIDS-specific policy addresses the issues associated with AIDS directly and enhances employee education about the disease. Employees have little question about where the company stands on this single topic and can act accordingly. Experience indicates that companies are less likely to follow this approach, however, and usually opt for the more general life-threatening illness policy.

Suppose, however, that a company chooses to avoid publishing any policy. This approach may reflect management's view that HIV infection and AIDS could not occur in its employees. It may also indicate that frank talk about transmission methods is too sensitive an issue, that open talk about AIDS will condone the methods by which it is transmitted, or that discussions of AIDS will indicate that someone in the company has the disease. Such views are generally cast in a negative light, and management is often seen as burying its head in the sand until the problem goes away; the so-called **ostrich mentality** (**Figure 10.6**). Workers usually appreciate insights into contemporary

FIGURE 10.6

Some companies choose to avoid developing an AIDS policy, hoping that the epidemic will not affect their businesses. Such an attitude is like the ostrich's practice of burying its head in the sand until danger passes.

issues such as AIDS and welcome a firm stand taken by the company. To be productive, employees need to know the rules of the workplace, including the rules on AIDS.

The CDC's encouragement that U.S. worksites implement AIDS-related policies bore fruit. By the end of the 1990s, a survey indicated that over half of worksites had AIDS policies in place and nearly 20% were offering AIDS education programs. On July 13, 2010, the White House released a plan that contains one of President Barack Obama's top HIV/AIDS policy priorities known as the **National HIV/AIDS Strategy (NHAS)**. It is the nation's first comprehensive plan with clear and measurable targets with three primary goals:

1. Prevent new HIV infections.
2. Increase access to care and optimize health outcomes.
3. Release HIV-related health disparities (reduce stigma and discrimination against people living with HIV).

The overall vision of the NHAS is that, "The United States will become a place where new HIV infections are rare and when they do occur, every person, regardless of age, gender, race/ethnicity, sexual orientation, gender-identity, or socioeconomic circumstance, will have unfettered access to high-quality, life-extending care, free from stigma and discrimination."

Legal Issues

In 1987, the U.S. Supreme Court rendered a landmark decision in the case of *School Board of Nassau County v. Arline*. The suit involved a teacher named Gene Arline who was dismissed from her teaching job in Nassau County, Florida because she had tuberculosis. Arline contended that by dismissing her, the school board violated her rights under the 1973 Federal Rehabilitation Act. This statute protects handicapped persons against discrimination based on, among other things, the erroneous perceptions of employers. The Supreme Court, in a seven to two decision, agreed with Arline and ruled that she was a handicapped person. It also emphasized that employers cannot distinguish between physical impairment and contagiousness because they both stem from the same handicap. In so doing, the justices reaffirmed and protected the employment rights of any individual who has had, or presently has, a contagious disease. Although the decision applied to tuberculosis, it also applied by inference to HIV infection and AIDS.

A second decision of importance was rendered in 1988 in the case of *Chalk v. U.S. District Court*. That year, the U.S. Supreme Court judged that a person suffering from AIDS was a handicapped person under the Federal Rehabilitation Act. The suit was brought by Vincent Chalk, a California teacher assigned to a nonteaching position when school officials learned he had AIDS (**Figure 10.7**). Finding in Chalk's favor, the Supreme Court reaffirmed the notion that the decision in *Nassau vs. Arline* applies to AIDS as well as tuberculosis. The court also concluded that an employer cannot defend against a claim by asserting that employees do not want to work with persons with AIDS. An easily transmitted secondary infection would be the only defense considered in such an instance.

A further clarification came in 1997. That year, the U.S. Supreme Court ruled that a person with HIV infection should be afforded the same rights as one with AIDS. The case was brought by an HIV-positive woman named Sidney Abbott who was refused treatment for a cavity at the office of a Maine dentist (the dentist insisted that the treatment be performed at a maximum-containment facility at a local hospital; the dentist's services would be free, but the hospital charges would be the

FIGURE 10.7

Vincent Chalk, the California teacher whose suit affirmed the U.S. Supreme Court's decision that persons with AIDS are to be treated the same as and to have the same rights as any handicapped persons. © L. Bakker/ Dreamstime.com.

patient's responsibility). Abbott's lawyer argued that she was protected from discrimination under provisions of the 1990 Americans with Disabilities Act, and the Court agreed. Justice Anthony Kennedy, writing for the majority, indicated that HIV infection satisfies the statutory and regulatory definition of physical impairment, especially because having the diseases impairs the reproductive process (Abbott testified that she feared passing the disease to her child if she conceived) and because HIV infection commonly leads to terminal disease. The decision has far-reaching ramifications as it suggests that a person with HIV infection has a disability even though no one can see signs of the disease.

The Supreme Court decisions in the three cases mandate that AIDS be treated as a handicap, not only by public institutions but also by private businesses that receive federal assistance. Such private businesses include those having substantial contracts with the federal government and those receiving financial assistance from federal agencies. They also encompass institutions receiving Medicaid or Medicare money. Any such institutions or businesses cannot discharge an employee unless the employee's disease is contagious and poses an actual danger to others and unless a reasonable accommodation would fail to eliminate the transmission danger (**Figure 10.8**).

The issue of "reasonable accommodation" for AIDS patients has stimulated a certain level of discussion, because having AIDS is synonymous with being handicapped and federal law requires these accommodations. For the AIDS patient,

FIGURE 10.8

The challenge of managing AIDS in the workplace should be met by a consistent, well-thought-out policy involving input from representatives from all over the company. One legal issue involved is the establishment of reasonable accommodations for AIDS patients. © BananaStock/Alamy Images.

reasonable accommodations include such things as released time for doctor's appointments or other medical care, rest periods during the workday if medically necessary, physical restructuring of the work area (e.g., ramps for wheelchair access), and the development of part-time work schedules to meet the individual's needs. According to law, the accommodations need not impose undue financial and administrative burdens on the employer, but to avoid extending such accommodations, the employer must show that they would create an imminent and substantial risk to the safety or health of the employee or others. Added expense, a change of schedule, or hiring additional employees are not considered undue burdens. In essence, the law has made it clear that the employer must cooperate as long as the AIDS patient can do the job.

Although the issue of discrimination has been addressed in federal law, it is spoken to primarily in state law. In New York, for example, the state Division of Human Rights handles all discrimination complaints from AIDS patients arising from violations of human rights law. Most complaints involve employment, housing, commercial space, and credit. After investigating a case, the division may recommend a hearing, while processing AIDS cases as soon as possible and preserving testimony on tape. If employment discrimination is proven, individuals are entitled to reinstatement of employment, back pay, admission to health care, and monetary compensation. Punitive damages and reimbursement for attorney's fees may also be awarded.

Confidentiality is essential when dealing with AIDS, particularly with respect to HIV testing. The individual states are largely responsible for confidentiality laws. In 1989, for example, the New York State legislature enacted one of the more comprehensive confidentiality laws. The law requires the written consent of the participant for any HIV-related testing. (The consent must be on a state authorized form, and pre-counseling and post-counseling are required.) It also specifies that anyone wishing to be tested must be informed that anonymous testing is available. Finally, it restricts disclosure of test results when providing health and social services (but permits disclosure in certain specified categories).

Confidentiality laws, however, can sometimes appear too restrictive. In 1996, for instance, an individual in New York was identified as the source of HIV infection in 20 or more women he named as sex partners. The state legislature immediately investigated the effects of New York's confidentiality laws. Two years later, it established a state registry and began requiring physicians to report the names of HIV-positive individuals. Trained health workers now ask (but do not require) the individuals listed in the registry to name their sexual and/or needle-sharing partners, and they notify the named contacts that they may have been exposed to HIV. Although confidentiality is a key provision of the law, AIDS activists voice the concern that the list could fall into the hands of insurance companies, employers, or others who could discriminate against the infected patient. Supporters of the law maintain that if they are located individuals can be treated at an early stage of the disease and increase their chances of survival.

Other legal issues arising from the AIDS epidemic involve laboratory and hospital care. It is important that managers take precautions to protect themselves against legal action for negligence, libel and slander, failure to obtain consent, breach of confidentiality, failure to warn, or failure to maintain a safe workplace. Negligence can be charged, for example, if a hospital mistakenly supplies units of HIV-positive blood for transfusion. A slander case could arise if a patient receives a positive result from an AIDS test that later is revealed to be false. (The patient could prove that the test was unreliable or improperly performed.) Failure to obtain consent may violate a state law if a patient is tested covertly.

Result reporting may be the basis for other legal problems. Positive test results should be reported to the physician whose duty is to inform the patient and health authorities according to state law (**Figure 10.9**). The reporting must be done according to standard protocols established by local authorities. Breach of confidentiality can be charged if the patient's consent forms are not filled out. Informing other people in a patient's life is also a delicate matter that could result in a breach of confidentiality by the physician. What position should the physician take, for example, if the HIV-positive patient refuses to inform a regular sex partner of the test results? What of the doctors, nurses, and laboratory technicians who are not warned that a particular patient may have AIDS? Can they cite the manager for failure to maintain a safe work environment?

Ultimately, the managers of laboratory and hospital facilities must ensure that their policies fall in the mainstream of acceptable procedures. Currently, most institutions use the CDC's guidelines for preventing HIV transmission as their blueprint. Furthermore, institutions generally have legal staffs to advise management and to keep it abreast of federal, state, and local laws as well as the standards for health care in sister institutions.

Economic Issues

In 1984, Congress made the first annual appropriation of funds to study and deal with AIDS. The amount was $61 million. By 1989, the annual appropriation had grown to $1.3 billion (an astonishing increase of 2000%), and by 2001, it was over $2.5 billion. In 2004, President Bush's Emergency Plan for AIDS Relief was funded to African and Caribbean nations most afflicted with the pandemic, for prevention programs as well as care of those with AIDS and children orphaned by AIDS. According to the most recent (2009) UNAIDS estimates available, more

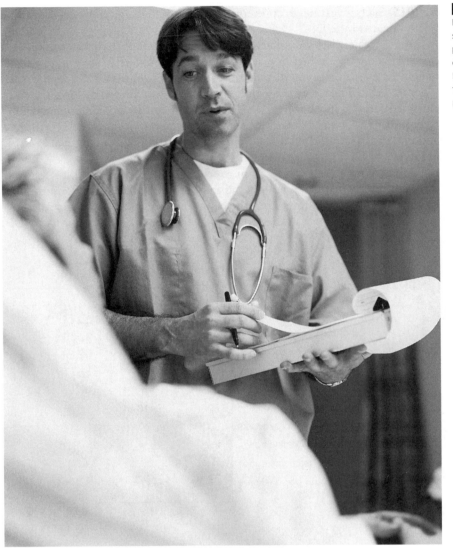

FIGURE 10.9
Part of the physician's responsibility is to inform a patient of positive HIV test results in a dignified and confidential manner. Failure to do so may be the basis for a legal judgment against the physician. © Photos.com.

than 14.9 million children under the age of 18 in Sub-Saharan Africa lost one or both parents due to HIV/AIDS. The total world estimate of children losing one or both parents to HIV/AIDS was 16.6 million.

There is little doubt that the AIDS epidemic has had a heavy impact on the world's economy. The costs can be direct as well as indirect. The U.S. 2012 annual federal expenditure of more than $20 billion is an example of a direct cost, as is the estimated $600,000 to care for an AIDS patient over the course of the illness. Indirect costs arise from such things as implementing universal precautions in hospitals (e.g., disposable gloves, masks, and protective wear), employing infection control procedures, mandating blood-screening, improving laboratory precautions, and establishing educational campaigns. For the "worried well" patient, indirect costs arise from counseling expenses, testing procedures. World leaders at the 2011 United Nations General Assembly Meeting on AIDS called for providing HAART for 15 million people in low- and middle-income countries by 2015. An estimated $28 to $33 billion annually is needed by 2015. Such investments could prevent 12 million infections and 7.4 million AIDS-related deaths by 2020.

Some economists would argue that spending more than $20 billion in federal funds annually is too high a price to pay for AIDS. The United States spent approximately $2.6 trillion on medical care (**Figure 10.10**) in 2010 compared to $256 billion spent in 1980. The rising costs of healthcare are attributed to longer lifespans and chronic illnesses, including AIDS. In 2003, the President's Emergency Plan for AIDS Relief (PEPFAR) was created. It brought significant attention to HIV, tuberculosis, and malaria. Total PEPFAR funding requested to Congress for 2004–2013 is shown in **Figure 10.11**.

Another reason offered for appropriating the necessary funds for AIDS is the disease's concentration in young and middle-aged adults. These individuals usually fund the healthcare system (and the economy in general), in contrast to children and elders who seek assistance from it. Moreover, in many nations, AIDS robs the

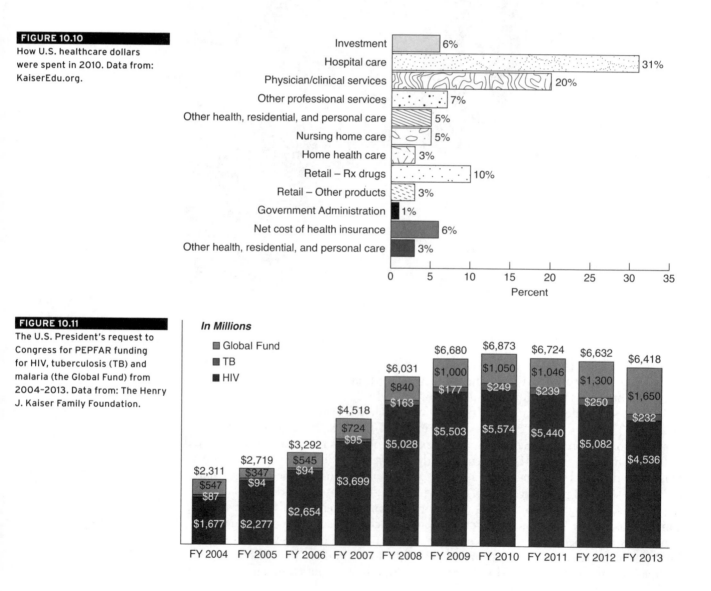

FIGURE 10.10

How U.S. healthcare dollars were spent in 2010. Data from: KaiserEdu.org.

FIGURE 10.11

The U.S. President's request to Congress for PEPFAR funding for HIV, tuberculosis (TB) and malaria (the Global Fund) from 2004-2013. Data from: The Henry J. Kaiser Family Foundation.

AIDS in Perspective

TABLE 10.1	Donor Government Funding for AIDS in 2011
Country	**Funding $US**
United States	$4.5 billion
United Kingdom	$971 million
Netherlands	$322 million
Germany	$304 million
Denmark	$189 million
Sweden	$164 million

Adapted from the Kaiser Family Foundation, 2012. "HIV/AIDS: The state of the epidemic after 3 decades." *JAMA* 308:330.

country of its young, educated, professional class of individuals who form the key underpinning of the society. Such a loss further weakens an economy that could already be fragile. Proponents of funding also point out that knowledge generated from AIDS research has many corollary benefits. Learning about HIV, for example, helps us understand how certain other retroviruses cause leukemia, and understanding the details of Kaposi's sarcoma gives insight on how other cancers may be initiated in the body.

On a global basis, it is imperative that wealthy nations become involved in the plight of poorer nations because AIDS is a worldwide disease that is continuing to spread. At the end of 2011, for example, sub-Saharan Africa was home to an estimated 23.5 million of the world's 34.2 million people infected with HIV. **Table 10.1** lists donor government funding for AIDS in developing countries in 2011. Private donors and pharmaceutical companies that have joined the effort are The Bill and Melinda Gates Foundation and Merck & Company. Furthermore, Merck is selling two of its protease inhibitors to poor countries at about a tenth of the U.S. price. Bristol-Myers has a program for its drugs and has waived its patent rights in South Africa to stimulate the production of generic equivalents in that country. **Figure 10.12** illustrates the allocations of resources invested in the global AIDS response.

In 2001, the AIDS epidemic was consuming about $20 billion each year within the United States, including about $2.5 billion for antiretroviral drugs. In 2012, the AIDS epidemic was consuming $28.4 billion dollars in which 53% or $14.9 billion dollars was used in care and treatment of people living with HIV. The cost of antiretroviral monthly antiretroviral therapy is shown in **Table 10.2**. Outside the United States, overall spending is a fraction of the U.S. statistic.

AIDS Drug Assistance Programs (ADAP) are present in every state. They provide FDA-approved antiretroviral therapy to low-income patients. More than one third of all people living with HIV in the U.S. are enrolled in ADAPs. The **Ryan White Program** is the largest ADAP for people living with HIV in the United States. Ryan White was an American teenager from Kokoma, Indiana. He was a hemophiliac and contracted HIV after receiving contaminated blood clotting factors. He was diagnosed with HIV in 1984 and died of AIDS in 1990. Congress passed the Ryan White Care Act shortly after White's death. The Act was reauthorized four times and is up for reauthorization by Congress in 2013.

Focus on HIV
Federal funding for the Ryan White Program was $0.02 billion dollars in 1991 and has increased to $2.3 billion dollars in 2011.

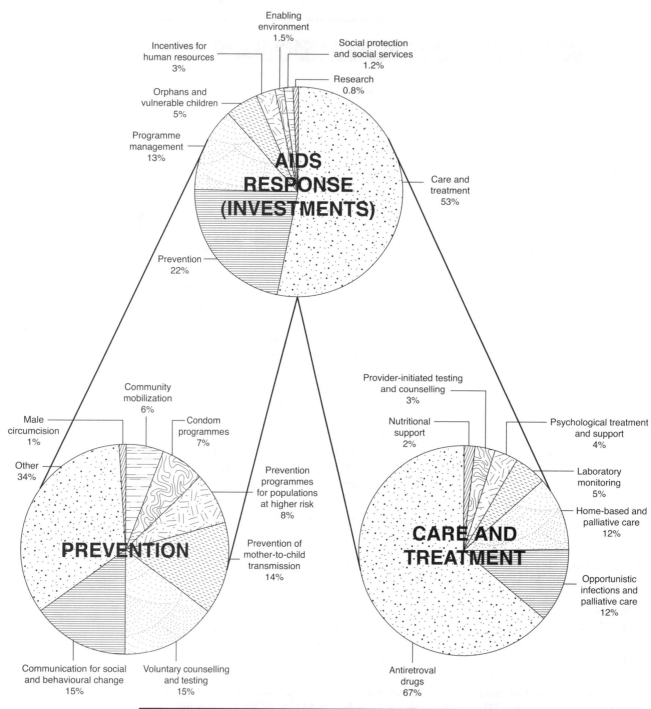

AIDS RESPONSE (INVESTMENTS)

Enabling environment 1.5%
Social protection and social services 1.2%
Research 0.8%
Incentives for human resources 3%
Orphans and vulnerable children 5%
Programme management 13%
Care and treatment 53%
Prevention 22%

PREVENTION

Community mobilization 6%
Condom programmes 7%
Male circumcision 1%
Other 34%
Prevention programmes for populations at higher risk 8%
Prevention of mother-to-child transmission 14%
Communication for social and behavioural change 15%
Voluntary counselling and testing 15%

CARE AND TREATMENT

Provider-initiated testing and counselling 3%
Nutritional support 2%
Psychological treatment and support 4%
Laboratory monitoring 5%
Home-based and palliative care 12%
Opportunistic infections and palliative care 12%
Antiretroval drugs 67%

FIGURE 10.12

Where the money/resources goes toward the AIDS response. Investment patterns vary from country to country.
Data from: UNAIDS, 2012.

10.2 Monthly Average Wholesale Price for the Leading Antiretroviral Drugs in the United States 2012

Brand	Dose	Monthly Cost
Nucleoside Reverse Transcriptase Inhibitors		
Zerit (d4T)	2 capsules/day	$493.38
stavudine (generic)	2 capsules/day	$411.16
Epivir (3TC)	1 tablet/day	$477.41
lamivudine (generic)	1 tablet/day	$429.66
Viread (tenofovir)	1 tablet/day	$873.28
Ziagen (abacavir)	2 tablets/day	$641.50
Retrovir (AZT)	2 tablets/day	$557.83
zidovudine (generic AZT)	2 tablets/day	$360.97
Videx (ddI)	1 capsule/day	$460.14
didanosine (generic)	1 capsule/day	$368.72
Emtriva (emtricitabine)	1 capsule/day	$504.37
Non-Nucleoside Reverse Transcriptase Inhibitors		
Viramune (nevirapine)	1 tablet/twice daily	$723.08
Sustiva (efavirenz)	3 capsules/3 times daily	$689.52
Rescriptor (delavirdine)	2 tablets/3 times daily	$365.45
Intelence (etravirine)	2 tablets per day	$978.64
Edurant (rilpivirine)	1 tablet/twice daily	$804.38
Protease Inhibitors		
Viracept (nelfinavir)	2 tablets/twice daily	$879.84
Crixivan (indinavir)	2 capsules/3 times daily	$548.12
Norvir (ritonavir)	1 tablet/twice daily	$617.20
Reyataz (atazanavir)	2 capsules/twice daily	$1,176.23
Prezista (darunavir)	2 tablets/twice daily	$1,230.20
Lexiva (fosamprenavir)	2 tablets/twice daily	$1,812.68
Aptivus (tipranavir)	2 capsules/twice daily	$1,335.14
Invirase (saquinavir)	2 tablets/twice daily	$1,088.84
Integrase inhibitors		
Isentress (raltegravir)	1 tablet/twice daily	$1,171.30
Fusion Inhibitor		
Fuzeon (enfuviritide)	1 injection/twice daily	$3,248.72
CCR5 Antagonist		
Selzentry (maraviroc)	1 tablet/twice daily	$1,148.16
Co-Formulated Combination Antiretroviral Drugs		
Epzicom (abacavir + lamivudine)	1 tablet daily	$1,118.90
Truvada (tenofovir + emtricitabine)	1 tablet daily	$1,391.45
Combivir (zidovudine + lamivudine)	1 tablet/twice daily	$1,035.12
(zidovudine + lamivudine) generic	1 tablet/twice daily	$931.61
Trizivir (abacavir + lamivudine + zidovudine)	1 tablet/twice daily	$1,676.62
Kaletra (lopinavir + ritonavir)	2 tablets/twice daily	$871.36
Complera (rilpivirine + tenofovir + emtricitabine)	1 tablet daily	$2,195.83
Atripla (efavirenz + tenofovir + emtricitabine)	1 tablet daily	$2,080.97

Data from: AIDSinfo Guidelines for the Use of Antiretroviral Agents in HIV-1 Infected Adults and Adolescents. Last updated March 27, 2012.

AIDS and Medical Professionals

Many Americans were shocked during the 1980s when several surgeons announced that they would no longer perform complex operations on patients who tested positive for HIV infection or had AIDS. In a day when vaccines and antibiotics control the transmission of nearly all of the infectious diseases, doctors had come to assume that both they and their patients were fully protected—that is, until the advent of the AIDS epidemic.

AIDS taxes to the limit the compassion and selflessness that should distinguish the health profession from other callings. It has imposed on the healing arts a number of other stressors that weigh heavily on the day-to-day practice of giving health care. Although the vast majority of health professionals have acted responsibly, it appears that some physicians and nurses have not shown the courage that society has come to anticipate from them.

Caring for an AIDS patient can be a stressful experience for many reasons. The AIDS patient is often young and desperate and may be about the same age as the healthcare provider. Such a patient, in contrast to an older patient, would not normally be expected to die soon, and it may be difficult to discuss diagnoses and prognoses with the patient. In the cases of homosexual men, there is usually no legal spouse, and the patient may wish a lover or friend to make responsible decisions in case of his incapacitation. Legal empowerment may also be a problem, and parents may insist on considerable input. Moreover, the healthcare professional may face the ethical dilemma of when to use or withhold life-sustaining treatment, especially in view of the patient's relative youth.

Having to observe universal precautions is another burden imposed by the AIDS epidemic (**Figure 10.13**). There are restrictions to observe (e.g., how to dispose of

FIGURE 10.13

The exposure to potentially infected blood places extraordinary stress on surgeons and other medical professionals. The need to observe universal precautions adds to that stress. © Photos.com.

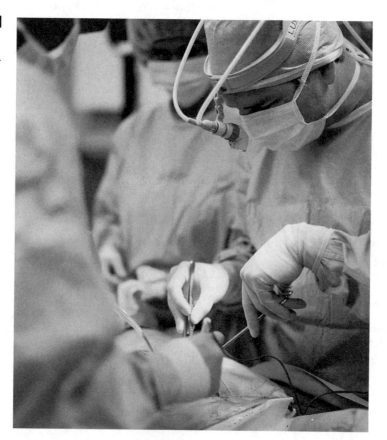

needles), barrier devices to be used (e.g., gloves, masks, goggles), and procedures for handling body fluids and tissues (e.g., blood). Added to these are the exaggerated fears arising from occupational exposures to infected tissues. Such fears are ill-founded because the overwhelming evidence indicates that health professionals have an extremely small chance of contracting HIV during their normal duties. Through June 2001, for instance, only 57 cases of HIV infection arising from occupational transmission had been documented in the United States (of more than 900,000 AIDS cases to that date).

There is also the ethical question of whether doctors and dentists who have HIV infection should inform their patients of their health condition. In 1991, both the American Medical Association (AMA) and the American Dental Association (ADA) suggested that their infected members should inform patients of their situation or cease practicing. The recommendations, which are nonbinding but influential, followed the revelation by the CDC that a Florida dentist had apparently transferred HIV to several patients during dental procedures. HIV-positive dentists were cautioned to avoid performing invasive techniques such as tooth removal, teeth cleaning, or orthodontia; however, compliance with the recommendations was voluntary and not all state health organizations agreed with the AMA and the ADA.

Practicing in a rural area can also pose an ethical dilemma for medical professionals. It may be uneconomical, for instance, for a local hospital to obtain the sophisticated machinery to test the T-lymphocyte counts of AIDS patients. A machine for doing such counts costs over $100,000, and a rural physician may be denied the diagnostic information available to urban counterparts. For rural patients, it may be difficult to obtain transportation to healthcare facilities. Support groups may be few and far between. Educational programs may be lacking, and confidentiality may be problematic since everyone in a small town is well-known.

Caring for the AIDS Patient

The AIDS patient requires special sensitivity from caregivers because of the unique aspects of the disease associated with its transmissibility and prevalence in certain groups in the population. Typically, an AIDS patient will experience multiple hospitalizations for the treatment of opportunistic diseases. In between hospitalizations, however, the AIDS patient may be cared for at home.

On these bases, two broad aspects of care emerge. In the hospital setting, medical care for the AIDS patient will depend on the type and severity of the opportunistic disease present. Another type of care, emotional care, can be equally important because a potentially fatal disease has been contracted, often in the patient's prime of life (**Figure 10.14**). Moreover, the patient may be aware that transmission could have occurred to a loved one, and there is the possibility that the patient may feel stigmatized because he or she practices a homosexual lifestyle or is a drug addict. Added to these issues is the possibility that the patient may need to be isolated because of the presence of a potentially transmissible opportunistic disease such as tuberculosis. In such a case, other patients in the hospital, especially those with compromised immune systems, may be susceptible to the opportunistic disease. If the patient does not understand the need for isolation, the feelings of being an outcast will exacerbate the emotional turmoil.

It is therefore important that the medical professional provide social and environmental stimulation to supplement medical care. Frequent office visits, for example, will assure the patient that healthcare providers have not withdrawn from fear and will

FIGURE 10.14

For AIDS patients, emotional care may be as important as medical care. The patient is often in the prime of life, and the disease is potentially fatal. Caregivers must therefore be prepared to deal with mental trauma as well as physical trauma. © LiquidLibrary.

increase the patient's self-esteem. Talking as well as touching can communicate acceptance and personal concern. If a patient is ambulatory, a conscious effort should be made to encourage use of common patient areas. During times of confinement, an abundance of reading material and easy access to radio and television will increase contact with the outside world and help alleviate the feeling of isolation. Visits from professional counselors and spiritual advisers help support the patient's emotional needs as well.

For the AIDS patient, there are minute-by-minute vacillations of mood and disposition. On any given day, emotions may include guilt, anger, self-pity, sorrow, determination, hope, and faith. Meditation, prayer, or other spiritual constructs can help keep the patient more centered and help reduce stress. Some patients keep themselves busy by participating in AIDS-related service projects, such as visiting schools to talk with young people about HIV infection. One patient calls it her way of "hanging onto life."

Education and Terminology

The medical professional must also assume the role of educator. Not only does the patient need authoritative information, but the patient's family and friends must know exactly what is taking place to be able to cope. It is surprising how ignorant people can be of a situation until it "hits home." In this respect, AIDS is like any other disease. Fear can quickly add to the stress of the situation, and the healthcare provider should be ready to correct misinformation while providing useful insights within the limits of what is known.

Sometimes terminology can make an important difference. For example, it is good to remember that persons with AIDS are "AIDS patients" rather than "AIDS victims." No crime has been committed, and thus, the term *victim* should be avoided. Furthermore, there are no "innocent victims" of AIDS because this term would imply that there are "guilty," and there are none. Neither is there any "blame" to be assessed because there was no intention to harm. There are only sick men, women, and children, all of whom need help.

Home Care

In the home setting, the AIDS patient needs compassionate care between episodes of opportunistic disease, during times of illness not appropriate for hospital treatment, and in the terminal stages of the disease if the patient desires to remain at home. Under these circumstances, the healthcare provider can help the AIDS patient avoid exposure to opportunistic diseases, maintain a program of exercise and nutrition, and alleviate the concerns of family and friends. In addition, the caregiver can provide emotional support and assist the patient in the final days.

In day-to-day activities, the AIDS patient needs to follow good hygienic practices to preclude exposure to opportunistic microorganisms. Frequent hand washing with antimicrobial soap is an example of such a hygienic practice. Avoiding certain animal pets, staying away from crowds, following disinfection practices when house cleaning, and cooking foods thoroughly are other examples of healthful practices (although different circumstances apply to each). Consulting with a nutritionist to devise a high-quality diet to maintain good health is well advised. Avoiding ill persons is also recommended because the AIDS patient has a compromised immune system.

The issue of sexual activity raises practical, social, and ethical dilemmas that can best be solved through love and understanding. Although sexual intercourse can transmit HIV, barrier protections such as condoms can substantially reduce the risk of infection and should be given serious consideration.

The emotional stresses in home care can increase when household members harbor fears about the risks of caring for the AIDS patient. Professional assistance by trained counselors and support groups can help those affected express and alleviate their fears. It is comforting to know, for example, that virtually no cases of HIV infection or AIDS have occurred in persons who gave home care to an AIDS patient and did not engage in risky behaviors. Once again, complete authoritative information is essential to understanding.

Providing quality care and comfort to the AIDS patient at home demands nothing more than love and common sense. The patient may need to be bathed, helped with walking or eating, or have dressings changed. Visiting nurses and healthcare professionals can guide caregivers who are unsure of themselves. Perhaps the patient will appreciate transportation to a support group of peers, or desire to have a prescription filled, or simply enjoy having someone to talk to. Filling these needs can be as valuable as injecting an antibiotic.

When the end is near, the AIDS patient will appreciate help putting things in order. An attentive ear will help the patient talk out whether medical intervention is desired during the final hours and, if so, how much. The physician should be involved in these discussions so that no uncertainty lingers. Patients will also

Mapping HIV

South and Southeast Asia

- In 2009, there were 4.1 million adults and children living with HIV in south and southeast Asia.

- More than half of the people in this region living with HIV are in China.

- The epidemic is concentrated-among people that inject drugs, sex workers and their clients, and men who have sex with men.

- In Myanmar, up to 38% of the people who inject drugs are HIV-positive.

- In Vietnam, between 32% and 58% of the people who inject drugs are living with HIV.

- In China, it is estimated that 7% to 13% of individuals who inject drugs are living with HIV.

- In southern India, up to 15% of sex workers are living with HIV.

- Women account for the growing proportion of HIV/AIDS from 21% in 1990 to 35% in 2009.

- HIV infections in Indonesia's Papua province are 15 times higher than the national average.

Data from: *UNAIDS Report on the Global AIDS Epidemic*, 2010 and *HIV/AIDS Epi Update* Public Health Agency of Canada, Centre for Communicable Diseases and Infection Control—July, 2010.

be thankful for help in completing final tasks and specifying funeral rites. Perhaps the greatest kindness that one can perform is helping to bring a friend's life to a dignified end.

Future of AIDS

In the June 25, 1990, issue of *Newsweek* magazine, a reporter wrote, "A sense of crisis is hard to sustain. It thrives on earthquakes and tornadoes, plane crashes and terrorist bombings. But forces that kill people one at a time have a way of fading into the psychic landscape."

So it has been with AIDS. Although the 1980s witnessed screaming headlines about the crisis wrought by HIV, the new century appears to be a time in which news about the AIDS epidemic is usually buried somewhere in the middle of the newspaper or discussed on the late news just prior to the sports and weather. Even the familiar red ribbons, worn to express concern about AIDS, have become scarcer (**Figure 10.15**). The fickle beacon of prime-time publicity that highlights certain causes seems ready to move on. According to the Henry J. Kaiser Family Foundation analysis, this trend is also depicted by survey information (**Figure 10.16**).

Challenges remain in obtaining sufficient funding for HIV-related interventions. However, if these challenges can be met globally, the goals toward an "AIDS-free generation" may be met one day and/or illness and death associated with AIDS will become increasingly rare.

FIGURE 10.15
Although fading from media head-lines, the AIDS epidemic is far from over. For example, 34.2 million people worldwide were living with HIV and nearly 30 million have died of AIDS as of 2012.

Percent of US Public Naming HIV/AIDS as the Most Urgent Health Problem

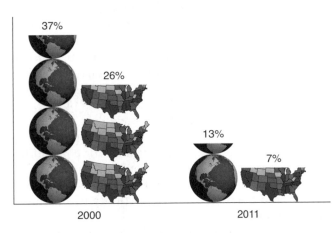

37%

26%

13%

7%

2000

2011

FIGURE 10.16
Percentage of the U.S. public naming HIV/AIDS as the most urgent problem. Adapted from the Kaiser Family Foundation, 2012. "HIV/AIDS: The state of the epidemic after 3 decades." *JAMA* 308:330.

Complexity of AIDS

AIDS was easily the most complex disease of the twentieth century. Ironically, the century opened with a strikingly similar disease, tuberculosis. In the early 1900s, tuberculosis was responsible for one death in seven from all causes. The disease primarily affected young adults, and it often incubated in the patient long before symptoms manifested themselves. Tuberculosis seemed to be "someone else's disease" because it occurred in underprivileged classes and people from lower socioeconomic groups, but it occasionally affected the better-known; it claimed the lives of the composer Chopin and the poet Keats. In recent decades, drug therapy has brought tuberculosis under a measure of control. In many cases, the tubercle bacilli remain in the body, but their multiplication is retarded by the drug and the severest effects of the disease are not experienced.

Many epidemiologists foresee a similar pattern for HIV infection and AIDS. Rather than remaining an acute disease with clear symptoms of illness, AIDS is undergoing an evolution to a chronic disease, that is, a disease persisting over a long period of time with few symptoms and no crisis period. According to this model, HIV would remain in the body, kept under control by drugs, and the infected person could look forward to a reasonably normal life. Diabetes is another chronic disease in today's society. How people live with tuberculosis and diabetes is how they may one day live with AIDS.

Before AIDS joins the ranks of chronic diseases, however, we can expect many more bits of knowledge that add to our understanding of the disease. For example, Kaposi's sarcoma is now thought to be a disease linked to AIDS only by coincidence. A separate virus, a herpesvirus, may be the cause. We can also expect new insights into the wasting syndrome that often characterizes AIDS, and we may possibly see a connection between the wasting and a specific parasitic infection influenced by a deteriorating immune system. Researchers will also continue to unlock the secrets of the immune system and may discover chemical substances that retard HIV replication.

Other advances in HIV and AIDS research concern the modalities used in treatment. We can look back to 1996 and the emergence of HAART treatment regimen for an example. The multidrug therapies, despite their drawbacks, slowed the growth of the U.S. AIDS epidemic as the death rate in patients declined sharply. Development of the protease inhibitors exemplified the potency of molecular medicine: Beginning some years earlier, researchers deciphered the structure of protease, analyzed the dynamics of its action, and developed drugs targeted against it. The successful use of rational treatments bodes well for the future of investigator-driven basic science and the ability of applied and pure scientists to work in harmony. Of course, many challenges remain, including how to simplify regimens, how to reduce costs, how to cope with drug resistance, and how to target other weak points of HIV.

The development and use of new drugs reinforces the need for early diagnosis because early intervention is essential for precluding the development to AIDS. It is clear that the window of opportunity for controlling HIV infection is a relatively narrow one and that therapy should begin early after exposure. New and improved diagnostic tests focusing on HIV antigens can also be expected. These tests will expand the window of opportunity by making an earlier diagnosis of HIV infection possible.

We can also look ahead to vigorous new efforts to develop an AIDS vaccine. The commitment of significantly increased resources raised hopes of developing a vaccine before the end of the first decade of the twenty-first century, and this occurred. Unfortunately, however, it now appears that a vaccine having modest efficacy will be the goal for the next decade, considering the many setbacks already encountered. Nevertheless, having a less than perfect vaccine may be preferred to having no vaccine; it can serve as a jumping-off point for research on an improved preparation. Scientists can then continue to seek answers to the nagging problems of eliciting a completely protective immune response while addressing the difficult issues surrounding testing of candidate vaccines and trying to understand the body's unusual reaction to HIV.

Furthermore, the future holds new treatment modalities that focus on varying aspects of the patient and the disease. For cancer therapy, it is not unusual for a person to undergo radiation treatment, chemotherapy, and even surgery. For HIV infection and AIDS, a patient may undergo chemotherapy for controlling HIV, immune-enhancing treatments to restore the immune system, behavioral therapy to alter high-risk behavior patterns, and psychological therapy to restore the "thinking well" pattern.

Accompanying these advances, however, are a number of lingering questions. For example, how do HIV proviruses integrate into human DNA, and once there, how do they force the host cells to follow their commands? Why are there apparently no individuals who are naturally immune to HIV? Why do infection rates vary so substantially from country to country? Why is HIV-1 so much more dangerous than HIV-2? What roles do sexually transmitted diseases play in AIDS development? What are the precise functions of HIV's genes? Which is the ideal drug therapy, and when is the most effective time to use it? Will we ever have an AIDS vaccine? The daunting tasks of finding answers to these questions will require the best minds and huge investments.

Lessons of AIDS

We will not really know what the AIDS epidemic is like until current events become history. Our descendants will someday pass judgment on where our society stood tall and where it erred, but we need not wait that long to understand some of the lessons the AIDS epidemic has taught us. It has revealed, for instance, that scientists do not have an answer for everything, nor are they as all-knowing as we sometimes perceive them to be. Perhaps it is well to remember that scientists seeking research funding are somewhat like a set of wheels in which the squeaky wheel gets the oil of public support. Retrovirologists did not begin "squeaking" until the early 1980s, and thus, there were no quick fixes for HIV when it arrived in our midst.

That observation, however, leads to another lesson because we have seen how scientists can gear up for action when a crisis presents itself. Once it was clear that an epidemic of infectious disease was sweeping across the world, scientists moved quickly to identify the responsible virus, develop a diagnostic test, and make available a set of useful therapies. Although some may argue that the pace of discovery was painfully slow, others will note that it was relatively rapid. For comparison's sake, they will point to the bacteria. These microbes were identified as causes of disease in the 1880s, but antibiotics did not become available until the 1940s, some 60 years later. In contrast, the human immunodeficiency virus was identified in 1984, and AZT was already in use by 1987, only three years afterward.

The AIDS epidemic has also demonstrated that humans can act with compassion, intelligence, and selflessness to overcome the fear that accompanies the influx of a new disease. How dangerous and insidious this fear can be is illustrated by the following story adapted from an ancient legend:

> One day on a road in the English countryside, a clergyman happened to meet Plague. "Where are you bound?" asked the clergyman. "To London," responded Plague, "to kill a thousand." They chatted for a few moments, and then went their separate ways. Some weeks later, the two chanced to meet again on the same road. As they conversed, the clergyman inquired, "I may have misunderstood, but I recall you were going to kill a thousand. How is it that two thousand died?" "Ah yes," replied Plague. "I killed only a thousand. Fear killed the rest."

In the early and mid-1980s, a climate of fear marked the emerging AIDS epidemic. There was fear of the virus, fear of the disease, fear of the AIDS patient, and a general fear of the unknown. Mean-spirited and irrational responses often could be traced to these fears. The late 1980s and the 1990s, in contrast, were a time of understanding and compassion, of knowledge and coping, of wisdom and generosity. Leaders have come forward to wage the fight against HIV; adequate financial resources have been made available by government and industry; legal protection has been extended where necessary; and accurate, timely surveillance has been implemented to track the epidemic and project its future course, and fear has waned.

Probably the primary reason for the passing of fear has been education (**Figure 10.17**). Educational campaigns launched in the 1980s have generally been well funded and well managed, and they have usually reached the populations at which they were directed. The concern in future years is that AIDS will primarily affect groups removed from education (such as injection drug users). In general, however, education has done much to set society on a course where it can effectively deal with the AIDS epidemic. In 1920, in *The Outline of History,* the noted author H. G. Wells wrote, "Human history becomes more and more a race between education and catastrophe." Indeed, education prompted a certain author to write this book.

The epidemic of HIV infection and AIDS has affected contemporary society profoundly, touching everyone in the world in some way. It has compelled individuals to make new judgments or has reinforced old outlooks on many practices, while stimulating many debates and bringing the positive side of the human spirit to the fore.

In the workplace, the AIDS epidemic can be managed successfully if a policy is developed and promulgated by employers. The policy could be AIDS specific or more general in scope, each type having advantages. Workers appreciate knowing the position an organization takes. According to federal and state laws, HIV infection and AIDS are treated as handicaps, and the rights of handicapped workers are set forth, including the employer's responsibility to provide reasonable accommodations. Discrimination and confidentiality are legal issues arising from the AIDS epidemic, as are issues that involve hospital and home care.

The economic burdens of HIV infection and AIDS are substantial, in both direct and indirect forms. In the United States, federal expenditures for AIDS care and research now exceed $28 billion annually. Globally, 34 million people were living with HIV at the end of 2011. The AIDS epidemic is exacting a heavy toll in Sub-Saharan Africa, where 69% of the people were living with HIV in 2011.

The AIDS epidemic presents ethical dilemmas for medical professionals who do not wish to expose themselves to HIV-contaminated tissues. Moreover, there are special concerns in caring for AIDS patients, such as patients' legal empowerment and, in many cases, their relative youth.

The AIDS epidemic is expected to decrease during the twenty-first century, because in the United States and certain other parts of the world, HIV infection and AIDS are increasingly coming under control and are being viewed as chronic rather than acute diseases. In the future, the disease could be managed much as tuberculosis and diabetes are managed today. As the years pass, more knowledge will come to the fore about AIDS. New treatments will be devised, and a vaccine will become available. Education will continue to be the chief factor in allaying fear of HIV infection and AIDS.

Healthline Q & A

Q1 What should I do if I feel I am the subject of discrimination at work because of my HIV infection?

A The issue of AIDS discrimination is handled primarily at the local and state levels. If you believe you have been discriminated against, you should contact a lawyer or the state office of human rights. At the latter, your complaint will be handled by a person trained in dealing with human rights violations. You may be entitled to relief from the source of the discrimination as well as monetary compensation for any damages arising from the discrimination.

Q2 I'm somewhat confused about Medicare and Medicaid. Exactly what is Medicare?

A In the United States, Medicare is a federally sponsored healthcare program for persons over 65 years of age. It is part of the Social Security system. Hospital and physician costs are covered, and no limits exist on the financial assets of the recipient.

Q3 What about Medicaid?

A Medicaid is also a federally sponsored program under the auspices of the Social Security Administration. In contrast to Medicare, however, Medicaid is for persons of all ages, as long as their income is below a certain level. Hospital and physician costs can be covered under Medicaid, depending on what a person applies for. It is necessary, however, that a person's available assets be expended before Medicaid benefits will be available.

Having finished this chapter, you should feel confident in discussing some of the implications of the AIDS epidemic in society. To test your knowledge, place a T to the left of the statement if it is true or an F if it is false. The correct answers can be found elsewhere in the text.

_____ **1.** It is generally advantageous for a business to have an unstated policy regarding AIDS.

_____ **2.** According to decisions by the U.S. Supreme Court, a person suffering from AIDS is considered to be handicapped.

_____ **3.** The guidelines issued by the CDC are rarely used by laboratory or hospital facilities in the development of safety policies.

_____ **4.** The 2001 federal expenditure for AIDS research and care of patients was less than $10,000 per year, and it is becoming less each year.

_____ **5.** Federal law requires that employers make reasonable accommodations for AIDS patients, including time off for doctor's appointments and rest periods during the work day.

_____ **6.** Observing universal precautions is one of the multiple burdens imposed on members of the healthcare system by the AIDS epidemic.

_____ **7.** An AIDS-specific policy developed by businesses for use in the workplace projects the idea that AIDS is no different from other major illnesses such as cancer or heart disease.

_____ **8.** In future years, it is very likely that all persons infected with HIV will develop AIDS and can expect to die.

_____ **9.** HIV infection and AIDS are generally not a problem in the workplace because most people affected are not in their economically productive years.

_____ **10.** By 1996, the AIDS epidemic was virtually over.

_____ **11.** One of the ethical dilemmas that medical professionals must consider is whether to perform surgery on a patient whose HIV status is uncertain.

_____ **12.** The knowledge generated from AIDS research has no application in any other field of medicine.

_____ **13.** It is illegal to inform others of a patient's HIV status without the patient's consent.

_____ **14.** The development of new drugs for treating HIV infection and AIDS will require that diagnosis be performed as soon as possible because early intervention can preclude disease development.

_____ **15.** One of the lessons of the AIDS epidemic is that scientists do not have a ready answer for every medical problem that may face society.

FOR ADDITIONAL READING

Aggleton, P., Homans, H., eds., 1988. *Social Aspects of AIDS*. New York: Falmer Press.

Beardsley, T., 1998. "Coping with HIV's ethical dilemma." *Scientific American* 279:106–107.

Deeks, S. G., et al., 2012. "Towards an HIV cure: A global scientific strategy." *Nature Reviews Immunology* 12:607–614.

Fauci, A. S., 2011. "AIDS: Let science inform policy." *Science* 333:13.

Fauci, A. S., Folkers, G. K., 2012. "The world must build on three decades of scientific advances to enable a new generation to live free of HIV/AIDS." *Health Affairs* 7:1529–1536.

Kaiser Family Foundation, 2012. "HIV/AIDS: The state of the epidemic after 3 decades." *Journal of the American Medical Association* 308:330.

Kramer, L. C. 1990. "Legal and ethical issues affect conduct toward AIDS sufferers." *Occupational Health and Safety* 59:49–50, 57.

Kuritzkes, D. R., Hirsch, M. S., 2012. "The future of HIV treatment." *Journal of Acquired Immune Deficiency Syndrome* 60(Suppl 2):S39–S40.

Mermin, J., Fenton, K. A., 2012. "The future of HIV prevention in the United States." *Journal of the American Medical Association* 308:347–348.

Sesser, S., 1994. "Hidden death." *The New Yorker*, November 19:62–63.

Answers to Review Questions

Chapter 1

1. F (1981); 2. F (Kaposi's sarcoma); 3. T; 4. F (blood); 5. T (but not exclusively); 6. F (Luc Montagnier); 7. T; 8. F (decreased); 9. F (pandemic); 10. F (100,000); 11. T; 12. F (monkeys); 13. F (milder); 14. T; 15. F (Africa)

Chapter 2

1. e; 2. e; 3. b; 4. a; 5. c; 6. d; 7. a; 8. d; 9. b; 10. e; 11. d; 12. c; 13. a; 14. c; 15. a.

Chapter 3

1. K; 2. J; 3. N; 4. O; 5. O; 6. H; 7. L; 8. M; 9. D; 10. P; 11. Q; 12. B; 13. C; 14. M; 15. G; 16. H; 17. A; 18. O; 19. E; 20. H; 21. C; 22. I; 23. C; 24. F; 25. N.

Chapter 4

1. AIDS-dementia complex; 2. Walter Reed; 3. Kaposi's sarcoma; 4. cytomegalovirus; 5. *Pneumocystis jirovecii*; 6. 10 years; 7. HIV infection; 8. lymphadenopathy; 9. 430,000 in 2008 and 330,000 in 2011 (Note more than 90% of cases occur in Sub-Saharan Africa.); 10. 800; 11. *Candida albicans*; 12. gastrointestinal system; 13. HIV wasting syndrome; 14. pentamidine isethionate; 15. cats.

Chapter 5

1. b; 2. c; 3. d; 4. a; 5. d; 6. a; 7. b; 8. a; 9. b.

Chapter 6

1. F; 2. T; 3. T; 4. F; 5. T; 6. F; 7. T; 8. F; 9. T; 10. T; 11. F; 12. F; 13. F; 14. T; 15. T.

Chapter 7

1. ELISA; 2. false negative; 3. electrophoresis; 4. T-lymphocytes; 5. viral load; 6. serological; 7. Western blot analysis; 8. false positive; 9. antigens; 10. polymerase chain reaction; 11. Food and Drug Administration; 12. radioimmunoprecipitation assay; 13. serum; 14. mother; 15. gene probe; 16. 6 to 10 weeks; 17 .99 percent; 18. primer DNA; 19. ELISA; 20. military.

Chapter 8

1. L; 2. D; 3. F; 4. K; 5. C; 6. H; 7. D; 8. G; 9. F; 10. O; 11. M; 12. D; 13. B; 14. A; 15. O; 16. E; 17. F; 18. I; 19. G; 20. J; 21. N.

Chapter 9

1. c; 2. b; 3. a; 4. d; 5. a; 6. a; 7. d; 8. a; 9. a; 10. a; 11. b; 12. a; 13. d; 14. b; 15. b.

Chapter 10

1. F; 2. T; 3. F; 4. T; 5. T; 6. T; 7. F; 8. F; 9. F; 10. F; 11. T; 12. F; 13. T; 14. T; 15. T.

B A Suggested Format for a 10-Session Course: The Biological Basis of AIDS

A ten-session program highlighting the biological basis of AIDS can be developed using this book as a text. The ten sessions may be offered over a period of ten weeks or at shorter intervals. Each session could encompass 60 to 90 minutes of instruction and discussion, and sessions could be supplemented by a myriad of videos, brochures, pamphlets, and other educational materials available from numerous sources. A suggested format for such a course follows.

Session Number	Subject Matter	Chapter Number
1	Introduction to HIV infection and AIDS, including overview of the current magnitude of the pandemic and where it is likely to spread in the future	1
2	Structure and replication of HIV relative to other viruses; significance of key functional components of HIV; significance among the different strains of HIV by groups	2
3	Development and operation of the human immune system; mechanisms by which HIV depresses the functions of the immune system by depressing T-cell activity; role of CD4 and co-receptors essential for HIV entry into host cells; research development towards a "cure" for HIV infection	3
4	Case definition of HIV/AIDS; pathology of HIV infection and AIDS; signs and symptoms; stages of disease and descriptions of opportunistic diseases; pediatric AIDS, AIDS dementia, HIV wasting syndrome, and HIV nephropathy	4

(continued)

Session Number	Subject Matter	Chapter Number
5	Epidemiology and transmission of HIV by various modes; high risk behaviors and groups affected; patterns of spread throughout the world; HIV mother-to-child transmission	5
6	Methods for avoiding exposure to HIV, including methods for personal protection, protection of health care workers, and protection of public safety workers; AIDS education in schools; rare routes of HIV transmission; myths about HIV transmission	6
7	Diagnosis of HIV infection and AIDS; indirect antibody tests and direct antigen tests; HIV viral load testing; protein and T-lymphocyte tests; implications of mandatory, voluntary, including home HIV testing	7
8	Treatments for HIV infection and AIDS; classes of anti-HIV drugs and their mechanisms of action; therapeutic methods for opportunistic infections; drug development and testing	8
9	Hopes and hurdles towards an HIV cure or vaccine; different strategies for vaccine development; vaccine trials	9
10	AIDS in broad perspective; the workplace, policies, assistance programs; future expectations and why there is a clear path toward "an AIDS-free generation"	10

Pronunciation Guide

Chapter 1

amyl   a'mil
cytomegalovirus   si'to-meg"ah-lo-vi'rus
dengue   den'ge
Kaposi's sarcoma   kap'o-seez sar-ko'mah
lymphadenopathy   lim-fad"e-nop'ah-thee
mitogen   mi'to-jen
Montagnier   mon-tan-ya'
Pneumocystis carinii   nu"mo-sis' tis car-in'e-e
Pneumocystis jirovecii   nu"mo-sis' tis yee row vet zee
pneumocystosis   nu"mo-sis-to'sis

Chapter 2

acyclovir   a-si'klo-veer
attenuated   ah-ten'u-a-ted
azidothymidine   az'i-do-thi'mid-een
deoxyribonucleic   de-ox"e-ri"bo-nu-kla'ik
encephalopathy   en-sef"ah-lop'ah-thee
genome   je'nome
glycoprotein   gli"ko-pro'teen
ischohedron   i-ko"sah-heed'ron
polymerase   pol-im'er-ase
protease   pro'te-ase
syncytium   sin-sish'e-um

Chapter 3

apoptosis   a-pop'to-sis
Candida albicans   kan'dee-dah al'bi-kanz
Cryptococcus neoformans   krip"to-kok'us ne-o-form'anz
cytomegalovirus   si'to-meg"ah-lo-vi'rus
lymphocyte   lim'fo-site
lymphokinase   lim'fo-ki-nase
lymphopoietic   lim"fo-poi-et'ik

phagocytosis   fag"o-si-to'sis
Pneumocystis jirovecii   nu"mo-sis' tis yee row vet zee
syncytium   sin-sish'e-um
Toxoplasma gondii   toks"o-plaz'mah gon'de-e

Chapter 4

Candida albicans   kan'dee-dah al'bi-kanz
candidiasis   kan"de-di'ah-sis
Cryptococcus neoformans   krip"to-kok'us ne-o-form'anz
cryptosporidiosis   krip"to-spor-id'e-o'sis
Cryptosporidium coccidii
   krip"to-spor-id'e-um kok-sid'e-e
cryptococcosis   krip"to-kok-o'sis
cytomegalovirus   si'to-meg"ah-lo-vi'rus
dementia   de-men'she-ah
encephalopathy   en-sef"ah-lop'ah-thee
Histoplasma capsulatum   his"to-plaz'mah cap-su-lat'um
Isospora belli   i-sos'po-rah bel'li
Kaposi's sarcoma   kap'o-seez sar-ko'mah
lymphadenopathy   lim-fad"e-nop'ah-thee
Mycobacterium avium-intracellulare
   mi"ko-bak-te're-um ave'e-um in"tra-sel-u-lar'e
opportunistic   op"or-tu-nis'tik
pentamidine isethionate   pen-tam'i-deen i-se-thi'o-nate
Pneumocystis jirovecii   nu"mo-sis' tis yee row vet zee
Toxoplasma gondii   toks"o-plaz'mah gon'de-e
toxoplasmosis   toks"o-plaz-mo'sis

Chapter 5

chancroid   shank'roid
chlamydia   klah-mid'e-ah
clitoris   klit'o-ris
columnar   co-lum'nar

Cryptosporidium   krip"to-spor-id'e-um
cytomegalovirus   si'to-meg"ah-lo-vi'rus
epidemiologist   ep'i-de"me-ol'o-jist
Isospora   i-sos'po-rah
Kaposi's sarcoma   kap'o-seez sar-ko'mah
Legionella pneumophilia   lee-jen-el'ah mu-mof'I-lah
pentamidine isethionate   pen-tam'i-deen i-se-thi'o-nate
proboscis   pro-bahs'kis
urethra   yu-re'thra

Chapter 6

candidiasis   kan"de-di'ah-sis
chlamydia   klah-mid'e-ah
lymphogranuloma   lim"for-gran-yu-lo'mah
nonoxynol   no-noks'i-nol
paraphernalia   par"ah-fer-nal'e-ah
trichomoniasis   trik"o-mo-ni'ah-sis

Chapter 7

electrophoresis   e-lek"tro-for-e'sis
immunofluorescence   im"yu-no-floor-es'ens
isothiocyanate   i"so-thi'o-si'ah-nate
polymerase   pol-im'er-ase
radioimmunoprecipitation
  ra'de-o-im"yu-no-pre-sip"i-ta'shun

Chapter 8

acyclovir   a-si'klo-vir
amantadine   ah-man'tah-deen
atovaquone   a-tov'a-quon
azidothymidine   a'zi-do-thi'mi-deen
bisheteroarylpiperazine   bis-het-er-o-aryl-pip-er-a-zine
castanospermine   kas"ta-no-sper'meen
clotrimazole   clo-tri'mah-zole
Crixivan   cri-vax'in
cytomegalovirus   si'to-meg"ah-lo-vi'rus
delavirdine   del-a-vir'deen
deoxyadenosine   de-ox"e-ah-den'o-seen
deoxycytidine   de-ox"e-ci'ti-deen
deoxyguanosine   de-ox"e-gwan'o-seen
deoxythymidine   de-ox"e-thi'mi-deen
didanosine   di-dan'o-seen
dinitrochlorobenzene   di-ni-tro-klor"o-ben'zeen
efavirenz   eh-fa-vir'enz
erythropoietin   e-rith"ro-poi'e-tin

foscarnet   fos-car'net
fluconizole   flu-kon'ah-zole
fullerens   ful'er-eenz
ganciclovir   gan-si'klo-vir
heteropolyanion   het"er-o-pol"e-an-i'on
hydroxynaphthoquinon   hi-drok'si-naf"tho-qwin-on
hypericin   hy"per-i'cin
Indinavir   in-din'a-vir
interferon   in"ter-fer'on
Invirase   in'vir-ase
isoniazid   i"so-ni'ah-zid
ketoconazole   ke"to-kon'ah-zole
lamivudine   lam-iv'u-deen
leucovorin   loo"ko-vor'in
miconazole   mi-kon'ah-zole
nevirapine   ne-vir'a-peen
Norvir   nor'vir
pentamidine isethionate
  pen-tam'i-deen i-se-thi'o-nate
phosphonoformate   fos"fo-no-for'mate
placebo   plah-se'bo
Pneumocystis jirovecii   nu"mo-sis' tis yee row vet zee
protease   pro'te-ase
ribavirin   ri"bah-vi'rin
rifabutin   rif-a-bu'tin
ritonavir   ri-ton'a-vir
saquinavir   sa-quin'a-vir
stavudive   sta"vyu-deen
syncytium   sin-sish'e-um
thalidomide   thal-id'o-mide
thimosins   thi'mo-sinz
thrombocytopenia   throm"bo-si'to-pe'ne-ah
trimethoprom-sulfamethoxazole
  tri-meth'o-prim sul'fah-meth-oks'ah-zol
trimetrexate   tri-meh-trex'at
zalcitabine   zal-cit'a-been
zidovudine   zi-do'vyu-deen

Chapter 9

adjuvant   ad'ju-vant
attenuated   ah-ten"yu-a'ted
azidothymidine   a'zi-do-thi'mi-deen
genome   je'nome
idiotype   id'e-o-tipe
macaque   ma-cak'
vaccinia   vak-sin'e-ah

Glossary

This glossary contains concise definitions of approximately 200 terms related to the biology of AIDS. The number in the parentheses refers to the chapter in which the term is discussed most completely.

accelerated approval regulation   process used by the FDA to allow certain drugs be allowed as treatment for "serious" illnesses that fill an unmet need.

acquired immune deficiency syndrome (AIDS)   as defined by the CDC, a disabling or life-threatening illness caused by the human immunodeficiency virus (HIV) and characterized by HIV encephalopathy, HIV wasting syndrome, or certain diseases due to immunodeficiency in a person with laboratory evidence for HIV infection or without certain other causes of immunodeficiency.

active (phase of viral production)   phase of viral production in which genomes, capsid proteins, and enzyme molecules are assembled together within an infected cell.

acyclovir   an antiviral drug commonly used to treat herpes simplex infection.

adjuvant   a substance such as aluminum sulfate that is chemically bound to subunits in a vaccine to make the subunits more attractive to phagocytes and thereby enhance the immune process.

adsorption   the first step in viral replication involving the highly specific union between the virus and its host cell.

AIDS-dementia complex (ADC)   a severe brain infection caused by HIV and accompanied by mental and physical deterioration (4).

AIDS-related complex (ARC)   a pre-AIDS condition in which the patient experiences lymphadenopathy, constant fever, fatigue and night sweats, extensive diarrhea, and other severe symptoms that accompany the destruction of T-lymphocytes (4).

alpha helix   the spiral structure of a polypeptide consisting of amino acids stabilized by hydrogen bonds.

alpha interferon   a form of interferon produced by genetic engineering and used against Kaposi's sarcoma.

amino acid   an organic acid containing one or more amino groups; the monomers that build proteins in all living cells.

amino group (NH$_2$)   the amine functional group that is located on one end (terminus) of a peptide or protein.

amphotericin B   an antifungal drug used to treat serious fungal diseases such as cryptococcosis .

ampligen   a drug consisting of double-stranded RNA that may stimulate immune system activity to destroy HIV.

anal intercourse   sexual activity in which the penis of the insertive partner is placed into the rectum of the receptive partner.

anemia   a condition in which a person's number of healthy red blood cells is too low to support adequate oxygen to tissues causing an individual feel fatigued.

antagonist (drug)   a drug that blocks the viral host cell co-receptor (e.g. CCR-5 or CXCR4) preventing the interaction between the HIV gp120 and its human co-receptor, preventing HIV entry into host cells.

antibodies   proteins derived from B-lymphocytes and plasma cells that combine chemically with antigens and microorganisms containing antigens to effect neutralization; antibodies are an essential feature of specific resistance against infectious disease.

antibody-mediated immunity (AMI)   immunity in which antibodies provide the actual mode of defense, in comparison with cell-mediated immunity in which cells function as defenders.

anticodon   a three-base sequence on the tRNA molecule that binds to the codon on the mRNA molecule during translation.

antigen a substance, usually a large protein or polysaccharide, that stimulates the activity of the immune system; antigens are interpreted as nonself in the immune process.

antisense molecule an anti-HIV synthetic RNA molecule used as a drug; the molecule unites with and neutralizes the messenger RNA molecule coded by proviral DNA following entry of HIV into the host cell.

antiseptic a chemical used to kill pathogenic microorganisms on a living object, such as the surface of the human body.

apoptosis the natural mechanism by which a cell dies at the end of its life cycle.

aspergillosis a fungal infection caused by an *Aspergillus* species which can cause a lung/pulmonary illness in patients with advanced HIV disease.

atom the smallest portion into which an element can be divided and still enter into a chemical reaction.

attenuated viruses viruses able to multiply in the body but at a rate so low that disease is not established; used in first-generation vaccines.

autoclave a laboratory instrument used to sterilize instruments or materials by means of pressurized steam.

autoimmune condition a pathologic condition arising from the production of antibodies against "self" antigens.

azidothymidine (AZT) an anti-HIV compound closely related to deoxythymidine that interferes with the production of proviral DNA by taking the place of deoxythymidine in the nucleic acid and bringing chain elongation to an end; also known as zidovudine and Retrovir.

bacteriophage a type of virus that attacks and replicates within bacteria.

Berlin patient Timothy Ray Brown, a man infected with HIV from San Francisco. While he was residing in Germany, he underwent a treatment for leukemia that included a bone marrow transplant from a donor who had cells naturally resistant to HIV infection. Medical experts concluded that this treatment cured him of HIV and he has not been using antiviral medication because his viral loads are nonexistent.

beta (β-pleated sheet) a type of secondary structure of proteins that arise from the primary sequence.

blood-brain barrier a semipermeable interface between the bloodstream and central nervous system tissues that protects the brain from potentially toxic substances or microbes.

blood supply banked blood obtained through donations for individuals who need transfusions.

B-lymphocyte, (also, B-cell) an essential cell of the immune system that dominates antibody-mediated immunity; when stimulated by lymphokines, B-lymphocytes convert to antibody-producing plasma cells.

bone marrow spongy tissue found inside of the hollow interior of long bones. In adults, the marrow in the large bones produces new red blood cells.

bone marrow transplant a procedure in which diseased bone marrow is replaced with healthy bone marrow from a donor. The diseased bone marrow is deliberately killed by high doses of chemotherapy or radiation prior to replacement with healthy donor bone marrow.

budding the process by which HIV passes through the membrane of a host cell at the completion of the replication process; while passing through, the HIV nucleocapsid takes on a portion of the cell membrane as an envelope.

bursa of Fabricius the lymphoid organ of the gastrointestinal tract of the embryonic chick where B-lymphocytes are formed.

canarypox a weakened poxvirus used to immunize canaries and other birds against canarypox. Canarypox is being engineered to carry genes for HIV antigens as a potential HIV vaccine. Canarypox cannot replicate in human cells, making it a safe candidate vaccine.

Candida albicans a yeastlike fungus that lives benignly in many individuals but causes candidasis (yeast infections), thrush, and esophageal infection in AIDS patients.

candidate drug or vaccine a candidate drug or vaccine is one in development in order to define its efficacy, safety, and immunogenicity (for a vaccine) *in vitro* and *in vivo* before it is licensed by the FDA.

candidiasis an opportunistic disease in AIDS patients in which *Candida albicans* multiplies in the oral cavity and esophagus causing tissue erosion, inflammation, and difficulty in swallowing and eating; also known as thrush.

capsid the layer of protein that surrounds the nucleic core of a virus such as HIV.

capsomeres the protein subunits of the capsid of a virus.

carbohydrate an organic compound consisting of carbon, hydrogen, and oxygen that is an important source of carbon and energy for all organisms; examples include sugars, starch, and cellulose.

carboxyl group (COOH) functional group located at the terminus of a peptide or protein.

case-control studies a type of epidemiological study in which two groups are observed in order to identify factors that may contribute to a disease/illness.

case definition a standard set of criteria used to identify which person has the disease being studied.

castanospermine an anti-HIV drug that apparently prevents syncytium formation by infected host cells.

CCR-5 (chemokine receptor) a co-receptor present on T-cells, dendritic cells, and macrophages that allows HIV to enter them.

CD4 or CD4 receptor   a protein molecule on the surface of a T-lymphocyte or brain cell that serves as the attachment point for HIV; each CD4 molecule contains 433 amino acids.

cell-mediated immunity (CMI)   immunity in which cells provide the actual mode of defense, in comparison with antibodymediated immunity, in which antibodies function as defenders.

cell membrane   a thin bilayer of phospholipids and proteins that surrounds the prokaryotic cell cytoplasm. *See also* plasma membrane.

Centers for Disease Control and Prevention (CDC)   a major agency of the United States Public Health Service charged with protecting the health of the U.S. populace by providing leadership and direction in the prevention and control of infectious disease and other preventable conditions.

chancroid   an STD caused by the bacterium *Haemophilus ducreyi* that is characterized by painful ulcers or sores on the genitals. The disease is common in sub-Saharan Africa among men who have frequent contact with prostitutes. Chancroid increases the risk of HIV transmission. It is treatable with antibiotics.

chemical bond   a force between two or more atoms that tends to bind those atoms together.

chemical element   any substance that cannot be broken down into a simpler one by a chemical reaction.

chemokine (cytokines)   a small hormone-like protein that is secreted by certain body cells such as phagocytic cells and that acts as a chemical messenger.

chemokine receptor proteins (also CCR-5 and CXCR-4 or fusins)   molecules involved in the immune system that act as receptors for chemokines. HIV may enter and infect host cells when gp120 binds to CD4 and a chemokine co-receptor.

chernaya   pre-filled used syringes containing homemade heroin for sale. The syringes are picked off the streets in Eastern Europe and Central Asia that are contaminated with blood (the blood may be from an HIV-infected person).

cholera   an infectious disease caused by the bacterium *Vibrio cholerae*. Cholera is characterized by profuse, watery diarrhea, vomiting, cramping, life-threatening dehydration, and depletion of electrolytes.

clinical trial   involves human volunteers in order to evaluate the effectiveness and safety of medications, medical devices, and vaccines.

codon   a three-base sequence on the mRNA molecule that specifies a particular amino acid insertion in a polypeptide.

community-based clinical trials   are clinical trials being conducted through doctors and clinics (the community) rather than academic research facilities.

competitive inhibition   a mechanism of enzyme inhibition in which an inhibitor is bound to the active site on an enzyme preventing blocking of the substrate and vice versa (e.g. the nucleoside analog AZT binding to the active site of HIV reverse transcriptase instead of thymidine).

condom   a soft, stretchable sheath placed over the penis during sexual activity to form a barrier between semen and the internal tissues of the sexual partner.

control group   a group of subjects in an experiment who are not receiving the factor/treatment being tested.

coreceptors   protein molecules involved in viral binding to cells.

cryptococcosis   an opportunistic disease in AIDS patients in which *Cryptococcus neoformans* multiplies in the lungs and brain causing headaches, stiff neck, and paralysis.

Cryptococcus neoformans   a fungal parasite often located in the lungs as a benign inhabitant but that causes severe lung and brain infections in AIDS patients.

cryptosporidiosis   an opportunistic disease in AIDS patients in which *Cryptosporidium coccidi* multiplies in the intestinal tissue and causes severe diarrhea, dehydration, and emaciation.

Cryptosporidium coccidi   a protozoal parasite that is a benign inhabitant of the intestine in many individuals but causes severe diarrhea in AIDS patients.

CXCR-4 (chemokine receptor aka"fusin")   are proteins embedded in the membrane of a cells that are chemokine receptors. HIV is able to bind to CD4 and CXCR-4 to facilitate binding and entry into T-cells.

cytomegalovirus (CMV)   a virus that exists benignly in many individuals but causes disease of many internal tissues in AIDS patients.

cytomegalovirus disease   an opportunistic disease in AIDS patients in which the cytomegalovirus multiplies aggressively in lung, liver, kidney, and other tissues, causing multiple forms of disease.

cytoplasm   the complex of chemicals and structures within a cell; in plant and animal cells excluding the nucleus.

cytotoxic T-lymphocyte   a T-lymphocyte that is activated by the lymphokines of helper T-lymphocytes; once activated, the cells enter the circulation and move to the antigen site, where they attack antigen-infected cells; the cytotoxic T-lymphocyte is a host cell for HIV.

dehydration synthesis   chemical reaction in which the components of water are removed from adjacent amino acid molecules to form a bond between the organic acid group of one amino acid and the amino group of the second amino acid.

deoxyribonucleic acid (DNA)   the genetic material of all cells and many viruses.

deoxyribose   the pentose sugar obtained by the hydrolysis of DNA.

dextran sulfate   an anti-HIV agent that apparently blocks viral binding to host cells and prevents syncytium formation.

dideoxycytidine (ddC)   an anti-HIV drug similar to AZT but containing cytosine where AZT contains thymine; the mode of action of ddC is similar to that of AZT, but the side effects are believed less severe.

dideoxyinosine (ddI)   an anti-HIV drug similar to AZT but containing inosine where AZT contains thymine; the mode of action of ddI is similar to that of AZT, but the side effects are believed less severe.

dideoxynucleosides   are synthetic nucleoside analogs of which several are used as antiretroviral drugs (e.g. AZT, ddI, ddC).

disinfectant   a chemical used to kill pathogenic microorganisms on a lifeless object such as a tabletop.

disulfiram (Antabuse)   is a drug being used experimentally to break HIV latency, reducing HIV sleeper cells (reservoirs) *in vitro* and *in vivo*. It was originally used as a drug to treat alcoholism by causing individuals to feel an aversion to alcohol. Disulfiram inhibits alcohol metabolism, resulting in the build up of high concentrations of acetaldehyde in the body.

DNA   *See* Deoxyribonucleic acid

DNA polymerase   an enzyme that catalyzes DNA replication by combining complementary nucleotides to an existing strand.

DNA replication   the process of copying the genetic material.

DNA vaccine   a noninfectious vaccine comprising plasmids modified to carry one or more protein-encoding genes.

donor   is someone who supplies or donates living tissues or fluids (e.g. organs or plasma or blood transfusion) to be used in another body.

dose   the quantity to be administered at one time e.g. a specific amount of drug, vaccine or radiation.

double helix   the structure of DNA, in which the two complementary strands are connected by hydrogen bonds between complementary nitrogenous bases and wound in opposing spirals.

electron   a negatively charged particle with a small mass that moves around the nucleus of an atom.

electrophoresis   a laboratory procedure in which an electrical current is used to separate a mixture of proteins in a gel.

emtricitabine (FTC, or trade name Emtriva)   a nucleoside reverse transcriptase analog used in HAART therapy to treat HIV infection

empty sink model   is an alternative theory that HIV interferes with the production of new cells (the disease "turns down the tap" rather than empties the sink). The theory is based on the study of telomeres that shorten at the ends each time a cell divides.

entry inhibitors   molecules that prevent HIV infection by preventing formation of cell receptor sites so that the virus particle can not enter the cell.

env gene   the gene that codes for HIV envelope proteins.

envelope   a flexible lipid-bilayer membrane that surrounds the capsid of many viruses; the envelope can contain projections called spikes.

enzyme-linked immunosorbent assay (ELISA)   a serological test in which enzyme-linked substances are used to test for the presence of HIV antibodies; the test is used to indirectly identify HIV.

epidemic   an unusually high number of cases in excess of normal expectation of a particular illness in a population, community, or region.

epidemiologist   one who discovers that an epidemic is in progress, defines the circumstances under which it spreads, and makes recommendations on controlling the epidemic in a population.

epidemiology   the study of the various factors that influence the frequency and distribution of diseases in a community.

erythropoietin   a protein hormone produced synthetically and by kidney cells, used to promote red blood cell production in the bone marrow of AIDS patients.

eukaryote   an organism whose cells contain a cell nucleus with multiple chromosomes, a nuclear envelope, and membrane-bound compartments; see also prokaryote.

eukaryotic   referring to a cell or organism containing a cell nucleus with multiple chromosomes, a nuclear envelope, and membrane-bound compartments.

European Agency for the Evaluation of Medicinal Products   is the agency established in 1995 to scientifically evaluate applications for medicinal products derived from biotechnology for European marketing authorization.

exon   any region of a gene that is transcribed and retained in the final messenger RNA product, in contrast to an intron, which is spliced out from the transcribed RNA molecule.

false negative   a result of a laboratory test that suggests that HIV is not present in the body when in fact it is present.

false positive   a result of a laboratory test that suggests that HIV is present in the body when in fact it is not present.

fast-track approval   an FDA procedure in which a treatment investigational new drug (IND) is released for patient use during a Phase II trial.

first-generation vaccine   a vaccine composed of whole microorganisms or their toxins.

flow cytometry   is a method that was traditionally used to count T lymphocytes in AIDS patients in which blood cells flow in a single file through a measuring device where they scatter laser light, allowing a count to be taken. Only one blood sample is measured at a time in a flow cytometer, and samples of blood must be no more than 2 days old. In 1995, a different test replaced this method.

follicular dendritic cells are found within the germinal centers of the lymph nodes (follicles) and have tree-like branches (they are dendritic). These cells remove HIV from the macrophages and "trap" the viruses.

Food and Drug Administration (FDA) an agency of the U.S. federal government that tests drugs to ensure their safety and effectiveness.

fusins are CXCR-4 chemokine receptors that also act as a co-receptor for HIV entry.

fusion inhibitors a class of drugs designed to interfere with the fusion of a virus with its host cell.

gag gene one of the nine HIV genes; codes for the viral capsid proteins.

ganciclovir an antiviral drug commonly used to treat cytomegalovirus (CMV) infections of the eye and other organs.

gay-related immune deficiency (GRID) was the original name given to AIDS by the CDC in 1981.

gender identity is a person's defined sexual orientation of oneself as male or female (or rarely both or neither).

gel electrophoresis is a technique used to separate biological molecules such as DNA, RNA, or proteins according to size and electrical charge by applying an electric current to them when they are in a gel matrix.

gene a segment of a DNA molecule that provides the biochemical information for a function product.

gene probe a single-stranded molecule of DNA that can recognize and bind to a complementary DNA segment on a large DNA molecule.

gene probe test a laboratory test in which a gene probe is used to bind with proviral DNA from HIV if the proviral DNA is present; the test is used to directly identify HIV.

gene therapy is an experimental treatment that involves the introduction of genes into a person's cells to replace or compensate for defective genes in a person's body that are responsible for a disease or medical problem; genes can be carried and delivered by customizing vectors such as viruses.

genetic code the specific order of nucleotide sequences in DNA or RNA that encode specific amino acids for protein synthesis.

genome the complete set of genes in a virus or an organism.

genotypic tests are used to determine if an HIV-infected person contains strains of HIV resistant to antiviral therapy based on specific mutations present within the viral genomes of the patient undergoing HAART.

glycoprotein any protein molecule containing one or more carbohydrate groups.

glycoprotein (gp) 41 a glycoprotein molecule having a molecular weight of 41 kilodaltons and found in the envelope spikes of HIV; the gp41 molecule makes the connection between HIV and host cell after the reaction between gp120 and CD4 has taken place.

glycoprotein (gp) 120 a glycoprotein molecule having a molecular weight of 120 kilodaltons and found in the envelope spikes of HIV; the gp120 molecule unites with the CD4 molecule of the host cell during the adsorption phase of viral replication.

glycoprotein (gp) 160 a forerunner of gp120 and gp41 molecules in HIV that can stimulate antibody production when used in an AIDS vaccine.

gonorrhea is a sexually transmitted disease caused by the bacterium *Neisseria gonorrhoeae*.

half-life the period of time required for half the original amount of drug in the blood to disappear.

HAART (highly active antiretroviral therapy) or HAART regime the administration of at least three different compounds to jointly block viral replication by inhibiting the viral protease and/or reverse transcriptase of HIV; was a possible approach to eliminate the virus from the body altogether.

helical shape of a virus in which the virus particle has a long-cylindrical shaped capsid.

helix the tightly wound coil form taken by many viruses, including the virus of rabies.

helper T-lymphocyte a T-lymphocyte that releases lymphokines after being activated by an antigen-bearing macrophage; lymphokines stimulate other cells of the immune system; the helper T-lymphocyte is a host cell for HIV; also called CD4 or T4 cell.

hematopoietic cells are found in the bone marrow where many leukocytes, including B-cells, reach maturity and are prepared to fight infections.

hemophiliac a person whose blood system lacks a certain clotting factor and, therefore, has difficulty in forming a blood clot; hemophiliacs have been exposed to HIV via infusions of clotting factors derived from human blood that was infected.

heterosexual a male or female who prefers a sexual partner of the opposite gender.

histone deacetylase (HDAC) is an enzyme in human cells that acetylates histones, altering the structure of chromatin and regulating the transcription of cellular genes.

histone deacetylase (HDAC) inhibitors such as vorinostat block the unregulated growth of cancer cells by regulating HDAC activity with little or no toxicity to normal cells.

Histoplasma capsulatum a fungus that lives benignly in many individuals but causes severe lung disease (histoplasmosis) in AIDS patients.

Hit-it-early, hit-it-hard approach is Dr. David Ho's approach to treating HIV infection by suggesting that a cocktail of antivirals should be used to treat HIV infected people to maximize suppression of the virus and minimize damage to the immune system.

HIV-1 worldwide predominant strain or type of HIV in the U.S. that causes AIDS.

HIV-2 predominant strain or type of HIV in West Africa that is rarely found elsewhere. It is less easily transmitted and the period between initial infection and illness is longer.

HIV antibodies are antibodies directed against one or more proteins or antigens on the surface of HIV.

HIV-associated neurocognitive disorder (HAND) term used to describe the full spectrum of neurological disease observed in HIV patients.

HIV-associated nephropathy usually begins with loss of blood proteins in the urine (proteinuria), and within one year, the patient usually develops what is known as end-stage renal failure. With end-stage renal failure, the patient's kidneys are not functioning. Dialysis can prolong life; however, the best treatment is a kidney transplant.

HIV encephalopathy a condition due to HIV and characterized by disabling cognitive and/or motor dysfunction interfering with activities of daily living or loss of behavioral developmental milestones in a child, progressing over weeks to months, in the absence of concurrent illness or condition other than HIV infection that could explain the findings.

HIV infection the condition wherein HIV exists as a provirus in the nucleus of its host cells and the infected person has mild nonspecific symptoms such as fatigue, mild fever, and swollen lymph nodes.

HIV wasting syndrome a condition due to HIV characterized by involuntary loss of more than 10 percent of baseline body weight plus either chronic diarrhea or chronic weakness and documented fever in the absence of concurrent illness or condition other than HIV infection that could explain the findings.

HLA (human leukocyte antigen system) is the name of the major histocompatibility complex (MHC) in humans.

HLA receptor gene codes for HLA receptors.

HLA receptors are proteins that act as a warning sign to neighboring cells of the body. During HIV infection, the HLA receptors attract T-helper and cytotoxic T-cells to destroy HIV-1 infected cells.

homosexual a male who prefers a sexual partner of the same gender; a female homosexual is referred to as a lesbian.

hormone is a protein synthesized by one type of cell that travels through the bloodstream to affect and/or direct the function of another type of cell.

HTLV-III/LAV the name used for the AIDS virus before 1986 and before the name human immunodeficiency virus (HIV) was introduced; the acronym stands for human T-cell lymphotropic virus type III/lymphadenopathyassociated virus.

human immunodeficiency virus (HIV) the RNA-containing particle of nucleic acid and protein that is regarded as the cause of HIV infection and AIDS; the virus replicates inside host cells of the immune system and brain and brings about their destruction.

humanized mouse model is a special breed of mouse born with a severe combined immune deficiency (SCID) that is genetically altered to include elements of the human immune system so that it can be used as a model for HIV research in pre-clinical trials (animal studies).

hydrogen bond a weak chemical bond between a positively charged hydrogen atom (covalently bonded to oxygen or nitrogen) and a covalently bonded, negatively charged oxygen or nitrogen atom in the same or separate molecules.

hydrophilic referring to a substance that dissolves in or mixes easily with water; *see also* hydrophobic.

hydrophobic referring to a substance that does not dissolve in or mixing easily with water; see also hydrophilic.

hypericin an antiviral drug for possible use against HIV.

hypogonadism a low level of the sex hormone testosterone which is a hormonal abnormality associated with wasting syndrome.

icosahedron a geometric shape with 20 triangular faces; the form taken by many viruses, including, for a time, HIV.

idiotype the region of an antibody molecule that binds to an antigen molecule in an antigen-antibody reaction; currently under study for use in an AIDS vaccine.

immune system a complex set of cells, chemical factors, and processes in which blood cells called lymphocytes respond to and eliminate foreign agents or substances in the body's tissues.

immunogens a molecule that stimulates an immune response in the host that is exposed to them.

immunosuppressant drugs sometimes referred to as anti-rejection drugs; a drug that lowers the body's immune response often used to prevent organ or tissue rejection in transplant recipients or they are used to treat inflammatory and autoimmune diseases.

inactivated viruses or inactivated viral vaccines viruses unable to multiply in host cells because of some chemical or physical treatment; used in certain vaccines.

incubation period the time in which an individual is infected with an infectious agent but has not yet displayed signs and symptoms of illness.

indinavir (Crixivan) an anti-HIV compound that acts as a protease inhibitor to block the final steps of HIV synthesis in the host cells.

indirect immunofluorescence assay (IFA) a laboratory test for HIV antibodies in which fluorescent-tagged substances unite with HIV antibodies if the latter are present on the surface of a carrier particle.

infection-control procedures measures practiced by healthcare personnel in healthcare facilities to decrease transmission of infectious diseases. Measures include handwashing, disinfection, practicing universal precautions, isolation of infected individuals, etc.

insertive partner during sexual intercourse, the insertive partner's penis is inserted into the partner's vagina, anus, or mouth.

integrase an enzyme that incorporates proviral DNA in the DNA of the host T-lymphocyte.

integrase inhibitors are a class of antiviral drugs used to treat people infected with HIV. They block the action of integrase, an HIV enzyme involved in inserting the viral genome into the DNA of the HIV-infected host cell.

interferon an antiviral substance produced by human cells on exposure to viruses.

interleukin-2 a lymphokine that activates natural killer cells.

investigational new drug (IND) program within the FDA that allows a pharmaceutical company permission to use an experimental drug in clinical trials.

investigational new drug for treatment (a so-called treatment IND) can be used as treatments for serious or life-threatening conditions for one person or a group of people.

isoniazid a chemotherapeutic agent effective against tuberculosis.

Kaposi's sarcoma a cancerous condition often associated with AIDS, in which slow-growing tumors in the blood vessel linings cause red to violet patches on the skin surface that eventually become purplish-brown nodules; in AIDS patients, nodules also form in the internal organs.

latent time period when HIV remains inactive (is not expressing viral mRNA/ translating proteins) because a copy of its genome is integrated into the host genome and replicates silently with the host genome until the cells are reactivated into a productive HIV infection.

latent reservoirs of HIV are cells that contain HIV proviral DNA.

Legionnaires' disease a new disease that emerged during July 1976 at the Bellevue-Stratford Hotel in Philadelphia. This disease is used as an analogy for the HIV epidemic that insidiously emerged as a new infection during the 1980s.

leukocytes white blood cells that develop in the bone marrow and thymus whose function is to traffic through the lymphatic system and to communicate the presence of infection, protecting the body against microbial infections

lipid a nonpolar organic compound composed of carbon, hydrogen, and oxygen; examples include triglycerides and phospholipids.

liposomes artificial membranes composed of lipidlike substances and used as adjuvants in vaccines.

live (attenuated) vaccines are traditional vaccines that stimulate the host to produce antibodies against a highly weaked, yet live, virus.

long-term nonprogressors a subgroup of individuals infected with HIV who have not previously taken or received antiretroviral therapy and show no sign of disease for long period (decades) of time. Bob Massie is a long-term nonprogressor.

lymphadenopathy swelling of the lymph nodes, adenoids, and other tissues of the immune system.

Lymphadenopathy-associated virus (LAV) name suggested by French researcher/virologist Luc Montagnier for HIV.

lymphatic system the series of interconnected and interdependent spaces and channels between the organs and body tissues that circulate lymph (including infection fighting cells [lymphocytes]), removing cell debris and foreign matter from the tissues and transferring them to the blood; part of the immune defense.

lymph nodes pockets of white blood cells prevalent in the neck, armpits, groin, and other body regions; cells of the immune system inhabit the lymph nodes.

lymphocyte a type of leukocyte that functions in cell mediated immunity. (e.g. T-cells and B-cells)

lymphogranuloma is a sexually transmitted disease caused by *Chlamydia*. It usually affects the lymph glands in the genital area.

lymphokines a series of reactive proteins that stimulate either B-cells and/or cytotoxic T-cells, depending on the nature of the antigen that began the process.

lymphopoietic cells forerunner cells of the immune system that can become B-lymphocytes or T-lymphocytes.

lysogeny the phenomenon in which a virus remains in the cell cytoplasm as a fragment of DNA and fails to replicate in or destroy the cell.

lytic cycle the process wherein a virus replicates within a host cell and destroys the host cell during the process.

M group (of HIV-1) one of three major groups of HIV-1 viruses, it causes 99 percent of the world's AIDS cases.

macrophage a large amoeboid cell that phagocytizes antigens and sets the immune process into motion.

major histocompatibility (MHC) genes code for MHC proteins involved in transplant rejection.

major histocompatibility (MHC) proteins proteins present on the surface of body cells that are unique for that particular individual and act as recognition sites during the immune process.

messenger RNA (mRNA)   an RNA transcript containing the information for synthesizing a specific polypeptide.

microbiocide   a substance (e.g. gel or cream) used to kill or destroy HIV or other infectious agents.

microglia   the brain's macrophages, which can be infected by HIV and cause encephalopathy.

microorganism   a microscopic form of life, including bacteria, viruses, fungi, protozoa, and some multicellular parasites.

Millenium Village Project sites   established in 2010 through a partnership of WHO and UNAIDS to create HIV mother-to-child transmission (MTCT)-free zones located in 10 countries across sub-Saharan Africa.

modified vaccinia ankara (**MVA**)   a highly attenuated vaccinia virus strain that is missing 10% of the vaccinia genome and is unable to replicate in primate or human cells.

monotherapy   is treating a disease (e.g. HIV) with only one drug.

Morbidity and Mortality Weekly Reports (MMWR)   a weekly publication of the Centers for Disease Control that summarizes contemporary health problems in the United States and presents cumulative statistics on certain infectious diseases.

mother-to-child-transmission (MTCT)   HIV transmission to a newborn child via an HIV-infected mother (also referred to as pediatric AIDS).

Mycobacterium aviam **complex (MAC)**   infections caused by *Mycobacterium avium* and *Mycobacterium intracellulare* that can be a serious threat to those with HIV.

Mycobacterium avium-intracellulare   a small bacterial rod that causes severe lung disease in AIDS patients.

Mycobacterium tuberculosis   a small bacterial rod that causes tuberculosis.

mutation   is a change in the characteristic organism or virus arising from a permanent alteration of a DNA or RNA sequence.

naive T-lymphocytes   a new T-cell which is ready to process and contain new antigens recognized by the immune system.

N group (of HIV-1)   one of three major groups of HIV-1 viruses.

nanometer   a unit of measurement equivalent to one billionth of a meter; the unit is abbreviated as nm and is used to measure the dimensions of viruses.

National Condom Awareness Day   is February 14th (the same day as Valentine's Day) and is aimed to create HIV/AIDS awareness.

National HIV/AIDS strategy   is one of U.S. President Barack Obama's top policy priorities in reducing HIV incidence, increasing access to care and optimizing health outcomes, and reducing HIV-related health disparities in the U.S.

natural killer   cell a type of T-lymphocyte that attacks tumor cells.

needlestick   the pricking of one's finger with a contaminated needle while performing a medical procedure.

nef **gene**   the "negative regulatory factor gene," which codes for a protein that enhances the movement of genetic messages in the cytoplasm of the host cell.

new molecular entities (NMEs)   An NME is an active ingredient that has never been marketed in the U.S. in any form.

nonnucleoside reverse transcriptase inhibitors (NNRTIs) or analogs   a group of drugs that target and react with reverse transcriptase.

nonreactive   testing negative in a serological test.

nonspecific resistance   protection against all foreign organisms, not just specific ones.

nucleases   are enzymes that break the phosphodiester bonds of nucleic acids (e.g. DNases break those bonds in DNA or RNases break those bonds in RNA).

nucleic acid   a high-molecular-weight molecule consisting of nucleotide chains that convey genetic information and are found in all living cells and viruses; *see* DNA and RNA.

nucleic acid testing (NATS)   methods used to detect HIV nucleic acids in order to diagnose and manage patients suffering from HIV/AIDS.

nucleocapsid   the combination of genome and capsid of a virus.

nucleoside   a molecule consisting of deoxyribose and a nitrogenous base; nucleosides are building blocks of DNA and RNA.

nucleoside reverse transcriptase inhibitors (NRTIs)   are compounds that inhibit the action of the HIV reverse transcriptase enzyme, thus preventing its genome from being replicated.

nucleotide   a component of a nucleic acid consisting of a carbohydrate molecule, a phosphate group, and a nitrogenous base.

NYC Condom Finder smartphone application   was developed as a smartphone application with GPS technology to locate free distribution of condoms in New York City. The app was launched and made freely available on Valentines Day (February 14th) 2011.

O group (of HIV-1)   one of three major groups of HIV-1 viruses.

Okazaki fragment   a segment of DNA resulting from discontinuous DNA replication.

opportunistic disease or opportunistic infection   a disease or infection caused by a virus or microorganism that does not cause disease but that can become pathogenic or life-threatening if the host is immune suppressed (e.g. AIDS patients).

opportunistic organism an organism that may exist in the body but cause no harm because the body's natural defense keeps it under control; when the defense is compromised, the organism seizes the "opportunity" to infect (e.g. a patient suffering from AIDS experiences many opportunistic infections).

Ora-Quick Advance Rapid HIV-1/2 Antibody test an over-the counter in-home rapid HIV test kit approved by the FDA in 2012.

p24 antigen makes up most of the viral core of HIV particles. Serum concentrations of HIV p24 are high before an infected individual seroconverts. The American Red Cross screens blood donations for HIV p24 antigen.

ostrich mentality is the mentality in which some people "bury their head in the sand" to pretend that all is well in the world. In other words, it is avoidance or lacking confrontation to a situation etc. such as addressing HIV/AIDS issues.

pandemic an epidemic of worldwide scope.

parallel-track procedure one in which patients are enrolled in a clinical trial even though they are ineligible for controlled clinical trials of experimental drugs.

parasite a type of heterotrophic organism that feeds on live organic matter such as another organism.

parasitism the relationship between organisms in which one lives as a parasite on another (e.g. HIV parasitizes human T-cells).

pediatric AIDS AIDS in a baby or a young child.

pentamidine isethionate a drug available in injectable and aerosolized form for use against *Pneumocystis jirvovecii* pneumonia.

peptide bond a linkage between the amino group on one amino acid and the carboxyl group on another amino acid.

peptide T a peptide consisting of eight amino acids and currently under study as an anti-HIV agent.

percutaneous injury is a puncture wound which exposes/draws blood e.g. needlestick injury.

phagocytic cells white blood cells capable of engulfing and destroying foreign materials or cells, including bacteria and viruses.

phagocytosis the engulfment and ingestion of foreign material (e.g. viruses, bacteria) by phagocytes.

Phase I trial involves testing relatively few individuals (perhaps 25) to determine the safety and the high and low dosing range of a vaccine or drug in humans.

Phase II trial involves testing several hundred individuals to assess the side effects and optimal dose/effectiveness as a therapeutic drug or vaccine.

Phase III trial involves testing thousands of volunteers to determine if the drug or vaccine is effective and safe.

phenotypic test is one that measures the behavior or phenotype of a patient's HIV response to particular antivirals (e.g. determining how well the virus replicates in the presence of antivirals as opposed to determining the mutations that cause drug resistance).

phosphonoformate an antiviral drug active against cytomegaloviruses; also known as Foscarnet.

placebo is a substance containing no medication or drug but prescribed or given to reinforce a patient's expectation to get well. Placebos are sometimes used clinical trials.

plasma cells highly active antibody-producing cells that are derived from B-lymphocytes.

Pneumocystis jirovecii a yeast-like fungal pathogen that is a benign inhabitant of the lungs in most individuals but causes a severe pneumonia in AIDS patients.

pneumocystosis opportunistic lung infection common in AIDS patients caused by *Pneumocystis jirvovecii*.

Pneumocystis pneumonia (PCP) an opportunistic disease in AIDS patients in which yeast-like fungi multiply furiously in the lungs, take up most of the air spaces, and severely reduce the ability of the body to exchange gases with the external environment.

pneumonia a microbial disease of the bronchial tubes and lungs; may be caused by various bacteria, viruses, fungi, and protozoa.

pol gene the gene that codes for HIV viral enzymes.

polylactide coglycolide is an adjuvant used in surgical sutures that boosts cell-mediated and antibody-mediated immunity in laboratory animals.

polymer a substance formed by combining smaller molecules into larger ones; see also monomer.

polymerase chain reaction (PCR) a procedure in which a small amount of DNA is amplified by encouraging copying of the DNA using specific enzymes, a piece of primer DNA, and a series of DNA building blocks; the gene probe test can then be utilized to identify the DNA.

postexposure prophylaxis (PEP) preventive measures taken after exposure to a causative agent.

preexposure prophlylaxis (PrEP) preventative measures taken before exposure to a causative agent (e.g. the MMR vaccine is given to children before they are exposed to measles, mumps, rubella viruses).

pre-mastication is the practice of pre-chewing of food for infants. HIV-positive mothers should not practice pre-mastication as it can result in mother-to-child transmission.

primary or acute HIV infection the initial infection caused by HIV which lasts for several weeks. It includes such symptoms as fever, swollen lymph nodes, muscle pain, sore throat, malaise, and mouth sores. Antibodies against are not detectable at this time of infection.

primary structure the sequence of amino acids in a polypeptide.

prime and boost trials are used when a single dose of an antigen is insufficient in providing effective protection/

vaccination against a pathogen. They involve using two different vaccine preparations in a series to "prime and boost" the immune response. There are HIV vaccine trials using this strategy e.g. using an HIV DNA vaccine followed by a genetically engineered canarypox HIV vaccine.

prophylaxis   measures used to prevent the development of a disease (e.g. residents of a nursing home taking Tamiflu to prevent an influenza outbreak at the facility or antibiotics taken by AIDS patients to prevent opportunistic bacterial infections).

protease   an enzyme that cleaves the proteins of potential HIV particles and prepares the proteins for union into the viral capsid.

protease inhibitor   a synthetic anti-HIV drug that unites with and neutralizes the enzyme protease, which is used in synthesis of the HIV capsid.

protein   a chain or chains of linked amino acids used as a structural material or enzyme in living cells.

protein (p) 17   a protein of the HIV capsid currently under study for use in a vaccine.

protein synthesis   the process of forming a polypeptide or protein through a series of chemical reactions involving amino acids.

provirus or proviral DNA   a double-stranded DNA molecule derived from the activity of reverse transcriptase and incorporated into the genetic material in the nucleus of a host cell.

proviral HIV DNA   is integrated into the chromosome/genome of cells using HIV integrase.

pseudovirions   are empty capsid or lipid outer shells of a virus that do not contain genetic material used as a potential vaccine to prevent viral infection.

purines   adenine or guanine nitrogen bases of DNA or adenine in RNA.

pyrimethamine   an antimicrobial drug used to treat infections by *Toxoplasma gondii*.

quarantine   is the segregation of infected individuals from the general population of healthy people who are not ill but have been exposed to an individual suffering from a communicable disease.

quaternary structure   complex structure of proteins characterized by the formation of large, multi-unit proteins by more than one of the polypeptides.

radioimmunoprecipitation assay (RIPA)   a laboratory test for HIV antibodies in which radioactive-tagged HIV antigens unite with HIV antibodies if the latter are present.

random coils   secondary structure of a protein that arises from primary structure.

rapid home (HIV) test   is a simple test to detect HIV; it can be an over-the-counter test (Ora-Quick) that provides the final results of the test at home or a test in which a "finger-prick" blood sample is applied to filter paper, packaged, and then mailed to a laboratory for NATs testing.

reactive   testing positive in a serological test.

rebound effect   occurs when a lower dose of antiviral drug is used to minimize side effects of the drug but the reduction incites a dangerous unexpected flare-up of symptoms.

receptive partner   in sexual intercourse, receives the partner's penis into the vagina, anus, or mouth.

receptors   on the surface of cells that allow receptor-binding protein or viral attachment proteins on viruses to adhere to a cell.

recipient   is a person who receives (e.g. organ or bone marrow transplant recipient).

recombinant vaccine   a purified antigenic component of a virus or other pathogen; safer than live vaccines.

replication fork   the point where complementary strands of DNA separate and new complementary strands are synthesized.

retinitis   is a sight-threatening complication of HIV/AIDS caused by cytomegalovirus infection. If untreated, it can cause a detached retina and blindness in 2–6 months.

retrovirus   a virus such as HIV whose RNA is used as a template for the synthesis of DNA mediated by the enzyme reverse transcriptase.

rev gene   "regulator of the expression of viral protein" gene, which shifts the balance from production of viral regulatory proteins to proteins that make up virus particles.

reverse transcriptase   the enzyme of certain viruses (such as HIV) that uses RNA as a template to synthesize a complementary molecule of DNA.

reverse transcriptase inhibitors   a chemical that prevents the action of reverse transcriptase.

ribonuclease   a portion of the reverse transcriptase molecule that destroys the RNA after it has been used as a template for DNA synthesis.

ribonucleic acid (RNA)   the nucleic acid involved in protein synthesis and gene control; also the genetic information in some viruses.

ribose   five carbon sugar unit found in RNA.

ribosomal RNA (rRNA)   an RNA transcript that forms part of the ribosome's structure.

ribosome   a cellular component of RNA and protein that participates in protein synthesis.

ritonavir (Norvir)   an anti-HIV compound that acts as a protease inhibitor to block the final steps of HIV synthesis in the host cells.

RNA   *See* Ribonucleic acid

RNA polymerase   the enzyme that synthesizes an RNA polynucleotide from a DNA template.

RNase (ribonuclease)   degrades ribonucleic acid (RNA).

Ryan White Program is the largest federal grant program first enacted in 1990 to work with various organizations at the city, local, and state level to provide HIV-related services in the U.S.

Sabin oral polio vaccine is a type of vaccine prepared with attenuated polioviruses and taken orally.

Salk injectable polio vaccine is a type of vaccine prepared with inactivated or killed polioviruses and injected into the body.

saquinavir (Invirase) an anti-HIV compound that acts as a protease inhibitor to block the final steps of HIV synthesis in host cells.

secondary structure the region of a polypeptide folded into an alpha helix or pleated sheet.

second-generation vaccine a vaccine composed of microbial subunits obtained from broken microorganisms.

self-antigens natural markers that are recognized by the human body.

semen the mixture of sperm cells and glandular fluids that is ejaculated from the male reproductive system during sexual activity.

semiconservative replication the DNA copying process where each parent (old) strand serves as a template for a new complementary strand.

sensitivity refers to the probability that a serological test such as ELISA will give a positive result when the serum contains the antibodies sought.

seroconvert or seroconversion to test positive in a serological test.

serological test any laboratory test used to detect antibodies in the serum.

seronegative lacks the antibodies sought in a serological test.

seropositive has the antibodies sought in a serological test.

serum the clear fluid portion of the blood that has no blood cells or clotting agents.

severe immune deficiency (SCID) a serious condition in which the person/animal is born with reduced numbers of T- and B-lymphocytes which impairs their immune systems, making them susceptible to severe infections and cancers.

sexually transmitted diseases (STDs) infections generally acquired by sexual contact (e.g. HIV, gonorrhea, syphilis, chlamydia).

sharps are any item with a thin cutting edge or fine point (e.g. syringe needles, scalpels, razor blades)

simian refers to a monkey or monkeylike animal such as a chimpanzee.

simian AIDS AIDS in monkeys or monkey-like animals.

single nucleotide polymorphisms (SNPs) DNA sequence variation occurring when a single nucleotide A, T, C, or G in the DNA differs between humans or paired chromosomes in humans.

soluble CD4 a synthetic drug consisting of free (soluble) CD4 molecules intended to unite with HIV in the bloodstream and prevent HIV attachment to host cells.

specific resistance resistance founded in the immune system and directed at neutralizing a specific microorganism or compound.

specificity refers to the probability that a serological test such as ELISA will give a negative result when the serum lacks the antibodies sought.

spikes projections of the viral envelope that assist the union of the virus with its host cell.

sterilization the removal of all life forms, with particular reference to bacterial spores.

sterilizing immunity the action of a vaccine that prevents infection.

steroid hormones (glucocorticoids) are a class of hormones that upregulate the expression of anti-inflammatory proteins.

substrate a chemical upon which an enzyme acts.

subunits concentrated suspensions of viral or bacterial fragments used in a vaccine to stimulate an immune response.

subunit or protein vaccine a vaccine containing parts of microorganisms such as capsid proteins or purified spike glycoproteins.

sulfamethoxazole a sulfur-containing drug that when combined with trimethoprim, can be used to treat *Pneumocystis jirovecii* pneumonia.

Super strain a multi-drug resistant strain of HIV that causes a rapid onset of AIDS.

suppressor T-lymphocyte a T-lymphocyte that dampens the activity of the immune system by affecting the cytotoxic T-lymphocytes.

syncytium a giant, multinucleated, functionless cell mass; the infection of host cells by HIV marks the cell surface with gp120 molecules that encourage cells to bind together and form a syncytium.

synergistic effect occurs when a combination of drugs work together to results in an additive or greater effect than if each drug was taken separately.

syphilis is a sexually transmitted disease caused by *Treponema pallidum*.

T-lymphocyte (also, T-cell) T-helper cell that promotes cell mediated immune responses.

T4 cell an alternative term for helper T-lymphocyte.

T8 cell an alternative term for suppressor T-lymphocyte.

Take Charge, Take Test (TCTT) a phase of the *Act Against AIDS* CDC campaign to increase HIV testing among black/African American women ages 18–34.

tat gene (transactivator gene)   a proviral gene that appears to be involved in the activation of HIV from the latent proviral state to the state of replication.

tat (protein)   HIV protein that increases the transcription of HIV RNAs.

telomere   DNA structures present at the ends of eukaryotic chromosomes that are not included in DNA replication; they protect the chromosome from fusion to other chromosomes degraded by nucleases.

tenofovir   an anti-HIV compound that is a nucleoside reverse transcriptase inhibitor. It is also part of the combination Truvada once-a-day dosing drug.

tertiary structure   the folding of a polypeptide back on itself.

The Magic Johnson Foundation   nonprofit organization founded by former NBA player Earvin "Magic" Johnson (who discovered he tested positive for HIV infection in 1991) to address the educational, health, and social needs of ethnically diverse, urban communities. Magic Johnson is an advocate for HIV/AIDS prevention and safe sex.

therapeutic immunity   the action of a vaccine that prevents disease.

third-generation vaccine   a vaccine composed of microbial subunits obtained synthetically by genetic engineering techniques.

thrombocytopenia   a side effect associated with AZT use in which an allergic reaction to the drug occurs and the body's blood platelet count and ability to form blood clots are reduced.

thrush   an infection of the mouth (and, possibly, esophagus) caused by the fungus *Candida albicans* and accompanied by milky white flakes on the tongue and other parts of the oral cavity; also known as candidiasis.

thymosins   are small proteins found in many tissues that were originally isolated from the thymus. Thymosins stimulate the pituitary gland via the hypothalamus to release hormones.

thymus   a flat, bilobed organ that lies in the neck below the thyroid, the T-lymphocyte is produced in this organ.

T-lymphocyte   an essential cell of the immune system that dominates cell-mediated immunity and functions as a helper, cytoxic, or suppressor T-lymphocyte; certain T-lymphocytes are host cells for HIV; also called T-cells.

trimethoprim-sulfamethoxazole (TMP-SMX)   is a drug that can be used to treat *Pneumocystis jirovecii* pneumonia.

Toxoplasma gondii   a protozoal parasite that is a benign inhabitant of the blood in many individuals but causes severe infection of the brain in AIDS patients; domestic housecats harbor *T. gondii*.

toxoplasmosis   an opportunistic disease in AIDS patients in which *T. gondii* multiplies in brain tissue, causing lesions, cerebral swelling, and seizures; in patients with out AIDS, toxoplasmosis is a mild mononucleosislike disease.

transcription   the biochemical process in which RNA is synthesized according to a code supplied by the bases of a gene in the DNA molecule.

transfer RNA (tRNA)   a molecule of RNA that unites with amino acids and transports them to the ribosome in protein synthesis.

transgender   a term for an individual who is identified or expresses a gender other than their biological one.

translation   the biochemical process in which the code on the mRNA molecule is translated into a sequence of amino acids in a polypeptide.

treatment IND   an investigational new drug used for treatment of a disease; often released to physicians while drug trials are ongoing.

trimethoprim   a drug that when combined with sulfamethoxazole, can be used to treat *Pneumocystis jirovecii* pneumonia.

trimetrexate   an anticancer drug also used to inhibit *Pneumocystis jirovecii*.

Truvada   a combination of tenofovir/emtricitabine antiretroviral drugs approved for HIV prevention (PrEP or pre-exposure prophylaxis) in healthy individuals at high-risk for acquiring an HIV infection.

tuberculosis   a bacterial disease of the lungs caused by *Mycobacterium tuberculosis* accompanied by progressive deterioration, difficult breathing, blood in the sputum, and eventual suffocation.

universal precautions   recommendations published by the CDC to protect health care workers from infection by patients, as well as the reverse, and to prevent the spread of HIV among patients through contaminated devices or surfaces.

urethra   the thin tube that carries semen and urine through the penis and out to the exterior.

V3 loop   a loop of HIV amino acid that help the virus to attach to the CD4 receptor site.

vaccinia   is a poxvirus used as a live vaccine to prevent smallpox.

vaginal intercourse   sexual activity in which the penis of the insertive male partner is placed into the vagina of the receptive female partner.

vectored immunoprophylaxis (VIP)   is a new approach to HIV prevention which involves inserting the gene for a neutralizing antibody against HIV into a viral vector such as adenovirus. The engineered adenovirus is injected into an animal (in preclinical studies) as a vaccine to prevent HIV infection.

vector vaccine   involves using a mild or avirulent virus carrying genes to HIV antigens as a vaccine. Examples of vector vaccines are modified vaccinia, adenovirus, and canarypox virus.

viable-vector vaccine   a vaccine consisting of nonpathogenic viruses whose DNA contains a segment that will code for HIV proteins.

***vif* gene**   regulatory gene necessary for the reverse transcription of RNA to DNA.

viral load   the measure of the number of HIV-1 RNA genome copies per milliliter of plasma; a predictor of the disease progression to AIDS.

virion   a completely assembled virus outside its host cell.

viruses   particles of nucleic acid (either DNA or RNA) surrounded by a protein sheath and, sometimes, a membranous envelope; neither prokaryotic nor eukaryotic; highly infectious.

virus-like particles (VLPs)   self-assembled viral proteins that make up the outer shell or coat of virus. These empty viruses or VLPs do not contain any genetic material and are unable to replicate or infect cells but can be used as a vaccine.

vorinostat   is a lymphoma cancer drug that is being used as a drug to break HIV viral latency in Phase I clinical trials.

***vpr* gene**   regulatory gene that encodes viral protein R, which assist transport of viral DNA into the host cell nucleus.

***vpu* gene**   regulatory gene that encodes viral protein U, which breaks down the CD4 receptor protein, a protein that prevents HIV from budding out of the cell.

Walter Reed classification system   a seven-stage classification system that charts the course of disease development from exposure to HIV through to AIDS.

wasting phenomenon   a condition accompanied by dramatic weight loss of more than 10% of a person's baseline body weight plus either chronic diarrhea lasting at least 10 days or chronic weakness and fever lasting 30 days. HIV wasting syndrome occurs in many AIDS patients.

Western blot analysis   a serological test in which standardized HIV proteins are used to test for the presence of HIV antibodies separated from one another by gel electrophoresis; the test is used to identify HIV indirectly.

whole-virus vaccine   a vaccine containing entire virus particles that have been killed.

window period   is the duration of time after HIV infection until antibodies against HIV are detectable by serologic methods.

World AIDS Day   (December 1) brings people together from around the world to raise awareness about HIV/AIDS.

World Health Organization (WHO)   a specialized agency of the United Nations that works to promote physical, mental, and social health in peoples of the world.

zidovudine   an alternate name for azidothymidine.

zinc finger   a region present on certain proteins responsible for binding to DNA.

zinc finger nucleases   are site-specific DNA editing proteins.

zinc finger proteins   are naturally occurring proteins that bind DNA during transcription.

Index

Note: Italicized page locators indicate photos/figures; tables are noted with *t*.

Antibiotics, 204, 288
  for sexually transmitted diseases, 147
Antibodies, 25, 56, 63, 66, 71, 73, 80, 180
  types of, 72
Antibody-mediated immunity, 68, 70–73, *71,*
  180, 259
Antibody response, viral vaccines and, 251
Antibody tests, HIV
  enzyme-linked immunosorbent assay, 180–181
  false positives and negatives, 185–186
  home and simplified tests, 186–188
  Western blot analysis, 182–185
Antigen-antibody reaction, 73
Antigens, 63, 66–67, 80
  antibody interactions with, mechanisms of, *72*
  HIV, 180
Anti-HIV therapies, developing, approaches to, *207*
Antiretroviral drugs
  development and therapy regimes, timeline
    of, *209*
  monthly average wholesale price for, U.S.,
    2012, 289*t*
  preventing maternal-child HIV transmission with,
    136, 241
  resistance to, 192
  side effects with, 222*t*
Antiretroviral therapy, 239, 241, 244
  cost of, 287
  incubation period and, 94
  in low- and middle-income population countries,
    2010, *138*
  public assistance programs and, 248
  in sub-Sahran Africa, 137
Antitoxins, *72*
Apoptosis, 78, 80
Aptivus, *209*
  monthly average wholesale price for, U.S.,
    2012, 289*t*
ARC. *See* AIDS-related complex
Arline, Gene, 281
Armed forces, HIV testing for, 193
Armstrong, Neil, 268
Artemether-lumefantrine, for malaria, 228*t*
Asia, south/southeast, HIV/AIDS in, 10, 294
Asians
  AIDS cases in children, 2002–2006 and
    cumulative, 134*t*
  AIDS in regions of U.S. and, 2010, *118*
  diagnosis of HIV infection attributed to injection
    drug use only in, 2010, U.S., 124*t*

HIV/AIDS diagnosed in, 115, 116*t*
HIV diagnoses of male adults or adolescents
  by high-risk heterosexual transmission, 2010,
  U.S., 119*t*
Aspergillosis, 105, 105*t*
Astrocytes, AIDS-related dementia and, *96, 96*–97
Atazanavir, *209*
  monthly average wholesale price for, U.S.,
    2012, 289*t*
Atovaquone/proquanil, expanded access enrollment
  related to HIV/AIDS and, 231*t*
Atripla, *209*
  monthly average wholesale price for, U.S., 2012, 289*t*
Attenuated vaccines, safety and efficacy of, *250*
Attenuated viruses
  producing, *249*
  in vaccines, 56
Australia
  HIV/AIDS in, 10, 193
  needle exchange programs in, 155
Autoimmune condition, 67
Autonomic nervous system, mental states,
  disease and, 233
Autopsy and embalming procedures and deceased
  HIV patients, precautions used during, 160–161
AVRS. *See* AIDS Vaccine Research Subcommittee
Azidothymidine. *See* AZT
Azithromycin, 228
AZT, 19, 20, 22*t*, 24, 52, 56, 122, 203, *207, 209,* 217,
  218, 219, 235, 297
  anemia and, 229
  benefits and uses for, 213–215, 215*t*
  community trials and, 232
  efficacy in newborns of HIV-positive pregnant
    women treated with, 214*t*
  expanded access enrollment related to HIV/AIDS
    and, 231*t*
  history behind, 211
  mode of action for, 211–213, *212*
  monthly average wholesale price for, U.S.,
    2012, 289*t*
  obstacles to use of, 215–216, 215*t*
  postexposure prophylaxis and, 226
  preventing prenatally acquired AIDS with, 135
  side effects with, 222*t*

B
Bacterial respiratory infections and HIV infection, drug
  treatment for, 228*t*
Bacterial vaginosis and STDs, 122, 147

Organ donors, HIV screening for, 171, 195
Ostrich mentality, AIDS in workplace and, 280, *280*
Outline of History, The (Wells), 298

P

Pacific Islanders
    AIDS cases in children, 2002–2006 and cumulative, 134*t*
    AIDS in regions of U.S. and, 2010, *118*
    HIV/AIDS diagnosed in, 115, 116*t*
    HIV diagnoses of male adults or adolescents by high-risk heterosexual transmission, 2010, U.S., 119*t*
Pan troglodytes troglodytes (chimpanzee), *12,* 12–13, 24
Papillomavirus infections, 147
Papua New Guinea, HIV/AIDS in, 193
Parallel-track procedure, experimental drugs and, 229, *230,* 231, 236
Paramedics
    blood exposure and, 115
    protecting, 167
Parasites/parasitism, 41, 67
Partner notification, diagnosis and, 179
Pasteur Institute, Paris, 7, 8, 9, 73
Pathogens, therapeutic agents and, 203–204
Patient care equipment, processing in health care settings, 163*t*
Patients in health care settings, CDC HIV screening recommendations for, 179*t*
PCP. *See Pneumocystis* pneumonia
PCR. *See* Polymerase chain reaction
Peace Corps, HIV testing for entrants into, 195
Pediatric AIDS, *106,* 106–108, 133, 135
    prenatal AZT treatment and, 213
    reporting in New Hampshire, 196
    in urban areas of U.S., 136
Pelvic inflammatory disease, 147
Penetration, 42–43, *43*
Penicillin, *56,* 204, *204*
Penile lesions, HIV-infected T-lymphocytes, heterosexual intercourse and, 129
Pentamidine isethionate, 100, 112, 227, 236
    aerosolized, 235
        approval of, 232
    expanded access enrollment related to HIV/AIDS and, 231*t*
PEP. *See* Postexposure prophylaxis

PEPFAR. *See* President's Emergency Plan for AIDS Relief
Peptide bond, 29
Peramivir, 228*t,* 231
Percutaneous injury, 164
Personal protection, 147–148
Personal protective equipment, for worker protection against HIV transmission in prehospital settings, 168*t*
Pertussis, 11
Peyer's patches, 64, *65*
Phagocytic cells, 67
Phagocytosis, 62, 67, *67, 72, 73,* 80
Phase I trial, *230,* 231, 232, 236, 265, 269
Phase II trial, *230,* 232, 236, 265, 269
Phase III trial, *230,* 231, 232, 236, 265, 269
Phenol, as antiviral chemical agent, *55, 56*
PI+NRTI, *209*
Placebos, 229, 232, 277
Plague, 73, 171, 276
Plasma donors, HIV testing for, 195, 199
Plasmids, 249, *250*
Pneumocystis jirovecii, 3, 77, 100, *101*
Pneumocystis jirovecii pneumonia, 3, 4, 22*t,* 100, *102,* 105*t,* 108, 109, 112, 139, 211, 227, 228*t,* 235, 236
Pneumocystosis, 3, 4, 19
Pneumonia
    recurrent, 108
    in seniors, 130
pol gene, 51, *53,* 57
Polio Surveillance Unit (CDC), 4
Polio vaccines, 252, 258, 269
    first-generation, 248
    HIV/AIDS hypothesis and, 15
Polyactide coglycolide, 259
Polymerase chain reaction, 189, 190, *191*
Polypeptides, 32
Positive ELISA HIV test, *181*
Postexposure prophylaxis, 226, 241
Postmortem procedures, precautions used during, 160–161
Poverty, AIDS and, 21, 23
Precautions policy, at workplace, 166
Preexposure prophylaxis, 226–227
Pregnancy
    HIV/AIDS and, 133, 141
    HIV status and, 122
    HIV therapy and, 135
    syphilis and, 147

STEP vaccine trials, 265
Sterilization procedures, in health care settings, 161, 163*t*
Sterilizing immunity, 252, 253, 270
Steroid hormones, 233
Stigma
    AIDS in seniors and fear of, 131
    HIV/AIDS in prisons and, 132
    women in India with HIV and, 141
Streptococcus pneumoniae, 105*t*, 249
    drug treatment for, 228*t*
Stribild, 209, 221
Substance abuse. *See also* Injection drug users
    transgender women with HIV infection and, 128
Substrate, 181
Subunit vaccines, 248
    AIDS, 253–256
    safety and efficacy of, *250*
Suicide bombers, postexposure prophylaxis and, 226
Sulfur, in amino acids, 28
Superstrain, of HIV, 208
Suppressor T-lymphocytes, 70, 77
Supreme Court, 281, 282
Surgical masks, for healthcare workers, 158
"Survival kits" for injection drug users, 153
Surviving AIDS (television broadcast), 78
Sustiva, *209, 218*
    monthly average wholesale price for, U.S., 2012, 289*t*
    side effects with, 222*t*
Sweden, donor government funding in, 2011, 287*t*
Switzerland, vaccine trials in, 265
Syncytium, 77
    deleterious effects of, 48
    formation of, *50*
Synergistic effect, 218
Syphilis, 11, 122, 126, 129, 139, 147, 194, 199
    congenital, 147
    trends in, 2010, *148*

T
Take Charge, Take Test, 122
Tamiflu, 231
tat gene, 51, *53*, 256
Tattoo needles and HIV transmission, 169
T-cells, 41, 66, 74
TCTT. *See* Take Charge, Take Test
Teenage pregnancy, AIDS epidemic and, 278
Teenagers, HIV risk and, 132, *133*
Telomeres, 74, 76

Tenofovir, *209, 217, 227, 235*
    monthly average wholesale price for, U.S., 2012, 289*t*
    vaginal gel, 241
Tenofovir + emtricitabine, monthly average wholesale price for, U.S., 2012, 289*t*
Tenofovir + emtricitabine + efavirenz, *209*
Tenovir, 221
Terrorism, postexposure prophylaxis and acts of, 226
Tertiary structure, of protein, 30, *31*
Testing. *See* HIV testing and diagnosis
Tetramer, 32
T4 cells, 77
Thailand, vaccine trials in, 265
T-helper cells, 74, *75*
Therapeutic drugs, objectives of, 203–205
Therapeutic immunity, 252–253, 270
Thinking well, disease progression and, 233–235, *234*, 297
Thoracic duct, *65*
Three-drug combination therapy, 217, 218
3TC, *207*, 216, 217, 218, 219, 235
    efficacy in newborns of HIV-positive pregnant women treated with, 214*t*
    expanded access enrollment related to HIV/AIDS and, 231*t*
    monthly average wholesale price for, U.S., 2012, 289*t*
    pregnant women and use of, 213
    side effects with, 222*t*
Thrombycytopenia, AZT and, 215
Thrush, 92, *93*, 102, 108
Thymosins, 233
Thymus gland, 64, *65, 66*, 233, 234
Ticks, 171
Tipranavir, *209*
    monthly average wholesale price for, U.S., 2012, 289*t*
Tissue donors, HIV testing for, 195
Tissue lymphatics, *65*
Tissue-specific HIV-associated diseases. *See* HIV-associated diseases, tissue-specific
T-lymphocyte counts, viral load determinations paired with, 192, *192*
T-lymphocytes, 41, 42, *43*, 57, 66, 69, 73, 80, 92, 97, 235
    anal intercourse, HIV transmission and, 125–126
    CD4, 73
    cytotoxic, 68, 69, *69*, 70, 80, 89, 251, 258